FRONTIER ORBITALS

FRONTIER ORBITALS
A PRACTICAL MANUAL

Nguyên Trong Anh

Formerly Research Director at CNRS
and Professor at the École Polytechnique, France

John Wiley & Sons, Ltd

Other Wiley Editorial Offices

John Wiley & Sons Inc., 111 River Street, Hoboken, NJ 07030, USA

Jossey-Bass, 989 Market Street, San Francisco, CA 94103-1741, USA

Wiley-VCH Verlag GmbH, Boschstr. 12, D-69469 Weinheim, Germany

John Wiley & Sons Australia Ltd, 42 McDougall Street, Milton, Queensland 4064, Australia

John Wiley & Sons (Asia) Pte Ltd, 2 Clementi Loop #02-01, Jin Xing Distripark, Singapore 129809

John Wiley & Sons Ltd, 6045 Freemont Blvd, Mississauga, Ontario L5R 4J3, Canada

Wiley also publishes its books in a variety of electronic formats. Some content that appears in print may not be available in
electronic books.

Anniversary Logo Design: Richard J. Pacifico

Library of Congress Cataloging-in-Publication Data

Anh, Nguyen.
  Frontier orbitals : a practical manual / Nguyen Anh.
    p. cm.
  Includes bibliographical references and index.
  ISBN: 978-0-471-97358-4 (cloth : alk. paper)
1. Molecular orbitals.   2. Chemistry, Physical organic.   I. Title.
  QD461.A65 2007
  541′.224–dc22   2006100392

British Library Cataloguing in Publication Data

A catalog record for this book is available from the British Library

ISBN 13 9780471973584 (Cloth) ISBN 13 9780471973591 (Paper)

Typeset in 10.5/12.5 pt Palatino by Thomson Digital

To Vân Nga

To Dao, Chuong and Nam

Acknowledgments

This book would probably never have been written without the friendly insistence of G. Bram and the help of O. Eisenstein, J. M. Lefour, A. Lubineau, Y. T. N'Guessan, P. Metzner, J. P. Pradère and A. Sevin. I have also benefited from the vast chemical knowledge which J. Boivin and S. Zard have regularly shared with me. Many thanks are due to D. Carmichael for the first English draft and for correcting a number of obscurities and numerical errors. Naturally, I am solely responsible for any mistakes in this book.

Contents

3 The Perturbation Method

4 Absolute and Relative Reactivities

5 Regioselectivity

Preface

Many chemical phenomena cannot be explained by classical physics. Examples include the covalent bond, the Walden inversion, the Hückel $4n + 2$ rule, pericyclic reactions, C-alkylation of enolates (despite the higher charge density at oxygen), axial attack at cyclohexanones (although the equatorial face is less hindered), the head-to-head cyclodimerization of acrolein (through atoms having the same charge), the anomeric effect, the *cis* configuration of certain enol ethers and many others. The list is lengthening by the day.

The reason is that chemical reactions occur at molecular level, so quantum mechanics is required to understand them.[1] However, computational chemistry is time consuming and solves problems one by one. For teaching purposes and everyday work, chemists need simple methods capable of giving *general predictions*. This book introduces the perturbation approach, the most valuable of these methods, and its simplified offspring, the *frontier orbital approximation*. It is based upon a course taught to Master's level students at the Université Paris Sud, and is aimed at experimentalists who are well versed in organic chemistry but have little or no understanding of quantum mechanics. The theoretical sections are succinct, the mathematics is kept to a strict minimum and, consequently, the explanations are not always totally rigorous. Greater emphasis is put on chemistry than on quantum mechanics, and the intelligent use of perturbation methods rather than their mathematical derivation. For example, the three-orbital perturbation equation is given without proof, but its limits and physical significance are detailed. The successes and limitations of the FMO method are discussed extensively; an understanding of when it is likely to fail is important, because valid results are obtained in only about 80% of cases. This is not an exceptional success rate, but I am unaware of any other simple method which is so versatile and effective.

This book is a *practical manual* and is intended for tutorial classes or self-studies. Being a *manual*, it should provide a firm enough background to allow the student to understand perturbation theory, rather than using it as a black box. The exercises found throughout the text are classified by symbols: **E** (easy), **M** (moderate), or **D** (difficult) to indicate their complexity. Full solutions are given in each case. These exercises must be considered an integral part of the course.

---

[1]'I had always felt – and of course still do – that the synthetic chemist would not go far unless he were to mobilize and apply, to the best of his ability – and within the limits set by the many other things he must know and do – the maximum in the way of principle and theory.' (R. B. Woodward, A. C. Cope Award address, 1973).

The organization of this manual reflects a desire to be *practical*. Hence, applications are not classified by reaction families, but rather by criteria used by the synthetic chemist: competition between reagents (relative reactivity), sites (regio- or chemoselectivity) or reaction trajectories (stereoselectivity). The steps involved in solving each problem, such as the choice of model, the calculation of molecular orbitals and the interpretation of results, are explained. At each stage, potential pitfalls are pointed out. Some are trivial, others more subtle (such as mathematically valid calculations which are physically absurd). Important points are highlighted in boxes, extended explanations are printed on a gray background and exercises stressing algebraic or numerical manipulations are marked with asterisks, to allow them to be skipped over upon first reading. The chapters dealing with applications begin with a box explaining the rules which they illustrate, and can be read independently. Cross-references are used to offset the potential disadvantages of such a compartmentalized structure. An Appendix containing the necessary MOs allows those without access to a computer to work out the exercises. The book is addressed more to students than specialists, so I have made no attempt to cover the literature exhaustively.

Chemistry underwent an explosive development in the second half of the 20th century. It is impossible to cover all of its aspects, so I have limited myself to organic chemistry. There remains much to be done; I hope that this book will provide the reader with the basics needed to do it.

Nguyên Trong Anh

n.t.anh@wanadoo.fr

1 What Can We Do With Frontier Orbitals?

1.1 The Advantages of the Perturbation Method

In essence, there are only two really important themes in chemistry: *structure* and *reactivity*. In structural problems, we usually compare the *relative stabilities* of two isomers (**1** and **2**) or conformers (**3** and **4**). Their energy differences are of the order of a few percent. Thus, benzene (**1**) is more stable than Dewar benzene (**2**) by $60 \, \text{kcal} \, \text{mol}^{-1}$, about 5% of its molecular energy ($\sim 1230 \, \text{kcal} \, \text{mol}^{-1}$).[1] Similarly, *trans*-butadiene (**3**) is more stable than *cis*-butadiene (**4**) by $2.7 \, \text{kcal} \, \text{mol}^{-1}$, or 3% of its energy of formation.

Reactivity is governed by two fundamental quantities: the *activation energy* $\Delta E^{\ddagger}$, given by the energy gap between the starting materials and the transition state, and the *reaction enthalpy* ΔH, which is the difference between the energies of the reagents and the products.[2] Again, these differences are small. For the electrocyclization of hexatriene, the energy of the system is $\sim 1300 \, \text{kcal} \, \text{mol}^{-1}$, the activation energy is $\sim 30 \, \text{kcal} \, \text{mol}^{-1}$ (2.5%) and the reaction enthalpy is $\sim 50 \, \text{kcal} \, \text{mol}^{-1}$ (4%).

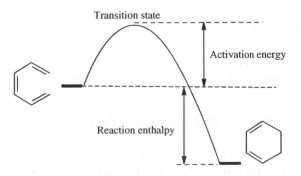

[1]Typical σ bond strength is approximately $90 \, \text{kcal} \, \text{mol}^{-1}$ and π bond strength approximately $50 \, \text{kcal} \, \text{mol}^{-1}$.
[2]For the moment, we will ignore the finer distinctions between E, H and G.

Frontier Orbitals   Nguyên Trong Anh
© 2007 John Wiley & Sons, Ltd

Therefore, the chemist is usually interested in small differences between large energies, which is why *perturbational approaches* are particularly useful. Suppose that we wish to calculate the activation energy of the hexatriene cyclization to an accuracy of 10 kcal. If $\Delta E^{\ddagger}$ is calculated by simple subtraction, it will be necessary to evaluate the energies of the reagent and of the transition state to a precision of 5 kcal. This means a margin of error of 0.33%, which is only possible using highly sophisticated techniques. Compare this with the case where we regard the transition state as a perturbed form of the initial system, and then calculate the energy of the perturbation. This gives the difference directly and requires a precision of only 33%, i.e. 100 times less. The perturbation method offers three advantages:

1. It provides chemically meaningful results with a minimum of effort. Simple Hückel calculations are sufficient in many cases.
2. By treating transition states as perturbations of the starting material, it allows us to avoid the (difficult) calculations of these unstable species.
3. It requires little equipment (these are `back of an envelope' calculations) and a minimum of theoretical knowledge (we only need to learn three perturbation schemes).

However, it does require some chemical intelligence!

1.2 The Uses of Frontier Orbitals

The frontier orbital approximation[3] is a special case of perturbation theory. It is very simple to use; we merely maximize the frontier orbital interactions. Its conclusions are correct in about 80% of cases, so it is not infallible. Nonetheless, to the best of the author's knowledge, no other simple theory applicable to a wide range of problems is any better. Furthermore, we can predict the cases where frontier orbital theory is likely to fail.

1.2.1 Five Standard Frontier Orbital Treatments of Reactivity

Absolute Reactivity

Question: Will A react with B?
Answer: Reaction is forbidden if their frontier orbital overlap is zero.

Relative Reactivity (Including Chemoselectivity)

Question: Will reagent A react preferentially with B_1 or B_2?
Answer: A reacts preferentially with the molecule whose frontier orbitals are closest in energy to its own. More precisely, if A is a nucleophile (electrophile), it will react

---

[3]Fukui K., Yonezawa T., Shingu H., *J. Chem. Phys.*, 1952, **20**, 722; Fukui K., Yonezawa T., Nagata C., Shingu H., *J. Chem. Phys.*, 1954, **22**, 1433.

with the electrophile (nucleophile) having the lowest lying LUMO (highest lying HOMO).

Regioselectivity

Question: Substrate B has two reactive sites. Which will be attacked preferentially by A?
Answer: If A is a nucleophile (electrophile), the attack will occur at the site having the largest LUMO (HOMO) coefficient.

Stereoselectivity

Question: Which is the best approach for A to attack a given site at B?
Answer: The preferred trajectory will have the best frontier orbital overlap.

Reversible and multistep reactions

Question: Is the initial reaction product unstable[4]?
Answer: The product will be unstable if it contains a bond which is unusually long. Such a bond is weak and can be broken easily. If this weak bond is formed during the reaction, then the process will be *reversible*. If not, the primary product will evolve to form a compound which is different from the starting material. We then have a *multistep reaction*.

1.2.2 Three Standard Frontier Orbital Treatments of Structural Problems

Stable Conformations

Question: Which are the most stable conformations?
Answer: If the molecule is formally divided into two fragments, the most stable conformations will be those having the smallest HOMO–HOMO interactions.

Reactive Conformations

Question: Which are the most reactive conformations?
Answer: Those having the highest lying HOMO and the lowest lying LUMO *in the transition state*.

Structural Anomalies

Question: When might structural anomalies occur?
Answer: A bond will shorten (lengthen) if bonding electron density increases (decreases) and/or antibonding electron density decreases (increases) between the extremities. If

---

[4]This question is related to the problem of structural anomalies.

the molecule is formally divided into fragments, angular deformations occur when they produce better interactions between the fragment frontier orbitals.

These abbreviated answers will be of little use to those who are not already familiar with the subject. They will be expanded in Chapters 4–7. Before considering them in detail, we will look at the concepts and methods which are needed to use frontier orbitals efficiently. Since the molecular orbitals employed in perturbation theory are generally expressed as linear combinations of atomic orbitals (LCAOs), Chapter 2 will review atomic orbitals (AOs), outline molecular orbitals (MOs) and describe the Hückel method for calculating them. Chapter 3 will set out perturbation methods in a *practical* fashion, putting more emphasis on applications and physical interpretation than upon mathematical derivation.

2 Atomic and Molecular Orbitals

2.1 Atomic Orbitals

According to quantum mechanics, an electron bound to an atom cannot possess any arbitrary *energy* or occupy any *position* in space. These characteristics can be determined by solving the time-independent Schrödinger equation:

$$\mathbf{H}\varphi = E\varphi \qquad (2.1)$$

where $\mathbf{H}$ is the *Hamiltonian operator* of the atom. We obtain a set of functions φ, which are termed *atomic orbitals* (AOs). Their mathematical equations are shown in Table 2.1, for the 1s to the 3d orbitals inclusive. With each electron is associated an atomic orbital, whose equation allows the position (or more precisely the probability dP of finding the electron within a given volume dV) and the energy of the electron to be calculated:

$$dP = \varphi\varphi^* \, dV \qquad (2.2)$$

$$E = \int \varphi\mathbf{H}\varphi^* \, dV = \langle\varphi|\mathbf{H}|\varphi\rangle \qquad (2.3)$$

In the above equations, φ^* is the complex conjugate of φ. In the cases which we will cover, it is always possible to chose atomic orbitals which are mathematically real, so we will do this systematically.

Don't panic!

To *use* frontier orbital theory efficiently, we have to understand its approximations, which define its limitations. This is not really complicated and requires more common sense than mathematical skills. So, don't worry about words like *operator* or about maths that we do not need to use.[1] Just to prove how little maths is in fact required, let us re-examine the previous section point by point.

---

[1]Chemistry is like any other science, in that the more we understand maths, the better things are. This does not mean that we have to employ maths continually: after all, a computer is not necessary for a simple sum. Maths is only a tool which allows us to make complicated deductions in the same way that computers allow us to do long calculations: quickly and without mistakes. Remember, though, the computing adage: *garbage in, garbage out*. If a theory is chemically wrong, no amount of mathematics will put it right.

Frontier Orbitals   Nguyên Trong Anh
© 2007 John Wiley & Sons, Ltd

1. In this course, we do not need to know how to solve the Schrödinger equation. In fact, after this chapter, we shall not even use the equations in Table 2.1. Just remember that orbitals are mathematical functions – solutions of Equation (2.1) – which are *continuous* and *normalized* (i.e. the square of φ is 1 when integrated over all space).

Equation (2.1) cannot be solved exactly for a polyelectronic atom A because of complications resulting from interelectronic repulsions. We therefore use approximate solutions which are obtained by replacing A with a fictitious atom having the same nucleus but only one electron. For this reason, atomic orbitals are also called *hydrogen-like orbitals* and the orbital theory the *monoelectronic approximation*.

Table 2.1 Some real atomic orbitals: Z is the atomic number and a is the Bohr radius ($a = $ $h^2/4\pi^2me^2 = 0.53 \times 10^{-8}$ cm)

$$\psi_{1s} = \frac{1}{\sqrt{\pi}}\left(\frac{Z}{a}\right)^{\frac{3}{2}} e^{-Zr/a}$$

$$\psi_{2s} = \frac{1}{4\sqrt{2\pi}}\left(\frac{Z}{a}\right)^{\frac{3}{2}}\left(2 - \frac{Zr}{a}\right)e^{-Zr/2a}$$

$$\psi_{2p_z} = \frac{1}{4\sqrt{2\pi}}\left(\frac{Z}{a}\right)^{\frac{5}{2}} re^{-Zr/2a}\cos\theta$$

$$\psi_{2p_x} = \frac{1}{4\sqrt{2\pi}}\left(\frac{Z}{a}\right)^{\frac{5}{2}} re^{-Zr/2a}\sin\theta\cos\varphi$$

$$\psi_{2p_y} = \frac{1}{4\sqrt{2\pi}}\left(\frac{Z}{a}\right)^{\frac{5}{2}} re^{-Zr/2a}\sin\theta\sin\varphi$$

$$\psi_{3s} = \frac{1}{81\sqrt{3\pi}}\left(\frac{Z}{a}\right)^{\frac{3}{2}}\left(27 - 18\frac{Zr}{a} + 2\frac{Z^2r^2}{a^2}\right)re^{-Zr/3a}$$

$$\psi_{3p_z} = \frac{\sqrt{2}}{81\sqrt{\pi}}\left(\frac{Z}{a}\right)^{\frac{5}{2}}\left(6 - \frac{Zr}{a}\right)re^{-Zr/3a}\cos\theta$$

$$\psi_{3p_x} = \frac{\sqrt{2}}{81\sqrt{\pi}}\left(\frac{Z}{a}\right)^{\frac{5}{2}}\left(6 - \frac{Zr}{a}\right)re^{-Zr/3a}\sin\theta\cos\varphi$$

$$\psi_{3p_y} = \frac{\sqrt{2}}{81\sqrt{\pi}}\left(\frac{Z}{a}\right)^{\frac{5}{2}}\left(6 - \frac{Zr}{a}\right)re^{-Zr/3a}\sin\theta\sin\varphi$$

$$\psi_{3d_z^2} = \frac{1}{81\sqrt{6\pi}}\left(\frac{Z}{a}\right)^{\frac{7}{2}}\left(\frac{Z}{a}\right)r^2e^{-Zr/3a}\left(3\cos^2\theta - 1\right)$$

Table 2.1   (*Continued*)

$$\psi_{3d_{xz}} = \frac{\sqrt{2}}{81\sqrt{\pi}}\left(\frac{Z}{a}\right)^{\frac{7}{2}} r^2 e^{-Zr/3a} \sin\theta\cos\theta\cos\varphi$$

$$\psi_{3d_{yz}} = \frac{\sqrt{2}}{81\sqrt{\pi}}\left(\frac{Z}{a}\right)^{\frac{7}{2}} r^2 e^{-Zr/3a} \sin\theta\cos\theta\sin\varphi$$

$$\psi_{3d_{x^2-y^2}} = \frac{\sqrt{2}}{81\sqrt{2\pi}}\left(\frac{Z}{a}\right)^{\frac{7}{2}} r^2 e^{-Zr/3a} \sin^2\theta\cos 2\varphi$$

$$\psi_{3d_{xy}} = \frac{\sqrt{2}}{81\sqrt{2\pi}}\left(\frac{Z}{a}\right)^{\frac{7}{2}} r^2 e^{-Zr/3a} \sin^2\sin 2\varphi$$

2. An *operator* is merely a symbol which indicates that a mathematical operation must be carried out upon the expression which follows it. Thus:
   3 is the operator 'multiply by 3';
   d/dx is the operator 'total differentiation with respect to x'.

   Each quantum mechanical operator is related to one physical property. The Hamiltonian operator is associated with *energy* and allows the energy of an electron occupying orbital φ to be calculated [Equation (2.3)]. We will never need to perform such a calculation. In fact, in perturbation theory and the Hückel method, the mathematical expressions of the various operators are never given and calculations cannot be done. *Any expression containing an operator is treated merely as an empirical parameter.*

   If a is a number and x and y are variables, then an operator f is said to be *linear* if $f(ax) = af(x)$ and $f(x + y) = f(x) + f(y)$. We will often employ the linearity of integrals in Hückel and perturbation calculations because it allows us to rewrite the integral of a sum as a sum of integrals.

By extension, *atomic orbital* has also come to mean a volume, limited by an equiprobability surface, wherein we have a high probability (let us say a 90% chance) of finding an electron. Figure 2.1 depicts the shapes of some atomic orbitals and a scale showing their relative energies. It deserves a few comments:

1. The energy scale is approximate. We only need remember that for a *polyelectronic* atom, the orbital energy within a given shell increases in the order s, p, d and that the first three shells are well separated from each other. However, the 4s and 3d orbitals have very similar energies. As a consequence, the 3d, 4s and 4p levels in the first-row transition metals all function as valence orbitals. The p orbitals are *degenerate* (i.e. the three p AOs of the same shell all have the same energy), as are the five d orbitals.
2. The orbitals of the same shell have more or less the same size. However, size increases with the principal quantum number. Thus a 3p orbital is more diffuse than a 2p orbital.

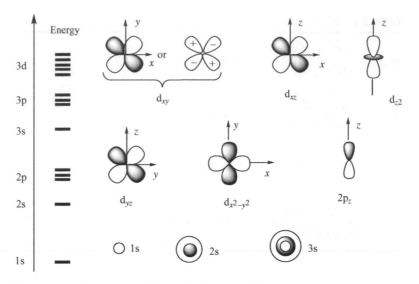

Figure 2.1 Shapes and approximate energies of some atomic orbitals.

3. The sign shown inside each orbital lobe is the sign of the function φ within that region of space. Taken on its own, this sign has no physical meaning, because the electron probability density is given by the square of φ [Equation (2.2)]. For this reason, we often distinguish between two different lobes by hatching or shading one of them, rather than using the symbols $+$ or $-$ (cf. the two representations of the d_{xy} orbital in Figure 2.1). However, we will see (p. 12) that the *relative signs* of two neighboring atomic orbitals *do* have an important physical significance.

    Let us now compare a 1s and a 2s orbital. If we start at the nucleus and move away, the 1s orbital always retains the same sign. The 2s orbital passes through a null point and changes sign afterwards (Figure 2.1). The surface on which the 2s orbital becomes zero is termed a *nodal surface*. The number of nodal surfaces increases with increasing energy: thus the 1s orbital has none, the 2s orbital has one, the 3s has two, etc.

4. Orbitals having the same azimuthal quantum number l have the same shape: all s orbitals have spherical symmetry and all p orbitals have cylindrical symmetry. The d_{z^2} orbital is drawn differently from the other d orbitals but, being a linear combination of $d_{z^2-x^2}$ and $d_{z^2-y^2}$ orbitals, it is perfectly equivalent to them. (This statement may be checked, using Table 2.1). The whole field of *stereochemistry* is founded upon the directional character of p and d orbitals.

5. Obviously, an orbital boundary surface defines an interior and an exterior. Outside the boundary, the function φ has very small values because its square, summed over all space from the boundary wall to infinity, has a value of only 0.1. Recognizing this fact allows the LCAO approximation to be interpreted in physical terms. When we say that a molecular orbital is a linear combination of AOs, we imply that it is almost indistinguishable from φ_k in the neighbourhood of atom k. This is because we are then inside the boundary of φ_k and outside the boundary of $\varphi_l (l \neq k)$, so that φ_k has finite values and contributions from φ_l are negligible. Therefore, an MO is broadly a series of AOs, the size of each AO being proportional to its LCAO coefficient.

Once the AOs are known, their occupancy is determined by:

1. The *Pauli exclusion principle*: each orbital can only contain one electron of any given spin.
2. The *Aufbau principle*: in the ground state (i.e. the lowest energy state), the lowest energy orbitals are occupied first.
3. *Hund's rules*: when *degenerate orbitals* (orbitals having the same energy) are available, as many of them as possible will be filled, using electrons of like spin.

Each electronic arrangement is known as a *configuration* and represents (more or less well) an *electronic state* of the atom.

2.2 Molecular Orbitals

All that we have just seen for atoms applies to molecules. Thus the *molecular orbitals* (MOs) of a given compound are the solutions of the Schrödinger equation for a fictitious molecule having the same nuclear configuration but only one electron. Once an MO's expression is known, the energy of an electron occupying it and the probability of finding this electron in any given position in space can be calculated. By extension, the term molecular orbital has also come to mean a volume of space wherein we have a 90% probability of finding an electron. Once the MOs are known, the electrons are distributed among them according to the Aufbau and Pauli principles and, eventually, Hund's rules. Each electronic configuration represents (more or less well) an electronic state of the molecule.[2]

The definitions above are rather abstract. Their meaning will be clarified in the examples given in the following sections. While working through these examples, we will be more concerned with the chemical implications of our results than with the detail of the calculations themselves. It would be a mistake to think that the diatomics we will study are theoreticians' molecules, too simple to be of any interest to an organic chemist. On the contrary, *the results in the next sections are important* because there is no significant conceptual difference between the interaction of two atoms to give a diatomic molecule and the interaction of two molecules to give a transition state, which may be regarded as a `supermolecule'. Formally, the equations are identical in both cases, and we can obtain the transition state MOs by just taking the diatomic MOs and replacing the atomic orbitals by the reactants' MOs, rather than having to start again from scratch. Hence the study of diatomic molecules provides an understanding of bimolecular reactions. Furthermore, the same general approaches can be used to investigate unimolecular reactions or conformations in isolated molecules. In these cases, it is only necessary to split the molecule into two appropriate fragments, and to treat their recombination as a bimolecular reaction.

---

[2]Electronic configurations are the MO equivalents of resonance structures. Sometimes a molecular state cannot adequately be represented by a single configuration, just as benzene or an enolate ion cannot be represented by only one Kekulé structure. The molecular state is then better described by a linear combination of several electronic configurations (*configuration interaction method*).

2.3 The MOs of a Homonuclear Diatomic Molecule

2.3.1 Calculations

Consider a homonuclear diatomic molecule A_2, whose two atoms A are identical. For the sake of simplicity, we will assume that each atom uses one (and only one) valence AO to form the bond. These interacting AOs, which we will call φ_1 and φ_2, are chosen so as to be mathematically *real*. The following procedure is used to calculate the resulting MOs:

1. The two nuclei are held at a certain fixed distance from each other (i.e. we apply the Born–Oppenheimer approximation).
2. The time-independent Schrödinger Equation (2.4) is written for the molecule, multiplied on the left-hand side by Ψ, and integrated over all space [Equation (2.5)]:

$$\mathbf{H}\Psi = E\Psi \tag{2.4}$$

$$\langle \Psi | \mathbf{H} | \Psi \rangle = E \langle \Psi | \Psi \rangle \tag{2.5}$$

3. Each MO is expressed as a linear combination of atomic orbitals (LCAOs):

$$\Psi = c_1 \varphi_1 + c_2 \varphi_2 \tag{2.6}$$

In Equation (2.6), we know φ_1 and φ_2. Calculating an MO Ψ_i therefore involves evaluating its associated energy E_i and the coefficients c_{i1} and c_{i2} of its LCAO expansion. Incorporating Equation (2.6) in Equation (2.5) gives

$$\langle c_1 \varphi_1 + c_2 \varphi_2 | \mathbf{H} | c_1 \varphi_1 + c_2 \varphi_2 \rangle = E \langle c_1 \varphi_1 + c_2 \varphi_{21} | c_1 \varphi_1 + c_2 \varphi_2 \rangle \tag{2.7}$$

The linearity of integrals (p. 7), allows the left-hand side of Equation (2.7) to be expressed as

$$\langle c_1 \varphi_1 + c_2 \varphi_2 | \mathbf{H} | c_1 \varphi_1 + c_2 \varphi_2 \rangle = \langle c_1 \varphi_1 | \mathbf{H} | c_1 \varphi_1 \rangle + \langle c_1 \varphi_1 | \mathbf{H} | c_2 \varphi_2 \rangle + \cdots$$

$$= c_1^2 \langle \varphi_1 | \mathbf{H} | \varphi_1 \rangle + c_2^2 \langle \varphi_2 | \mathbf{H} | \varphi_2 \rangle + \cdots$$

To express this more simply, let us set

$$\langle \varphi_i | \mathbf{H} | \varphi_i \rangle = \alpha_i$$

$$\langle \varphi_i | \mathbf{H} | \varphi_j \rangle = \beta_{ij}$$

$$\langle \varphi_i | \varphi_j \rangle = S_{ij}$$

where α_i is termed the *Coulomb integral*, β_{ij} the *resonance integral* and S_{ij} the *overlap integral*. We are using normalized AOs, so $S_{ii} = 1$. Furthermore, the two atoms are identical,[3] so

$$\alpha_1 = \alpha_2 \qquad \text{and} \qquad \beta_{12} = \beta_{21}$$

---

[3] In physical terms, $\beta_{12} = \beta_{21}$ simply means that the force binding atom 1 to atom 2 is the same as the force binding 2 to 1.

Thus, Equation (2.7) can be written as

$$(c_1^2 + c_2^2)\alpha + 2c_1c_2\beta - E(c_1^2 + c_2^2 + 2c_1c_2 S) = 0 \tag{2.8}$$

where α, β and S are parameters and c_1, c_2 and E are unknowns.

4. Let us now choose c_1 and c_2 so as to minimize E (*variational method*). To do this, we differentiate Equation (2.8), and set the partial derivatives to zero:

$$\frac{\partial E}{\partial c_1} = \frac{\partial E}{\partial c_2} = 0$$

thus obtaining the *secular equations*:

$$(\alpha - E)c_1 + (\beta - ES)c_2 = 0$$
$$(\beta - ES)c_1 + (\alpha - E)c_2 = 0 \tag{2.9}$$

These equations are homogeneous in c_i. They have a nontrivial solution if the *secular determinant* (i.e. the determinant of the coefficients of the secular equations) can be set to zero:

$$\begin{vmatrix} \alpha - E & \beta - ES \\ \beta - ES & \alpha - E \end{vmatrix} = (\alpha - E)^2 - (\beta - ES)^2 = 0 \tag{2.10}$$

The solutions to Equation (2.10) are

$$E_1 = \frac{\alpha + \beta}{1 + S} \quad \text{and} \quad E_2 = \frac{\alpha - \beta}{1 - S} \tag{2.11}$$

E_1 and E_2 are the only energies which an electron belonging to the diatomic molecule A_2 can have. Each energy level E_i is associated with a molecular orbital Ψ_i whose coefficients may be obtained by setting $E = E_i$ in Equation (2.9) and solving these equations, taking into account the normalization condition:

$$\langle \Psi_i | \Psi_i \rangle = c_{i1}^2 + c_{i2}^2 + 2c_{i1}c_{i2}S = 1 \tag{2.12}$$

The solutions are

$$\Psi_1 = \frac{1}{\sqrt{2(1+S)}}(\varphi_1 + \varphi_2) \quad \text{and} \quad \Psi_2 = \frac{1}{\sqrt{2(1+S)}}(\varphi_1 - \varphi_2) \tag{2.13}$$

Figure 2.2 gives a pictorial representation of Equation (2.11) and (2.13).

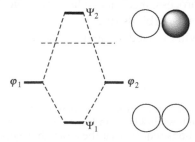

Figure 2.2 The MOs of the homonuclear diatomic A_2. φ_1 and φ_2 are arbitrarily drawn as s orbitals. Note that the destabilization of Ψ_2 is greater than the stabilization of Ψ_1.

2.3.2   A Physical Interpretation

Molecular Orbitals

As we can see from Figure 2.2, the approach of two atoms to form a molecule is accompanied by the mixing of their two AOs to form two MOs. One, Ψ_1, lies at lower energy than the isolated AOs whereas the other, Ψ_2, is at higher energy.

The destabilization of Ψ_2 with respect to the parent atomic orbitals is greater than the stabilization of Ψ_1, so the stability of the product will depend on the number of its electrons. When the molecule has one or two electrons, the Aufbau principle states that they will occupy Ψ_1, which has a lower energy than the orbitals in the separated atoms. Hence the molecule is stable with respect to the atoms. This analysis explains the phenomenon of *covalent bonds*.[4]

If the system contains three electrons, the two occupying Ψ_1 will be stabilized, and the other one, localized in Ψ_2, destabilized. Here, the stability of the molecule depends upon the relative energies of Ψ_1, Ψ_2 and the AOs: thus, HHe dissociates spontaneously, but the three-electron bond in He_2^+ is moderately robust. Note that, in contradiction with Lewis theory, a covalent bond may be formed with one or three electrons. Electron-deficient bonds (where there are fewer than two electrons per bond) are particularly prevalent amongst boron compounds.

If the system contains four electrons, two will be stabilized but the other two are destabilized to a greater extent. The molecule is then unstable with respect to the separated atoms. This is why the inert gases, where all the valence orbitals are doubly occupied, exist as atoms rather than behaving like hydrogen, oxygen or nitrogen and combining to give diatomic molecules. The mutual repulsion which occurs between filled shells is the MO description of *steric repulsion*.

Let us now turn to the LCAO expansions of Ψ_1 and Ψ_2. In Ψ_1, the AOs are in phase (they have the same sign). Thus, Ψ_1 has its greatest amplitude in the region between the two nuclei, where the AOs reinforce each other. An electron occupying Ψ_1 therefore has a high chance of being found in this internuclear region. Having a negative charge, it attracts the two (positive) nuclei and holds them together.[5] Hence, orbitals such as Ψ_1 are termed *bonding orbitals*.

In Ψ_2, the AOs have opposite phases, so Ψ_2 has different signs on A_1 and A_2. Ψ_2 is continuous, so it must pass through zero between A_1 and A_2. Consequently, an electron occupying Ψ_2 has only a small chance of being localized in the internuclear region where it can produce a bonding contribution. In fact, such an electron tends to break the bond: in the process, it can leave Ψ_2 for a lower lying AO. Hence the name *antibonding orbitals* is given to orbitals like Ψ_2.

These results will be used frequently in this book in the following form:

---

[4]Note that this kind of bond cannot be explained by classical physics. Two atoms will only form a bond if an attractive force holds them together. Newtonian gravitational forces are too weak, and Coulombian interactions require that the atoms have opposite charges, which is difficult to accept when the atoms are identical.
[5]Kinetic energy terms, which are more favorable in an MO than in an AO, also play a significant role in promoting bonding (Kutzelnigg W., *Angew. Chem. Int. Ed. Engl.*, 1973, **12**, 546).

> *An in-phase overlap is bonding and lowers the MO energy, whereas an out-of-phase overlap is antibonding and raises the MO energy.*

The Parameters

The Coulomb Integral α

To a first approximation, the Coulomb integral α_A gives the energy of an electron occupying the orbital φ_A in the isolated atom A. Therefore, its absolute value represents the energy required to remove an electron from φ_A and place it at an infinite distance from the nucleus where, by convention, its energy is zero. Consequently, α_A is always negative and its absolute value increases with the *electronegativity* of A.

The Resonance Integral β

The absolute value of the resonance integral gives a measure of the A_1A_2 *bond strength*.[6] It increases with increasing overlap. We will see that S_{12} measures the volume common to φ_1 and φ_2, which encloses the electrons shared by A_1 and A_2. Large values of S_{12} thus imply strong bonding between A_1 and A_2. When S_{12} is zero, β_{12} is also zero. It follows that two orthogonal orbitals cannot interact with each other. Conversely, the more two orbitals overlap, the more they interact. *Stereoelectronic control* results from this *principle of maximum overlap*: the best trajectory is that corresponding to the best overlap between the reagent and the substrate. The principle of maximum overlap is often expressed in terms of the *Mulliken approximation*:

$$\beta_{12} \approx kS_{12} \tag{2.14}$$

where the proportionality constant k is negative. Basis AOs are generally chosen with the same sign, so the overlap integrals are positive and the resonance integrals negative.

The Overlap Integral

Consider two overlapping orbitals φ_i and φ_j. They define four regions in space:

- Region **1** lies outside φ_i and φ_j, where both orbitals have small values. The product $\varphi_i\,\varphi_j$ is negligible.
- Region **2** (enclosed by φ_i but outside φ_j) and region **3** (enclosed by φ_j but outside φ_i) also have negligible values for $\varphi_i\,\varphi_j$: one component is appreciable, but the other is very small.
- Region **4**, where both φ_i and φ_j are finite. The value of S_{ij} comes almost exclusively from this region where the two orbitals overlap (hence the term 'overlap integral').

---

[6] β_{12} is sometimes said to represent the coupling of φ_1 with φ_2. This originates in the mathematical analogy between the interaction of two AOs and the coupling of two pendulums. The term resonance integral has similar roots (Coulson C. A., *Valence*, Oxford University Press, Oxford, 2nd edn, p. 79).

Mulliken Analysis

The MOs in the diatomic molecules discussed above have only two coefficients, so their chemical interpretation poses few problems. The situation becomes slightly more complicated when the molecule is polyatomic or when each atom uses more than one AO. *Overlap population* and *net atomic charges* can then be used to give a rough idea of the electronic distribution in the molecule.

Overlap Population

Consider an electron occupying Ψ_1. Its probability density can best be visualized as a cloud carrying an overall charge of one electron. To obtain the shape of this cloud, we calculate the square of Ψ_1:

$$\langle \Psi_1 | \Psi_1 \rangle = c_{11}^{2} \langle \varphi_1 | \varphi_1 \rangle + 2 c_{11} c_{12} S_{12} + c_{12}^{2} \langle \varphi_2 | \varphi_2 \rangle = 1 \tag{2.15}$$

Equation (2.15) may be interpreted in the following way. Two portions of the cloud having charges of c_{11}^{2} and c_{12}^{2} are essentially localized within the orbitals φ_1 and φ_2 and 'belong' to A_1 and A_2, respectively. The remainder has a charge of $2 c_{11} 2 c_{12} S$ and is concentrated within the zone where the two orbitals overlap. Hence this last portion is termed the *overlap population* of $A_1 A_2$. It is positive when the AOs overlap in phase (as in Ψ_1) and negative when they are out of phase (as in Ψ_2). The overlap population gives the fraction of the electron cloud shared by A_1 and A_2. A positive overlap population strengthens a bond, whereas a negative one weakens it. We can therefore take $2 c_{11} c_{12} S$ as a rough measure[7] of the $A_1 A_2$ bond strength.

Net Atomic Charges

It is often useful to assign a net charge to an atom. This allows the nuclei and electron cloud to be replaced by an ensemble of point charges, from which the dipole moment of the molecule can be easily calculated. It also allows the reactive sites to be identified: positively charged atoms will be preferentially attacked by nucleophiles, whereas negatively charged atoms will be favored sites for electrophiles.[8]

The *net charge* on an atom is given by the algebraic sum of its nuclear charge q_n and its electronic charge q_e. The latter is usually evaluated using the Mulliken partition scheme, which provides a simple way of dividing the electron cloud among the atoms of the molecule. Consider an electron occupying the molecular orbital Ψ_1 of the diatomic $A_1 A_2$. The contribution of this electron to the electronic charge of A_1 is then c_{11}^{2} *plus half of the overlap population*. In the general case:

$$q_e(A) = \sum_{i,j} n_i c_{iA} c_{ij} S_{Aj} \tag{2.16}$$

---

[7]In a polyelectronic molecule, it is necessary to sum over all electrons and calculate the total overlap population to obtain a measure of the bond strength.
[8]This rule is not inviolable. See pp. 87, 96 and 175.

where S_{Aj} is the overlap integral of φ_A and φ_j, n_i is the number of electrons which occupy Ψ_i and c_{iA} and c_{ij} are the coefficients of φ_A and φ_j in the same MO. The summation takes in all of the MOs Ψ_i and all of the atoms j in the molecule.

2.4 MOs of a Heteronuclear Diatomic Molecule

2.4.1 Calculations

A heteronuclear diatomic molecule is comprised of two different atoms A and B. For simplicity, we will again assume that only one AO on each atom is used to form the bond between A and B. The two relevant AOs are then φ_A, of energy α_A and φ_B of energy α_B. The calculation is completely analogous to the case of the homonuclear diatomic given above. For a heteronuclear diatomic molecule AB, Equation (2.10) – where the secular determinant is set to zero – becomes

$$(\alpha_A - E)(\alpha_B - E) - (\beta - ES)^2 = 0 \tag{2.17}$$

Equation (2.17) is a second-order equation in E which can be solved exactly. However, the analogs of expressions Equation (2.11) and (2.13) are rather unwieldy. For qualitative applications, they can be approximated as follows:

$$E_1 \approx \alpha_A + \frac{(\beta - \alpha_A S)^2}{\alpha_A - \alpha_B} \qquad E_2 \approx \alpha_B + \frac{(\beta - \alpha_B S)^2}{\alpha_B - \alpha_A} \tag{2.18}$$

$$\Psi_1 \approx N_1 \left(\varphi_A + \frac{\beta - \alpha_A S}{\alpha_A - \alpha_B} \varphi_B \right) \qquad \Psi_2 \approx N_2 \left(\varphi_B + \frac{\beta - \alpha_B S}{\alpha_B - \alpha_A} \varphi_A \right) \tag{2.19}$$

where N_1 and N_2 are normalization coefficients. Equations (2.18) assume that E_1 and E_2 are not very different from α_A and α_B, respectively. Using this approximation, it is possible to rewrite Equation (2.17) in the form

$$\alpha_A - E_1 = \frac{(\beta - E_1 S)^2}{\alpha_B - E_1} \approx \frac{(\beta - \alpha_A S)^2}{\alpha_B - \alpha_A} \tag{2.20}$$

which is equivalent to Equations (2.18). Equations (2.18) and (2.19) are shown pictorially in Figure 2.3.

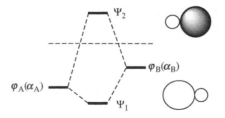

Figure 2.3 MOs of a heteronuclear diatomic molecule. φ_A and φ_B are arbitrarily shown as s orbitals.

2.4.2   A Physical Interpretation

Figure 2.3 shows that combination of the two AOs φ_A and φ_B (having energies $\alpha_A <$ α_B) produces two MOs: one, Ψ_1, has lower energy than α_A, whereas the other, Ψ_2, has higher energy than α_B. The destabilization of Ψ_2 with respect to α_B is always larger than the stabilization of Ψ_1 with respect to α_A. The bonding MO Ψ_1 comprises mainly φ_A, with a small contribution from an in-phase mixing with φ_B; the antibonding orbital Ψ_2 is mainly φ_B, with a small out-of-phase contribution from φ_A. Hence we can consider Ψ_1 as the φ_A orbital slightly perturbed by φ_B and Ψ_2 as the φ_B orbital perturbed by φ_A. This is the physical meaning of the right-hand side of Equations (2.18) and (2.19), which is why they appear as a main term and a correction. It is convenient to write the denominator of the correction in the form (energy of the perturbed orbital minus the energy of the perturbing orbital). The correction will then have a positive sign.

The stabilization of Ψ_1 with respect to α_A and the destabilization of Ψ_2 with respect to α_B increase as the $\alpha_A - \alpha_B$ energy gap decreases, the maximum being attained when the two AO's are degenerate ($\alpha_A = \alpha_B$), i.e. as in a homonuclear diatomic molecule. Comparison with Equations (2.11) and (2.13) shows that Equations (2.18) and (2.19) are only valid when

$$|\alpha_A - \alpha_B| > |\beta - \alpha_A S| \tag{2.21}$$

The physical meaning of this inequality is obvious: the correction can never be larger than the principal term. We will return to this point in the next chapter.

2.5 π MOs of Polyatomic Molecules

2.5.1   The Hückel Method for Polyatomic Molecules

In many exercises where only π systems are considered, we will employ Hückel calculations.[9] For polyenes, these simple calculations reproduce *ab initio* energies and coefficients fairly well.

The Hückel Method Applied to the Allyl System

We use the same approach as for diatomic molecules and begin with the Schrödinger Equation (2.22), which we multiply by Ψ on the left-hand side and integrate over all space [Equation (2.23)]. After replacing Ψ by its LCAOs [Equation (2.24)], we obtain Equation (2.25):

$$\mathbf{H}\Psi = E\Psi \tag{2.22}$$

$$\langle \Psi | H | \Psi \rangle = E \langle \Psi | \Psi \rangle \tag{2.23}$$

---

[9]For details on the different types of calculations (*ab initio*, semi-empirical, etc.), see Chapter 8.

$$\Psi = c_1\varphi_1 + c_2\varphi_2 + c_3\varphi_3 \tag{2.24}$$

$$\langle c_1\varphi_1 + c_2\varphi_2 + c_3\varphi_3 \mid \mathbf{H} \mid c_1\varphi_1 + c_2\varphi_2 + c_3\varphi_3 \rangle$$
$$= E\langle c_1\varphi_1 + c_2\varphi_2 + c_3\varphi_3 \mid c_1\varphi_1 + c_2\varphi_2 + c_3\varphi_3 \rangle \tag{2.25}$$

The Hückel treatment assumes that:[10]

(a) each Coulomb integral has the same value:
$$\alpha_1 = \alpha_2 = \alpha_3 = \alpha \tag{2.26}$$

(b) the resonance integral is the same for any two neighboring atoms and zero for any two atoms not directly bound to each other:
$$\beta_{13} = 0$$
$$\beta_{12} = \beta_{23} = \beta \tag{2.27}$$

(c) the overlap integrals S_{ij} are zero when $i \neq j$ and 1 when $i = j$:
$$S_{ij} = \delta_{ij} \quad (\delta_{ij} = \text{Kronecker symbol}) \tag{2.28}$$

Equation (2.25) then becomes
$$(\alpha - E)(c_1^2 + c_2^2 + c_3^2) + 2\beta(c_1c_2 + c_2c_3) = 0 \tag{2.29}$$

Differentiating Equation (2.29) and zeroing each partial derivative of E with respect to c_i, we obtain the secular equations:
$$(\alpha - E)c_1 + \beta c_2 = 0$$
$$\beta c_1 + (\alpha - E)c_2 + \beta c_3 = 0 \tag{2.30}$$
$$\beta c_2 + (\alpha - E)c_3 = 0$$

Writing $x = (\alpha - E) / \beta$ and setting the secular determinant to zero, this gives
$$\begin{vmatrix} x & 1 & 0 \\ 1 & x & 1 \\ 0 & 1 & x \end{vmatrix} = x(x^2 - 2) = 0 \tag{2.31}$$

whose roots are $x = 0$ and $x = \pm \sqrt{2}$. Hence an electron may have one of three possible energies:
$$E_1 = \alpha + \sqrt{2}\beta$$
$$E_2 = \alpha \tag{2.32}$$
$$E_3 = \alpha - \sqrt{2}\beta$$

which increase down the page. Substituting these energies into Equation (2.30) and normalizing according to

---

[10] The validity of these approximations is discussed in Anh N. T., *Introduction à la Chimie Moléculaire*, Ellipses, Paris, 1994, p. 200.

$$\langle \Psi_i \mid \Psi_i \rangle = c_{i1}{}^2 + c_{i2}{}^2 + c_{i3}{}^2 = 1 \tag{2.33}$$

we find that:

$$\Psi_1 = 0.5(\varphi_1 + \varphi_3) + 0.707\varphi_2$$
$$\Psi_2 = 0.707(\varphi_1 - \varphi_3) \tag{2.34}$$
$$\Psi_3 = 0.5(\varphi_1 + \varphi_3) - 0.707\varphi_2$$

Any electrons found in Ψ_2 have the same energy α as an electron in an isolated carbon atom. Hence they neither stabilize nor destabilize the allyl system. For this reason, Ψ_2 is termed a *nonbonding orbital*.

Coulson Formulae for Linear Polyenes

Linear polyenes are unbranched, open-chain conjugated hydrocarbons having the general formula $C_n H_{n+2}$. Coulson[11] has shown that the energy levels of a linear polyene having N atoms are given by Equation (2.35), with MOs labeled in order of increasing energy:

$$E_p = \alpha + 2\beta \cos\left(\frac{p\pi}{N+1}\right) \tag{2.35}$$

The coefficient c_{pk} of φ_k in the Ψ_p MO is given by

$$c_{pk} = \sqrt{\frac{2}{N+1}} \sin\left(\frac{pk\pi}{N+1}\right) \tag{2.36}$$

With respect to the median plane of a rectilinear polyene, atoms k and $N - k + 1$ are symmetrical. Now, the coefficient of atom $(N - k + 1)$ is given by

$$c_{p,N-k+1} = \sqrt{\frac{2}{N+1}} \sin\left[\frac{p(N-k+1)\pi}{N+1}\right] = \sqrt{\frac{2}{N+1}} \sin\left(p\pi - \frac{pk\pi}{N+1}\right)$$

Since

$$\sin(p\pi - x) = \sin x \quad \text{if } p \text{ is odd}$$

$$\sin(p\pi - x) = -\sin x \quad \text{if } p \text{ is even}$$

it follows that *all odd-numbered MOs are symmetrical*, i.e. the coefficients at C_1 and $C_{n'}$ at C_2 and C_{n-1}, etc., are identical. *All even-numbered MOs are antisymmetrical*, i.e. these coefficients are equal, but have opposite signs.

[11]Coulson C. A., *Proc. R. Soc. London*, 1939, **A169**, 413; Coulson C. A., Longuet-Higgins H. C., *Proc. Ry. Soc. London*, 1947, **A192**, 16; Coulson C. A., *Proc. Ry. Soc. London*, 1938, **A164**, 383.

We have just seen that coefficients at C_1 and C_n are either identical or opposite. According to formula Equation (2.36), they vary as

$$\sin\left(\frac{p\pi}{N+1}\right) \text{ with } p = 1, 2, 3, \ldots, N$$

Therefore, *the coefficients at the terminal atoms rise steadily, reaching a maximum in the HOMO and the LUMO, and then decline.* These properties will be useful for the derivation of the selection rules of pericyclic reactions.

Bond Orders and Net Charges

The overlap population is always zero in a Hückel calculation ($S_{ij} = 0$), so we employ a *bond order* p_{rs} to estimate the strength of a π bond between two atoms r and s. It is defined as

$$p_{rs} = \sum_j n_j c_{jr} c_{js} \tag{2.37}$$

where n_j represents the number of electrons and c_{jr} and c_{js} the coefficients of r and s, respectively, in Ψ_j. The summation includes all of the occupied orbitals (the vacant orbitals can be neglected, because $n_j = 0$). Therefore, the bond index p_{rs} is simply an overlap population obtained using Hückel coefficients and an arbitrary value of 0.5 for S_{rs}.[12] The electronic charge on the atom r is given by

$$q_e^{(r)} = \sum_j n_j c_{jr}^{\,2} \tag{2.38}$$

and its *net charge* is the sum of $q_e^{(r)}$ and its nuclear charge $q_n^{(r)}$.

Exercise 1 (E)[13]

(1) Use Coulson's equations to derive the π molecular orbitals of butadiene.
(2) Calculate the bond orders p_{12}, p_{23}, p_{34}. These results are a great success for Hückel theory. Why?

Answer

(1)
$$\Psi_1 = 0.37(\varphi_1 + \varphi_4) + 0.60(\varphi_2 + \varphi_3) \qquad E_1 = \alpha + 1.618\beta$$
$$\Psi_2 = 0.60(\varphi_1 - \varphi_4) + 0.37(\varphi_2 - \varphi_3) \qquad E_2 = \alpha + 0.618\beta$$
$$\Psi_3 = 0.60(\varphi_1 + \varphi_4) - 0.37(\varphi_2 + \varphi_3) \qquad E_3 = \alpha - 0.618\beta$$
$$\Psi_4 = 0.37(\varphi_1 - \varphi_4) - 0.60(\varphi_2 - \varphi_3) \qquad E_4 = \alpha - 1.618\beta$$

(2) In the ground state, only Ψ_1 and Ψ_2 are occupied. Each contains two electrons. Using formula Equation (2.37), we see that

---

[12] A bond order for two nonbonded atoms is meaningless, as S_{rs} is then zero.
[13] For the meaning of asterisks, **(E)**, **(M)**, etc., see the Preface.

$$p_{12} = p_{34} = 2(0.37 \times 0.60) + 2(0.60 \times 0.37) = 0.89$$
$$p_{23} = 2(0.60 \times 0.60) - 2(0.37 \times 0.37) = 0.45$$

The p_{23} index is smaller than the others, which suggests that the central bond is weaker. Thus the calculation reproduces the alternating single and double bonds, even though the same resonance integral was used for all of them.

Exercise 2 (M)

(1) Calculate the bond orders for ethylene in (a) the ground state and (b) the first excited state $(\pi \rightarrow \pi^*)$. What are the chemical consequences of these results?
(2) Introduce overlap [using Equation (2.13) and Figure 2.2]. What conformation would the ethylene excited state have if it were sufficiently long-lived to reach equilibrium?

Answer

(1) According to Coulson's equations, the π MOs of ethylene are:

$$\Psi_1 = 0.707(\varphi_1 + \varphi_2) \quad \text{with} \quad E_1 = \alpha + \beta$$
$$\Psi_2 = 0.707(\varphi_1 - \varphi_2) \quad \text{with} \quad E_2 = \alpha - \beta$$

In the ground state Ψ_1 contains two electrons. The bond order is given by

$$p_{12} = 2 \times 0.707^2 = 1$$

Ψ_1 and Ψ_2 both contain one electron in the excited state, so the bond order becomes

$$p_{12} = 0.707^2 - 0.707^2 = 0$$

and the π bond disappears. Since only a σ bond links the carbon atoms, they can rotate freely about the C–C axis. Hence alkenes can be isomerized by irradiation. It is worth remembering that one of the key steps in vision involves the photochemical isomerization of *cis*- to *trans*-rhodopsine.

(2) If overlap is neglected, the destabilization due to the antibonding electron is exactly equal to the stabilization conferred by the bonding electron. However, the destabilizing effects become greater when overlap is introduced [cf. Equations (2.11) and (2.14)]. When the p orbitals are orthogonal, the overlap is zero and the destabilization disappears. As a result, this conformation is adopted in the ethylene excited state.

* Exercise 3 (E)

Calculate the net atomic charges in the allyl cation.

Answer

In the allyl cation, the two electrons are both found in Ψ_1. The charges are:

$$q_1 = q_3 = 2 \times 0.5^2 = 0.5 \qquad \text{net charge: } 0.5$$
$$q_2 = 2 \times 0.707^2 = 1 \qquad \text{net charge: } 0$$

So, the positive charge is divided equally between the terminal atoms.

2.5.2 How to Calculate Hückel MOs

Why should we use Hückel calculations in some exercises, when it is now so easy to do semi-empirical or *ab initio* calculations? There are two reasons. First, experimentalists often need only rapid `back of an envelope' solutions, which can be readily obtained with Hückel calculations. Second, there is a close analogy between the formalisms of Hückel and perturbation methods. Understanding Hückel calculations will help you master perturbation theory.

Most modern Hückel programs will accept the molecular structure as the input. In older programs, the input requires the *kind of atoms* present in the molecule (characterized by their Coulomb integrals α_i) and the way in which they are *connected* (described by the resonance integrals β_{ij}). These are fed into the computer in the form of a *secular determinant*. Remember that the Coulomb and resonance integrals cannot be calculated (the mathematical expression of the Hückel Hamiltonian being unknown) and must be treated as empirical parameters.

Choosing the Parameters α and β

Heteroatoms

Theoreticians call any non-hydrogen atom a *heavy* atom, and any heavy atom other than carbon a *heteroatom*. In the Hückel model, all carbon atoms are assumed to be the same. Consequently, their Coulomb and resonance integrals never change from α and β, respectively. However, heteroatom X and carbon have different electronegativities, so we have to set $\alpha_X \neq \alpha$. Equally, the C–X and C–C bond strengths are different, so that $\beta_{CX} \neq \beta$. Thus, for heteroatoms, we employ the modified parameters

$$\alpha_X = \alpha + k\beta$$
$$\beta_{CX} = h\beta$$

(2.39)

When i and j are both heteroatoms, we can take $\beta_{ij} = h_i\,h_j\,\beta$. The recommended values for X = O, N, F, Cl, Br and Me are given in Table 2.2. The exact numerical values of these parameters are not crucially important but it is essential that values of α_i appear in the correct order of electronegativity and β_{ij} in the correct order of bond strength.[14]

Alkyl Substituents

Hückel calculations are very approximate, so it is pointless to use oversophisticated models. Therefore, all alkyl substituents can be treated as methyl groups.

The methyl group is represented as a doubly occupied orbital of energy $\alpha + \beta$ (Table 2.2). This may need some explanation. In a methyl group, the hydrogen s orbitals and the carbon valence orbitals combine to give seven three-dimensional `fragment orbitals', which are shown on p. 188. Only two of these, π'_{Me} and π'^*_{Me}, can conjugate with a neighboring π system: the others are orthogonal to it and cannot overlap. Hence, in

---

[14]Minot C., Anh N. T., *Tetrahedron*, 1977, **33**, 533.

Table 2.2   Some Hückel parameters for heteroatoms, after Streitwieser[15]

Atom or group	Coulomb integral	Resonance integral
Oxygen		
One electron	$\alpha_O = \alpha + \beta$	$\beta_{CO} = \beta$
Two electrons	$\alpha_O = \alpha + 2\beta$	$\beta_{CO} = 0.8\beta$
Nitrogen		
One electron	$\alpha_N = \alpha + 0.5\beta$	$\beta_{CN} = \beta$
Two electrons	$\alpha_N = \alpha + 1.5\beta$	$\beta_{CN} = 0.8\beta$
Fluorine	$\alpha_F = \alpha + 3\beta$	$\beta_{CF} = 0.7\beta$
Chlorine	$\alpha_{Cl} = \alpha + 2\beta$	$\beta_{CCl} = 0.4\beta$
Bromine	$\alpha_{Br} = \alpha + 1.5\beta$	$\beta_{CBr} = 0.3\beta$
Methyl	$\alpha_{Me} = \alpha + 2\beta$	$\beta_{CMe} = 0.7\beta$

calculations restricted to π orbitals, a methyl group can be represented *rigorously* by two orbitals: one bonding and doubly occupied the other antibonding and empty. The empty antibonding orbital is well removed from the α level, so it has little effect upon the system and can be ignored.

The Methyl Inductive Effect

Neglecting the π'^*_{Me} orbital amounts to assimilating the methyl group to an electron pair, in other words to consider that it has a pure π-donating effect. This is chemically reasonable.[16] In fact, a methyl is a *σ-attracting* and *π-donating* group.[17] This is the rea-son why, in the gas phase, the acidity order of amines increases with substitution as does *also their basicity order*: $Me_3N > Me_2HN > MeH_2N > H_3N$!

The nature of methyl inductive effect was the subject of a controversy in the 1960 and 1970s. However, a careful perusal of the literature shows in fact no contradiction, the criteria used being different with the authors. Those favoring an electron-donating effect based their arguments on the Markownikov rule, the Hammett equation and the acidity order of alcohols *in solution*. Authors advocating an electron-withdrawing effect justified their idea with NMR spectra, quantum mechanical calculations of atomic charges of molecules *in the gas phase* and acidity order of alcohols *in the gas phase*.

The inductive effect, as many other 'effects' in organic chemistry, is not an observable and cannot be defined precisely, in an objective manner. It is therefore not surprising that different criteria led to different conclusions. See Minot *et al.*[16] for a more detailed discussion.

---

[15]Streiwieser A., *Molecular Orbital Theory for Organic Chemists*, John Wiley & Sons, Inc., New York, 1961, p. 135

[16]Minot C., Eisenstein O., Hiberty P. C., Anh N. T., *Bull. Soc. Chim. Fr. II*, 1980, 119.

[17]A methyl is a true donor when borne by a cation, and is an apparent electron donor when borne by a double bond or an anion. By 'apparent donor', we mean that there is no real electron transfer to the double bond or the anion, but the HOMO energy is raised, compared with that of the parent unsubstituted system.

Writing the Secular Determinant

In some Hückel packages, the input (the atoms and their connectivities) must be introduced as a secular determinant. The latter can be written merely by looking at the structural formula. Let a_{ij} be the element in row i and column j, and set

$$x = \frac{\alpha - E}{\beta} \qquad \text{(in units of } \beta\text{)}$$

Using an arbitrary labeling scheme for the atoms, we then take:

- $a_{ii} = x$ if atom i is a carbon atom, $a_{ii} = x + k$ if i is a heteroatom [for the definition of h and k, see Equation (2.39)].
- $a_{ij} = 1$ if i and j are adjacent carbon atoms, and $a_{ij} = h$ if one of them is a heteroatom. If both are heteroatoms, we can use $a_{ij} = h_i\, h_j$ as a first approximation.
- $a_{ij} = 0$ if i and j are not adjacent to each other.

Checking the Calculations

Always check your calculations (your input may be erroneous). If your parameters are adequate, your calculations *must* reproduce the main chemical characteristics of your compound: the electronic charge should increase with the atom's electronegativity; the frontier orbitals of an electron-rich compound should be raised, etc.

Beware: Hückel calculations only recognize connectivities. So, for example, they are incapable of distinguishing between *cis*- and *trans*-butadiene. Care should also be taken over *degenerate orbitals*. Their *ensemble* must respect the molecular symmetries, but individual degenerate MOs may violate them. Many combinations of coefficients can be used to describe each pair of degenerate orbitals; some are more tractable than others. Thus, some program gives the following for the Ψ_2 and Ψ_3 MOs of the cyclopentadienyl radical:

$$\Psi_2 = 0.21\varphi_1 - 0.50\varphi_2 - 0.52\varphi_3 + 0.18\varphi_4 + 0.63\varphi_5$$
$$\Psi_3 = 0.60\varphi_1 + 0.38\varphi_2 - 0.36\varphi_3 - 0.61\varphi_4 - 0.01\varphi_5$$

All the coefficients are different. The MOs below are much more convenient to use:

$$\Psi_2 = 0.63\varphi_1 + 0.20\varphi_2 - 0.51\varphi_3 - 0.51\varphi_4 + 0.20\varphi_5$$
$$\Psi_3 = -0.60\varphi_1 - 0.37\varphi_2 + 0.37\varphi_4 + 0.60\varphi_5$$

The fivefold symmetry has been reduced to symmetry through a plane. These symmetry orbitals can be found easily, merely by redoing the calculations using slightly modified values for C_1 (e.g. 1.01β for its resonance integral).

* Exercise 4 (E)

Write the secular determinant for the following molecules:

1

2

3

4

Answer

1
$$\begin{vmatrix} x & 1 & 0 \\ 1 & x & 0.8 \\ 0 & 0.8 & x+2 \end{vmatrix}$$

2
$$\begin{vmatrix} x+1 & 1 & 0 & 0 \\ 1 & x & 1 & 0 \\ 0 & 1 & x & 1 \\ 0 & 0 & 1 & x \end{vmatrix}$$

3
$$\begin{vmatrix} x & 1 & 0 & 0 & 1 & 0 \\ 1 & x & 1 & 0 & 0 & 0 \\ 0 & 1 & x & 1 & 0 & 0 \\ 0 & 0 & 1 & x & 1 & 0 \\ 1 & 0 & 0 & 1 & x & 1 \\ 0 & 0 & 0 & 0 & 1 & x \end{vmatrix}$$

4
$$\begin{vmatrix} x+1.5 & 0.8 & 0 & 0 & 0.8 \\ 0.8 & x & 1 & 0 & 0 \\ 0 & 1 & x & 1 & 0 \\ 0 & 0 & 1 & x & 1 \\ 0.8 & 0 & 0 & 1 & x \end{vmatrix}$$

Electron Counting

An accurate electron count is necessary to determine which MOs are occupied in the ground state. Halogens always provide two electrons, because they interact with a conjugated system through their lone pairs. Oxygen and nitrogen may contribute one or two electrons according to the molecule in question. Lewis structures show that a heteroatom bound by a double bond provides one electron to the π system, whereas a singly bound heteroatom gives two. For example:

2.6 To Dig Deeper

Levine I. N., *Quantum Chemistry*, 4th edn, Prentice-Hall, Englewood Cliffs, NJ, 1991. Very lucid. Contains exercises with succinct answers. The reader is taken through the proof step-by-step, which is particularly agreeable for those who have forgotten their maths.

3 The Perturbation Method

3.1 Perturbations and Hückel Methods

A perturbation calculation requires a reference system, whose Hamiltonian $\mathbf{H}°$ and MOs $\Psi_i°$ (of energy $E_i°$) are known. The system we wish to study is closely related to it. In fact, it is assumed that the real system is a slightly perturbed version of the reference, so its Hamiltonian $\mathbf{H}$ can be written as

$$\mathbf{H} = \mathbf{H}° + \mathbf{P} \tag{3.1}$$

where $\mathbf{P}$, whose mathematical expression is *never given*, is the perturbation operator. The integrals

$$P_{ij} = \langle \Psi_i° | \mathbf{P} | \Psi_j° \rangle \tag{3.2}$$

being perturbations, are always small. The MOs of the perturbed system Ψ_i are expanded as linear combinations of the MOs of the reference system:

$$\Psi_i = \sum_j c_{ij} \Psi_j° \tag{3.3}$$

Therefore, perturbation calculations and Hückel calculations are very similar: (a) the Hamiltonian expression is not specified and (b) the required MOs are linear combinations of known orbitals. When Equations (3.1) and (3.3) are incorporated into the time-independent Schrödinger equation:

$$\mathbf{H}_i \Psi_i = E_i \Psi_i \tag{3.4}$$

and the latter is solved by the variation method, three types of integral appear:

$$\langle \Psi_i° | \mathbf{H} | \Psi_i° \rangle = E_i° + P_{ii} \tag{3.5}$$

$$\langle \Psi_i° | \mathbf{H} | \Psi_j° \rangle = P_{ij} \tag{3.6}$$

$$\langle \Psi_i° | \Psi_j° \rangle = S_{ij}° = 0 \tag{3.7}$$

These integrals are the analogs of α, β and S.

Frontier Orbitals   Nguyên Trong Anh
© 2007 John Wiley & Sons, Ltd

3.2 Study of Bimolecular Reactions Using Perturbation Methods

3.2.1 Two-orbital Systems

Consider a reaction between two molecules A and B. For simplicity, we assume that each molecule has only one MO (Ψ_A° of energy E_A° and Ψ_B° of energy E_B°, respectively). During the reaction, the reagents evolve to produce the 'supermolecule' (A $\cdots$ B). As we saw in the previous section, the MOs of (A $\cdots$ B) can be calculated by a perturbation approach which is entirely analogous to the Hückel treatment of a diatomic molecule. In fact, we only need to take the MOs of the diatomic and replace:

- the AOs φ by the MOs Ψ° of A and B
- α and β by the expressions in Equations (3.5) and (3.6).[1]

We should distinguish between the two cases below:

The MOs in the Starting Materials Are Degenerate

This system is the analog of a homonuclear diatomic, with $S = 0$. Equations (2.11) and (2.13) indicate that the mixing of two degenerate orbitals Ψ_A° and Ψ_B° gives two new ones:

$$\begin{aligned} \Psi_1 &= 0.707\,(\Psi_A^\circ + \Psi_B^\circ) \quad &\text{of energy} \quad & E_1 = E^\circ + P_{AB} \\ \Psi_2 &= 0.707\,(\Psi_A^\circ - \Psi_B^\circ) \quad &\text{of energy} \quad & E_2 = E^\circ - P_{AB} \end{aligned} \tag{3.8}$$

where P_{AB} represents the integral $\langle \Psi_A^\circ | \mathbf{P} | \Psi_A^\circ \rangle$.

The MOs in the Starting Materials Are Not Degenerate

This system is the analog of a heteronuclear diatomic (p. 15). Mixing of the two orbitals Ψ_A° and Ψ_B°, where $E_A^\circ < E_B^\circ$, will give two combinations, one of which is bonding (Ψ_A) and the other antibonding (Ψ_B):

$$\begin{aligned} \Psi_A &= N\left(\Psi_A^\circ + \frac{P_{AB}}{E_A^\circ - E_B^\circ}\Psi_B^\circ \right) \quad &\text{of energy} \quad & E_A = E_A^\circ + \frac{P_{AB}^{\ 2}}{E_A^\circ - E_B^\circ} \\[2em] \Psi_B &= N\left(\Psi_B^\circ + \frac{P_{AB}}{E_B^\circ - E_A^\circ}\Psi_A^\circ \right) \quad &\text{of energy} \quad & E_B = E_B^\circ + \frac{P_{AB}^{\ 2}}{E_B^\circ - E_A^\circ} \end{aligned} \tag{3.9}$$

---

[1] As neither $\mathbf{H}$ nor $\mathbf{P}$ is specified, this amounts to a mere change in notation! We only need to replace α_i by E_i° and β_{ij} by P_{ij}. The *intramolecular* perturbation of Ψ_i by itself, P_{ii}, may be neglected, because we will only study bimolecular reactions and will invariably use the (nonperturbed) frontier orbitals of the starting materials. This point is discussed on p. 51.

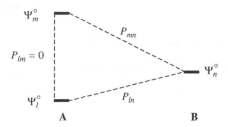

Figure 3.1 The three-orbital interaction diagram.

Remarks

1. If we ensure that the denominators are written in the form (energy of the perturbed orbital minus energy of the perturbing orbital), the correction terms in Equations (3.9) will always have a positive sign.

2. Since $E_A^\circ - E_B^\circ$ appears in the denominator, *Equations (3.9) can only be used when $E_A^\circ - E_B^\circ$ is greater than P_{AB}*. This is the analog of the constraint in Equation (2.21). Physically, this means that the correction term must be smaller than the principal term.

3. Because $P_{AB} < (E_A^\circ - E_B^\circ)$, the corrections in Equations (3.8) are greater than those in Equations (3.9). In other words, the interaction between degenerate orbitals is greater than between nondegenerate orbitals.

3.2.2 Systems Having More Than Two Orbitals

We now consider a more realistic case where molecules A and B each have several MOs, n_A and n_B, respectively. As a first approximation, we assume that each MO on A (or B) is perturbed by all the orbitals of B (or A), which act *independently* of each other. This amounts to treating $n_A n_B$ two-orbital problems. This number will be significantly reduced by employing the frontier orbital approximation (see below).

For the moment, suffice it to say that the *two-orbital perturbation* schemes give the *orbital energies* and the *sign* of the MO coefficients in the supermolecule (A $\cdots$ B) with a reasonable degree of precision. However, *three-orbital perturbations* are needed to determine the *relative sizes* of the coefficients.

What do we mean by this? Consider the interactions between two MOs, Ψ_l° and Ψ_m°, of molecule A and an MO Ψ_n° of molecule B (Figure 3.1). Ψ_l° and Ψ_m° belong to the same molecule so, *in the starting material*, they are orthogonal and cannot interact with each other. However, perturbation by B allows them to interact in the product. This interaction has little effect on the overall energy, but can markedly change the size of the MO coefficients. After reaction, orbital Ψ_l° is transformed into Ψ_l, which can be written as[2]

$$\Psi_l = N\left[\Psi_l^\circ + \frac{P_{ln}}{E_l^\circ - E_n^\circ}\Psi_n^\circ + \frac{P_{ln}P_{mn}}{(E_l^\circ - E_n^\circ)(E_l^\circ - E_m^\circ)}\Psi_m^\circ \right] \qquad (3.10)$$

The only difference between Equations (3.10) and (3.9) lies in the last term, which mixes Ψ_l° with Ψ_m°, and modifies the coefficients in the A component of the supermolecule.

---

[2] A proof may be found in Anh N. T., *Introduction à la Chimie Moléculaire*, Ellipses, Paris, 1994, p. 149.

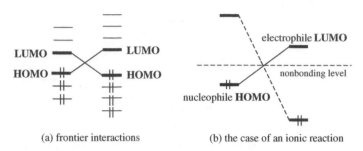

| (a) frontier interactions | (b) the case of an ionic reaction |

Figure 3.2   Frontier orbital interactions.

In physical terms, this mixing means that *the electron cloud of A is distorted by the approach of B*. The mixing coefficient may look daunting, but it is not very complicated. The numerator is the product of two resonance integrals. Only three resonance integrals, P_{lm}, P_{ln} and P_{mn}, can exist in a three-orbital system and the first, being zero (two orbitals of the same molecule), can be ignored. The denominator is the product of two energy differences: (the perturbed MO energy minus the first perturbing MO energy) multiplied by (the perturbed MO energy minus the second perturbing MO energy).

3.2.3   The Frontier Orbital Approximation

We saw above that $n_A n_B$ two-orbital interactions occur during the union of A and B. In 1952, Fukui introduced the bold approximation[3] that, of these, only the HOMO–LUMO[4] interactions significantly affect the outcome of the reaction (Figure 3.2a). These MOs are termed *frontier orbitals*, because they mark the border between occupied and unoccupied orbitals. The frontier orbital (FO) approximation means that we have only consider *two* interactions for reactions between neutral molecules, irrespective of the size and complexity of A and B.

Ionic reactions are simpler still: the *only* important interaction involves the HOMO of the nucleophile and the LUMO of the electrophile (Figure 3.2b). This is because a nucleophile (or any electron-rich compound) readily donates electrons, so it will react through its HOMO, where the highest energy electrons are localized. Conversely, an electrophile (or any electron-poor compound) accepts electrons easily. These electrons can only be put into vacant orbitals. Obviously, the lower the energy of the empty orbital, the more easily it accepts electrons. Thus an electrophile generally reacts through its LUMO.

3.2.4   Unimolecular Systems

Theoretically, Equations (3.8) and (3.9) apply only to bimolecular processes, so we employ a trick for *unimolecular reactions*: the molecule is formally divided into two fragments whose recombination is treated as a bimolecular reaction.[5] This technique is also very useful for treating *structural problems* (Chapter 7).

---

[3] This approximation is justified on p. 49. Its limitations will be discussed in Chapter 8.
[4] HOMO = highest occupied MO; LUMO = lowest unoccupied MO.
[5] The selection rules for sigmatropic rearrangements were deduced in this manner (Woodward R. B., Hoffmann R., *J. Am. Chem. Soc.*, 1965, **87**, 2511).

3.3 Perturbation Theory: The Practical Aspects

3.3.1 Numerical Calculations

Let us look at the MOs of an enol, which can be modeled naturally as the combination of an ethylene fragment and a hydroxyl group, i.e. a carbon skeleton and a substituent. The carbon AOs are denoted φ_1 and φ_2, the oxygen lone pair φ_3, the ethylene MOs π and π^* and the enol MOs Ψ_1, Ψ_2 and Ψ_3.

The interaction scheme is shown in Figure 3.3. Formally, the fragmentation process involves the breaking of one ij bond (in the present case, the C_2O_3 linkage). Care should be taken to employ the *same sign* for the coefficients of i and j in the fragment orbitals. For example, when φ_3 has a positive sign, the ethylene π^* orbital should be written as $0.707(-\varphi_1 + \varphi_2)$ and not $0.707(\varphi_1 - \varphi_2)$. We thus ensure that P_{ij} will be negative and all correction terms which appear in Equations (3.8)–(3.10) will then have a positive sign.

The interacting orbitals (π of energy $\alpha + \beta$, π^* of energy $\alpha - \beta$ and φ_3 of energy $\alpha + 2\beta$) are not degenerate, so we can evaluate the energies E_i of the enol MOs Ψ_i using the Equations (3.9):

$$E_3 = E(\pi^*) + \frac{P_{\varphi_3,\pi^*}^2}{E(\pi^*) - E(\varphi_3)}$$

with

$$P_{\varphi_3,\pi^*} = \langle \pi^* | \mathbf{P} | \varphi_3 \rangle = \langle 0.707(-\varphi_1 + \varphi_2) | \mathbf{P} | \varphi_3 \rangle$$
$$= 0.707 \langle -\varphi_1 | \mathbf{P} | \varphi_3 \rangle + 0.707 \langle \varphi_2 | \mathbf{P} | \varphi_3 \rangle$$

C_1 and O_3 are not bound directly to each other, so the $\langle -\varphi_1 | \mathbf{P} | \varphi_3 \rangle$ term is zero. $\langle \varphi_2 | \mathbf{P} | \varphi_3 \rangle$ measures the change in the C_2O_3 resonance integral during the recombination process. It is zero when the fragments are separated and 0.8β when bound (see the parameter Table 2.2). Thus:

$$P_{\varphi_3,\pi^*} = 0.707 \times 0.8\beta = 0.566\beta$$

$$E_3 = (\alpha - \beta) + \frac{(0.566\beta)^2}{(\alpha - \beta) - (\alpha + 2\beta)} = \alpha - 1.107\beta$$

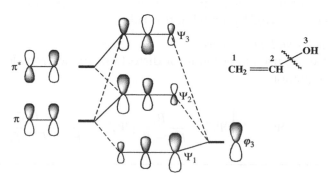

Figure 3.3 The MO diagram of an enol, built using a perturbation approach. The principal component of each MO is indicated by the unbroken line: Ψ_1 is derived from φ_3, Ψ_2 from π and Ψ_3 from π^*.

The three perturbation schemes

To recap: splitting a molecule into fragments (Section 2.4) allows unimolecular reactions to be treated as bimolecular processes. Only one or two frontier interactions have to be considered, irrespective of the problem (Section 2.3). Two cases can be distinguished:

The two interacting orbitals are degenerate
The product MOs are given by:

$$\Psi_1 = 0.707(\Psi_A{}^\circ + \Psi_B{}^\circ) \quad \text{with} \quad E_1 = E^\circ + P_{AB}$$
$$\Psi_2 = 0.707(\Psi_A{}^\circ - \Psi_B{}^\circ) \quad \text{with} \quad E_2 = E^\circ - P_{AB} \tag{3.8}$$

Equations (3.8) *must* be used when $E_A{}^\circ - E_B{}^\circ$ is smaller than P_{AB} (see Exercise 2, p. 33)

The two interacting orbitals are not degenerate
This is the more usual case. The product MOs are given by

$$\Psi_A = N\left(\Psi_A{}^\circ + \frac{P_{AB}}{E_A{}^\circ - E_B{}^\circ}\Psi_B{}^\circ\right) \quad \text{with} \quad E_A = E_A{}^\circ + \frac{P_{AB}{}^2}{E_A{}^\circ - E_B{}^\circ}$$

$$\Psi_B = N\left(\Psi_B{}^\circ + \frac{P_{AB}}{E_B{}^\circ - E_A{}^\circ}\Psi_A{}^\circ\right) \quad \text{with} \quad E_B = E_B{}^\circ + \frac{P_{AB}{}^2}{E_B{}^\circ - E_A{}^\circ} \tag{3.9}$$

These interactions are second order in P_{AB}, so they are weaker than those occurring between degenerate orbitals (first order in P_{AB}). The more stabilized Ψ_A, the easier is the reaction between A and B. To maximize this stabilization, the numerator $P_{AB}{}^2$ must be increased and/or the denominator $(E_A{}^\circ - E_B{}^\circ)$ decreased. Since Mulliken's approximation (p. 13) takes P_{AB} proportional to the overlap between $\Psi_A{}^\circ$ and $\Psi_B{}^\circ$, we can see that:

Rule. Reactions are facilitated when the frontier orbitals of the reagents are close in energy and when their overlap is large.

Except for some conformational studies, every example given in this course has been solved by applying this rule.

Three-orbital interactions
These are used to account for distortions in electron clouds. They are given by the equation

$$\Psi_1 = N\left[\Psi_1{}^\circ + \frac{P_{ln}}{E_l{}^\circ - E_n{}^\circ}\Psi_n{}^\circ + \frac{P_{ln}P_{mn}}{(E_l{}^\circ - E_n{}^\circ)(E_l{}^\circ - E_m{}^\circ)}\Psi_m{}^\circ\right] \tag{3.10}$$

The correct value of $\alpha - 1.108\beta$ is in excellent agreement with our E_3.

$$E_2 = (\alpha + \beta) + \frac{(0.566\beta)^2}{(\alpha + \beta) - (\alpha + 2\beta)} = \alpha + 0.680\beta$$

$$E_1 = (\alpha + 2\beta) + \frac{(0.566\beta)^2}{(\alpha + 2\beta) - (\alpha + \beta)} + \frac{(0.566\beta)^2}{(\alpha + 2\beta) - (\alpha - \beta)} = \alpha + 2.427\beta$$

The calculated values of E_2 and E_1 agree much less well with the correct values of $\alpha + 0.773\beta$ and $\alpha + 2.336\beta$ because we are approaching the point where second-order perturbation equations are no longer valid: the energy difference between π and φ_3 is β, while the corresponding resonance integral is 0.566β. Calculating the MOs using the two-orbital Equations (3.9) gives

$$\Psi_1 = N\left(\varphi_3 + \frac{P_{\varphi_3,\pi}}{E_{\varphi_3} - E_\pi}\pi + \frac{P_{\varphi_3,\pi^*}}{E_{\varphi_3} - E_{\pi^*}}\pi^*\right)$$
$$= N(\varphi_3 + 0.566\pi + 0.189\pi^*) = 0.23\varphi_1 + 0.46\varphi_2 + 0.86\varphi_3$$
correct solution: $0.16\varphi_1 + 0.36\varphi_2 + 0.91\varphi_3$

$$\Psi_2 = N\left(\pi + \frac{P_{\varphi_3,\pi}}{E_\pi - E_{\varphi_3}}\varphi_3\right) = 0.62(\varphi_1 + \varphi_2) - 0.49\varphi_3$$
correct solution: $0.74\varphi_1 + 0.57\varphi_2 - 0.37\varphi_3$

$$\Psi_3 = N\left(\pi^* + \frac{P_{\varphi_3,\pi^*}}{E_{\pi^*} - E_{\varphi_3}}\varphi_3\right) = 0.69(-\varphi_1 + \varphi_2) - 0.19\varphi_3$$
correct solution: $-0.66\varphi_1 + 0.73\varphi_2 - 0.19\varphi_3$

These results imply that the electron density on the central carbon is higher than that on the terminal carbon.[6] This makes no chemical sense and also disagrees with exact calculations. The error arises because we have ignored distortions of the molecular electron clouds. When the three-orbital correction term given in Equation (3.10) is added, Ψ_2 becomes

$$\Psi_2 = N\left[\pi + \frac{P_{\varphi_3,\pi}}{E_\pi - E_{\varphi_3}}\varphi_3 + \frac{P_{\varphi_3,\pi}P_{\varphi_3,\pi^*}}{(E_\pi - E_{\varphi_3})(E_\pi - E_{\pi^*})}\pi^*\right]$$
$$= N(\pi - 0.566\varphi_3 - 0.16\pi^*) = 0.71\varphi_1 + 0.51\varphi_2 - 0.49\varphi_3$$

Now the electron density is higher on C_1 than on C_2.

---

[6] The two lowest MOs are occupied. The C_2 coefficient is greater in Ψ_1, and the C_1 and C_2 coefficients are equal in Ψ'_2.

3.3.2 Qualitative Applications

The previous section was designed merely to provide some indication of the accuracy and limitations of perturbation methods. However, there is little point in doing approximate calculations by hand when the cheapest computers can give the correct results in seconds. The real utility of perturbation theory lies in qualitative applications. We will begin to explore its potential by asking, for example: (1) will an enol be more reactive exercise a given electrophile than ethylene? and (2) will an electrophile attack the enol at C_1 or at O_3?

Question (1) will be covered in detail in the next chapter, where we will look at relative reactivity. For the time being, we will merely state that the compound having the higher-lying HOMO will react preferentially. The HOMO of the enol is obviously Ψ_2, which is the ethylene HOMO (of energy $\alpha + \beta$) *destabilized* by the oxygen lone pair (of energy $\alpha + 2\beta$). We do not need to do calculations: when two orbitals interact, the lower is stabilized and the higher is destabilized. Thus, the enol is more reactive than ethylene. To answer question (2), we need to consider the coefficients of C_1 and O_3 in Ψ_2 (cf. pp. 96–102). Ψ_2 is an out-of-phase combination which mixes the π orbital with a *small* contribution from φ_3, so we can immediately conclude that the coefficient of C_1 is greater than that of O_3 and that they have opposed signs. Thus, the electrophile attacks preferentially at C_1.

Comparison of the Ψ_2 coefficients at C_1 and C_2 is only slightly more difficult. In the previous section, we saw that the difference between the coefficients at C_1 and C_2 is created by the admixture of π^* with π. Since

$$\pi = 0.707(\varphi_1 + \varphi_2) \quad \text{and} \quad \pi^* = 0.707(-\varphi_1 + \varphi_2)$$

and since the mixing coefficient of π with π^* is negative:

$$\frac{P_{\varphi_3,\pi} P_{\varphi_3,\pi^*}}{(E_\pi - E_{\varphi_3})(E_\pi - E_{\pi^*})} = \frac{(-)(-)}{(+)(-)} = -\mu \quad \text{with } \mu \text{ positive}$$

we can write

$$\Psi_2 = N\left[0.707(\varphi_1 + \varphi_2) - 0.707\mu(-\varphi_1 + \varphi_2) - \lambda\varphi_3\right]$$

So the mixture of π with π^* increases the coefficient at C_1 and diminishes it at C_2. Such simple arguments allow us to predict the results produced by computers, at least at a qualitative level. We can never evaluate the thousands, even millions, of integrals which are necessary for a Hartree–Fock calculation by hand, but it is satisfying to understand why the calculations *necessarily* give these results, rather than having to treat the computer simply as a black box.

Exercise 1 (E)

Which of the five drawings above represents the propene HOMO? The relative magnitudes of the coefficients are given by the symbols l, m and s for large, medium and small, respectively. Justify your choice *without* recourse to calculations.

Answer

In Hückel calculations, we represent the methyl group by a lone pair of energy $\alpha + \beta$. Propene contains therefore four electrons, so the HOMO is the second lowest in energy. Consequently, it must have a node and we can eliminate **B** immediately. Next, by regarding propene as a combination of a methyl group and ethylene, we can construct a perturbation scheme which is identical with Figure 3.3. Furthermore, Table 2.2 shows that the parameters for a methyl group and an oxygen lone pair are very similar, so the MOs of propene and an enol must be similar. Hence Ψ_2 should resemble π. This allows us to exclude **A** and **C** because they have large contributions from Me and π^*. We now have to choose only between **D** and **E**. The difference in the coefficients of the ethylene carbons results from the mixing of π with π^*. This increases the coefficient on the terminal atom and decreases it on the central atom, as we saw on p. 32. Hence **E** represents the propene HOMO.

* Exercise 2 (M) *For the relevant MOs, see p. 245*

Formulate fulvene as a combination of ethylene and butadiene and uses a perturbation approach to calculate its frontier orbitals.

Answer

Two disconnections are possible. However, (a) is clumsy, because it does not respect the symmetry of the system. Using scheme (b) saves a great deal of effort.

The atomic numbering scheme is given in (c). The ethylene MOs are denoted by π and π^* and those of butadiene and fulvene by Φ_i and Ψ_j, respectively. Orbitals Φ_2 and Φ_4 are antisymmetric with respect to the fulvene symmetry plane, whereas π, π^*, Φ_1 and Φ_3 are symmetrical. Consequently, the last four orbitals can interact during the union of the fragments, but the Φ_2 and Φ_4 orbitals will remain unchanged. The interaction diagram is shown below.

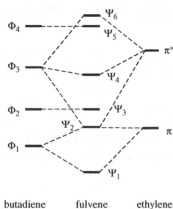

butadiene   fulvene   ethylene

Fulvene has six electrons, so the HOMO is the third from the bottom. The scheme suggests that this orbital is the butadiene Φ_2 MO. However, we need to ensure that Φ_1 does not perturb π enough to move it above Φ_2. Equation (3.9) shows that the change in the π orbital energy is

$$\Delta E = \frac{P_{\pi,\Phi_1}{}^2}{E_\pi - E_{\Phi_1}} + \frac{P_{\pi,\Phi_3}{}^2}{E_\pi - E_{\Phi_3}}$$

Since

$$P_{\pi,\Phi_1} = \langle \pi|\mathbf{P}|\Phi_1 \rangle = \langle 0.707(\varphi_5 + \varphi_6)|\mathbf{P}|0.37(\varphi_1 + \varphi_4) + 0.60(\varphi_2 + \varphi_3)\rangle$$
$$= 0.707 \times 0.37 \times (\langle\varphi_5|\mathbf{P}|\varphi_1\rangle + \langle\varphi_5|\mathbf{P}|\varphi_4\rangle) = 0.523\beta$$
$$P_{\pi,\Phi_3} = \langle \pi|\mathbf{P}|\Phi_3 \rangle = \langle 0.707(\varphi_5 + \varphi_6)|\mathbf{P}|0.60(\varphi_1 + \varphi_4) - 0.37(\varphi_2 + \varphi_3)\rangle$$
$$= 0.707 \times 0.60 \times (\langle\varphi_5|\mathbf{P}|\varphi_1\rangle + \langle\varphi_5|\mathbf{P}|\varphi_4\rangle) = 0.848\beta$$
$$\Delta E = \frac{0.523^2\beta^2}{(\alpha + \beta) - (\alpha + 1.618\beta)} + \frac{0.848^2\beta^2}{(\alpha + \beta) - (\alpha - 0.618\beta)} = 0.002\beta$$

correct value: 0

This confirms that the fulvene HOMO is indeed the butadiene Φ_2 orbital. Note that the destabilization of π by Φ_1 equals its stabilization by Φ_3. This seems surprising, given that the latter is further removed in energy, but there is a greater overlap with Φ_3 and the two effects cancel. Remember that two orbitals interact strongly if they are of similar energy *or* if their overlap is large. We can now turn to the LUMO, which is derived from Φ_3. If we take the energy of Φ_3 and correct for interaction with π and π^* using Equation (3.9), then

$$E(LU) = E(\Phi_3) + \frac{P_{\Phi_3,\pi}{}^2}{E(\Phi_3) - E(\pi)} + \frac{P_{\Phi_3,\pi^*}{}^2}{E(\Phi_3) - E(\pi^*)}$$

$$= (\alpha - 0.618\beta) + \frac{(0.848)^2}{(\alpha - 0.618\beta) - (\alpha + \beta)} + \frac{(0.848)^2}{(\alpha - 0.618\beta) - (\alpha - \beta)}$$

$$= (\alpha - 0.618\beta) - 0.444\beta + 1.882\beta = \alpha + 0.820\beta$$

This result is absurd because it places the LUMO at lower energy than the HOMO (energy $\alpha + 0.618\beta$). Obviously, the error comes from the enormous 'correction' term of 1.882β. Closer inspection shows that the energy differvence between Φ_3 and π^* is only 0.382β and their resonance integral is 0.848β, so we have violated the criterion given in remark 2 (p. 27). Hence the above second-order perturbation approach is not valid and we have to treat the orbitals Φ_3 and π^* as if they were degenerate. First-order perturbation equations give a much more reasonable result:

$$E(LU) = (\alpha - 0.618\beta) - 0.444\beta + 0.848\beta = \alpha - 0.214\beta$$

exact value: $\alpha - 0.254\beta$

We can now move on to calculate the LUMO coefficients. Equations (3.9) and (3.10) need correction for the quasi-degeneracy of orbitals Φ_3 and π^*. The resulting equation (shown below) can be analysed as follows. The first two terms are the principal components of the LUMO Ψ_4, the third the mixing of Φ_3 with π, the fourth the mixing of Φ_3 with Φ_1 through π^* (note how the mixing coefficient has been modified) and the last term its mixing with Φ_1 through π. Had we used disconnection (a), we would have to add four extra terms to describe the interactions of Φ_3 with Φ_2 and Φ_4 through π and π^*. The calculation would have been much more difficult and, being approximate, would not have reproduced the symmetry of fulvene (equality of the coefficients in 1 and 4, and in 2 and 3).

$$N\left[\Phi_3 + \pi^* + \frac{P_{\Phi_3,\pi}}{E_{\Phi_3} - E_\pi}\,\pi + \frac{0.848 P_{\Phi_1,\pi^*}}{E_{\Phi_3} - E_{\Phi_1}}\Phi_1 + \frac{P_{\Phi_3,\pi}P_{\Phi_1,\pi}}{(E_{\Phi_3} - E_{\Phi_1})(E_{\Phi_3} - E_\pi)}\Phi_1\right]$$

$$= N(\Phi_3 + \pi^* - 0.524\pi - 0.198\Phi_1 + 0.123\Phi_1)$$

$$= 0.37(\varphi_1 + \varphi_4) - 0.27(\varphi_2 + \varphi_3) + 0.23\varphi_5 - 0.73\varphi_6$$

The correct coefficients are 0.35 for 1 and 4, -0.28 for 2 and 3, 0.19 for 5 and -0.75 for 6. The orbital coefficients are now fairly correct.

3.4 The Dewar PMO Method

3.4.1 Alternant Hydrocarbons

Definition

*A conjugated hydrocarbon is **alternant** if its carbon atoms can be divided into two classes: **starred** and **non-starred**. No two atoms belonging to the same class can be linked directly.*

If the two classes contain different numbers of atoms, by convention the starred is the more numerous class. Compounds **1–4** are alternant, **5–7** are not. Note that:

- all linear polyenes are alternant;
- any compound containing a ring having an odd number of atoms is non-alternant.

Hückel MOs of alternant hydrocarbons have some noteworthy properties, outlined below.

The Pairing Theorem

The energy levels in an alternant hydrocarbon are symmetrical with respect to α. Therefore, any orbital Ψ_p having an energy $\alpha + x_p \beta$ has a corresponding orbital Ψ_{-p} of energy $\alpha - x_p \beta$. The coefficients of the starred atoms are identical in Ψ_p and Ψ_{-p}; for the non-starred atoms, they are equal but opposite in sign.

This theorem gives rise to some extremely useful corollaries:[7]

1. The nonbonding orbitals are paired with themselves, so the coefficients of their non-starred atoms are zero (they are their own opposites).
2. In any nonbonding molecular orbital (NBMO), the sum of the coefficients of the atoms s adjacent to a given atom r is zero. Hence the NBMO coefficients can be calculated very easily. The benzyl radical provides a nice example. All of the non-starred atoms have coefficients of zero. We give the *para* atom an arbitrary coefficient a. The sum of the coefficients of the atoms adjacent to the *meta* carbons must be zero, so the *ortho* coefficients are $-a$. For the sum of the coefficients around the *ipso* carbon to cancel, the benzylic carbon coefficient must be $2a$. The value for a is given by the normalization condition:

$$a^2 + a^2 + a^2 + 4a^2 = 1$$

3. In a neutral alternant, the π charge is 1 electron on every atom (i.e. the net charge is zero). In an ion, the charge is also 1 for non-starred atoms. For a starred atom r having NBMO coefficient c_{or}, the electronic charge is $(1 + c_{or}^2)$ in an anion and $(1 - c_{or}^2)$ in a cation.
4. The bond order p_{rs} between two atoms of the same class is always zero.

Exercise 3 (D)   *For the relevant MOs, see p. 246*

Why is azulene blue whereas naphthalene and anthracene are colorless?

Hints: The assumption that smaller HOMO–LUMO gaps will produce more strongly colored products is true to a first approximation, but not good enough for this problem. In the compounds under study, the HOMO–LUMO gap is largest in naphthalene, which could well explain its lack of color. However, the gap is very similar in anthracene and azulene. To understand their color differences, we need a much more accurate estimate of their excitation energies.

For the calculations, we start by formally removing an electron from the HOMO, which requires an energy equal to the first ionization potential. We then replace the electron in the LUMO, which releases an energy corresponding to the electron affinity of the ionized

---

[7] For a demonstration of these theorems, see: (a) Salem L., *The MO Theory of Conjugated Systems*, Benjamin, New York, 1966, p. 36; (b) Dewar M. J. S., *The MO Theory of Organic Chemistry*, McGraw-Hill, New York, 1969, p. 199.

molecule. This is always greater than the electron affinity of the neutral molecule, particularly when the HOMO and LUMO are not localized in the same regions of space, because the repulsion between the unpaired electrons is then smaller. This increased electron affinity may cause absorption in the visible rather than in the UV region.

Answer

The pairing theorem requires that the HOMOs and LUMOs of the alternant molecules anthracene and naphthalene are localized on the same atoms; this is not the case for the nonalternant azulene. For the detailed calculations concerning the color-determining interelectronic repulsion, see the paper by Michl and Thulstrup.[8]

Exercise 4 (M)

Find the NBMO coefficients in the following alternant hydrocarbons:

8   **9**   **10**

Answer

In **8**, we assign a value of a to the coefficient at atom 1. The sum of the coefficients around 2 is zero, so the coefficient at 3 is $-a$. The sum of the coefficients around 4 is again zero, but this does not allow us to define values for 5 and 9 unequivocally. Let us give 5 an auxiliary value b, which immediately imposes $-b$ for 7. If we zero around 4, the value of 9 is $a - b$; if we zero around 8, it is b. Equating these values gives $a = 2b$. Then, applying the normalization condition:

$$4b^2 + 4b^2 + b^2 + b^2 + b^2 = 1$$

we find that $b = 0.30$.

Note that if we start from an atom within the ring instead of position 1, the auxiliary b is not needed.

In **9**, the non-starred atom 9 has a coefficient of zero. The second corollary requires that the coefficient of 8 also be zero. Hence we can ignore the vinyl group completely, and are back to the case of the benzyl radical.

---

[8] Michl J., Thulstrup E. W., *Tetrahedron*, 1976, **32**, 205.

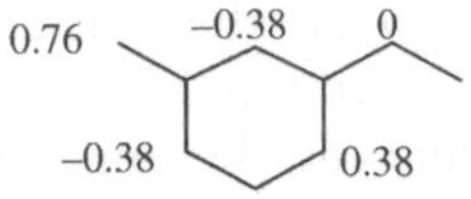

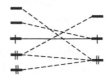

Compound **10** is another case where we need an auxiliary variable. If we start with atom 1, we can deduce the coefficient at 11. The variable b is assigned to the central carbon (not numbered in the diagram) to allow us to calculate the values for 9, 7, and 5. We then obtain the value for 3 by zeroing the coefficients around either 2 or 4. Equating these two values gives $b = 0$.

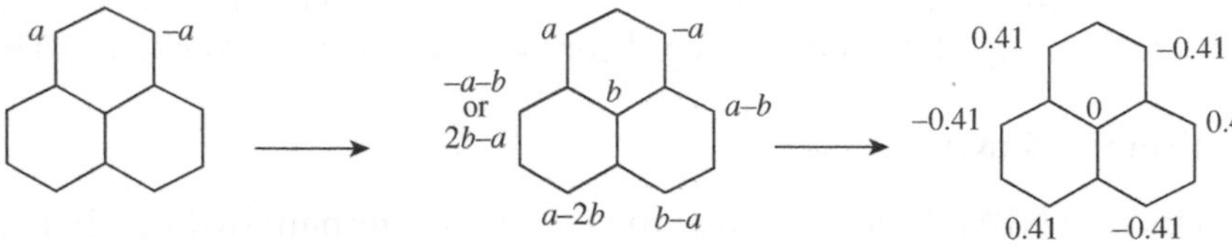

3.4.2 The Dewar PMO Method

The Dewar PMO (perturbational MO) method[9] avoids calculations by making clever use of alternants. The idea is to divide a molecule formally into two *alternant radicals,* whose recombination is studied. When the radicals are different, only their NBMOs lie at the same energy (Figure 3.4). Hence their interaction provides almost all of the recombination energy. The other interactions, which are second order in P_{AB}, can then be neglected. We will illustrate the method by deriving the *aromaticity rules.*

Figure 3.4 Interaction of two alternant radicals. Only the nonbonding orbitals give a first-order interaction (unbroken line).

Definition

An annulene is said to be aromatic, nonaromatic or antiaromatic if its π system is more stable, equally stable or less stable than the corresponding open chain polyene, respectively.

The first corollary on p. 36 states that the even-numbered atoms in a linear conjugated radical all have NBMO coefficients of zero. For odd-numbered atoms, the coefficients are the same in size, but alternate in sign. Thus, the coefficients at the termini are the same for radicals having $4n + 1$ atoms, but opposite in sign for radicals containing $4n - 1$ atoms.

A $4n + 2$ atom conjugated polyene and annulene can both be built from a monoatomic radical and a radical having $4n + 1$ atoms.[10] Diagrams **11** and **12** show the combinations which give the polyene and the annulene, respectively. The overlap in **12** is twice

---

[9] Dewar M. J. S., *J. Am. Chem. Soc.,* 1952, **74,** 3341, 3345, 3350, 3353, 3357. To be honest, these papers are rather indigestible. It is better to read instead Chapter VI of ref. 7b.

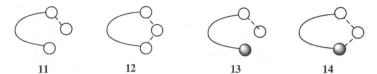

11	**12**	**13**	**14**

as large as in **11**, so the $4n + 2$ annulene is more stable than its polyene homolog. This annulene is thus aromatic. In diagrams **13** and **14**, we see the combinations of a monoatomic radical with the $4n - 1$ radical to give a $4n$ polyene and a $4n$ annulene, respectively. Here, the net overlaps are positive for **13** and zero for **14**, so we can see that the $4n$ annulene is antiaromatic. Finally, an annulene radical can be made by forming a bond between the two ends of the corresponding open-chain radical. Since these atoms are starred, their bond order is zero (corollary 4). Thus, cyclization has no effect on the π energy of the system and the annulene radical is nonaromatic.

Let us look at a radical having $4n + 1$ atoms. The termini have the same NBMO coefficients, so cyclization will cause them to overlap in-phase. The electron in the NBMO will then be stabilized. Now, let us pass from the $4n + 1$ electron radical to the corresponding $4n + 2$ electron anion by adding an electron. Such an electron, nonbonding in the open system, becomes bonding in the cyclized compound. Thus the anionic $4n + 2$ electron annulene is aromatic. The cationic $4n$ electron annulene, formed by removing a bonding electron from the nonaromatic annulene radical, is antiaromatic. A similar rationale starting with a $4n - 1$ atom chain shows that a $4n$ anionic annulene is antiaromatic and a $4n + 2$ cationic annulene is aromatic.

Hückel Rule

Annulenes are aromatic if they have $4n + 2$ electrons, irrespective of their charge. Those having $4n$ electrons are antiaromatic. Annulene radicals are nonaromatic.[11]

Let us now examine the stability of a $4n + 2$ annulene having the topology of a Möbius strip.[12] This annulene can be obtained by uniting a monoatomic radical with a $4n + 1$ conjugated radical which is given a half-turn (180°) before the union. This is equivalent to saying that, in this union, the $4n + 1$ radical reacts in an antarafacial manner (see arrows in the left drawing). In its nonbonding MO (NBMO), after the half-turn, the coefficients on the termini have coefficients with opposite signs (arrows on the right drawing):

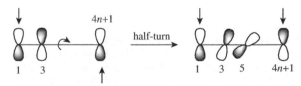

The overlap of this twisted radical with a monoatomic radical to give a twisted polyene is positive. It is zero for an annulene (Figure 3.5). A Möbius $4n + 2$ annulene is

---

[10] In a conjugated hydrocarbon, each carbon atom provides one π electron. Thus, for neutral systems, the number of atoms and the number of π electrons are identical.

[11] In 1930, when Hückel first derived his rule, he considered only aromatic annulenes. Antiaromatic and nonaromatic systems are extensions introduced by later authors, in particular by Dewar.

[12] Heilbronner E., *Tetrahedron Lett.*, 1964, 1923.

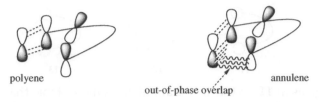

polyene   annulene

out-of-phase overlap

Figure 3.5   Formal unions giving a Möbius $4n + 2$ polyene and a Möbius $4n + 2$ annulene.

thus antiaromatic. A similar argument shows that a Möbius $4n$ annulene is aromatic. Obviously, if we give the conjugated radical an even (odd) number of half-turns, the termini in its NBMO have coefficients with the same (opposite) signs and the resulting $4n + 2$ annulene is aromatic (antiaromatic). Therefore, we can state the following.

Generalized Aromaticity Rules

A $4n + 2$ annulene is aromatic if it contains no or an even number of half-turns and antiaromatic if it contains an odd number of half-turns. A $4n$ annulene is aromatic if it contains an odd number of half-turns and antiaromatic if it contains no or an even number of half-turns.

Generalizing an idea first put forward by Evans,[13] Dewar and Zimmerman independently proposed a very elegant treatment of pericyclic reactions, based on aromaticity rules.[14] In a *pericyclic reaction*, all *intervening atoms*[15] form a single ring in the transition state and use only one AO[16] in the bond-making/bond-breaking process. If all intervening atoms are assimilated to carbons, then a pericyclic transition state resembles an annulene (the secular determinant is the same, only the β integrals may have different values).[17] Obviously, when the annulene is aromatic, the isoconjugate pericyclic transition state is stabilized and the reaction is allowed. When the annulene is antiaromatic, the isoconjugate transition state is high in energy and the reaction is forbidden.

Dewar–Zimmerman Rule

A thermal pericyclic reaction is allowed when its transition state is aromatic and forbidden when it is antiaromatic.

---

[13] Evans M. G., *Trans. Faraday Soc.*, 1939, **35**, 824.

[14] (a) Dewar M. J. S., *Tetrahedron Suppl.* 8, Part I, 1966, **75**; (b) Dewar M. J. S., Dougherty R. C., *The PMO Theory of Organic Chemistry*, Plenum Press, New York, 1975, pp. 106, 338; (c) Zimmerman H. E., *Angew. Chem. Int. Ed. Engl.*, 1969, **8**, 1; (d) Zimmerman H. E., in *Pericyclic Reactions*, Lehr R. E., Marchand A. P. (Eds), Academic Press, New York, 1977, Vol. 1, p. 53; see also ref. 2, p. 182 and ref. 7b.

[15] We define as *intervening atom* an atom extremity of a bond made or broken in the reaction.

[16] Thus hydroboration is *not* a pericyclic reaction, because the boron atom makes use of two AOs. Similarly, the reaction between a carbene and a double bond is *not* pericyclic because the carbon atom uses two AOs. Cheletropic reactions are not pericyclic either. Consider, for example, the fragmentation of 3-cyclopentenone to give CO and butadiene. Not only does the expelled carbon atom use two AOs to bond with its neighbours, but also the oxygen atom, which is an intervening atom (it was initially linked to the carbon atom by a double bond, which becomes a triple bond in CO) is exocyclic in the transition state.

[17] Two molecules having the same secular determinant (if all the atoms in conjugation are replaced by carbon atoms) are said to be '*isoconjugate*'.

Exercise 5 (D)

(1) Find the NBMO coefficients for methylenecyclobutadiene **15**.
(2) Hence, deduce that the reaction of X^+ with cyclobutadiene leads to an addition rather than a substitution. Prove the opposite for benzene.

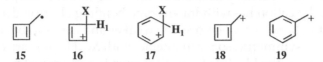

15   **16**   **17**   **18**   **19**

Hints: Either (a) compare the Wheland σ complexes **16** and **17** with the alternant cations **18** and **19** (by treating XH_1 as a single 'superatom') and then show that H_1^+ is extruded more easily from **19** than **18**; or (b) disconnect **16** and **17** into XC_1H_1 and an allyl or pentadienyl cation, respectively. The XC_1 and C_1H_1 bonds in the XC_1H_1 fragment give rise to four orbitals: $\sigma(XC_1)$, $\sigma^*(XC_1)$, $\sigma(C_1H_1)$ and $\sigma^*(C_1H_1)$. Then recombine the two fragments. The main interaction involves the LUMO of the cation and the HOMO of XC_1H_1, which is the antibonding combination **20** of $\sigma(XC_1)$ with $\sigma C_1H_1)$.[18] Note that **20** has the correct symmetry to interact with the π orbitals (cf. **21**). Finally, show that this interaction facilitates the extrusion of H_1^+ from **17** but not from **16**.

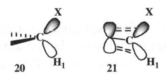

20 H_1 **21** H_1

Answer[19]

(1) Beginning with a starred ring atom, we find the coefficients easily:

with $a = 1\sqrt{2}$. Starting with the exocyclic atom requires an auxiliary variable:

(2) *First method*. The hydrogen H_1 will be lost (as a proton) much more easily if it is highly charged. The third property of alternant hydrocarbons (p. 36) indicates that the charge of the exocyclic atom in **18** is zero. Therefore, complex **16** can only lose a proton with difficulty, and additions are favored. The opposite occurs in **19** because the exocyclic atom has a large positive charge, so substitutions proceed readily.

---

[18] For more detailed explanations of how to build MOs from the orbitals of the fragments, see: Jorgensen W. L., Salem L., *The Organic Chemist's Book's of Orbitals*, Academic Press, New York, 1973; Jean Y., Volatron F., *An Introduction to Molecular Orbitals*, Oxford University Press, Oxford, 1993.
[19] Dixon W. T., *Chem. Commun.*, 1969, 559.

Second method. The interaction of the cation LUMO with the XC_1H_1 HOMO generates two orbitals. Only the bonding combination, which strongly resembles the HOMO but contains a small contribution from the LUMO, is occupied. Therefore, the two electrons which belonged to the HOMO of the XC_1H_1 fragment are now in an orbital which is partially delocalized on to the cationic portion in the molecule. In other words, the cation has withdrawn some bonding electron density from the XC_1H_1 fragment, the bonds of which become weaker and easier to break. The pentadienyl cation LUMO is symmetric, so it can interact with XC_1H_1 and assist the extrusion of H_1^+ from **17**. However, the LUMO of the allyl cation is antisymmetric; it cannot interact with the HOMO of XC_1H_1, so the C_1H_1 bond is not weakened.

Exercise 6 (E)

In the 1,5-sigmatropic transposition in cyclopentadiene, the experimentally observed migratory aptitude is $SiMe_3 \gg H > Me$. Explain.

Answer[20]

Due to the aromaticity of Cp^-, the transition state should not be modeled by two interacting radicals. A better model would be the ion pair $[X^+ \cdots Cp^-]$. The ease of migration reflects the capacity of X to accommodate a positive charge.

Exercise 7 (E)

Predict the stereochemistry of the following reaction:

Answer

This thermal pericyclic reaction involves six electrons and should takes place via a Hückel aromatic transition state. The following scheme shows that the ring closure must be disrotarory. Therefore, the phenyls are *cis* to each other.

---

[20] Kahn S. D., Hehre W. J., Rondan N. G., Houk K. N., *J. Am. Chem. Soc.*, 1985, **107**, 8291.

Exercise 8 (M)

The selection rules for sigmatropic rearrangements were derived for open-chain systems. *A priori*, it is not obvious that they also apply to cyclic systems. Using the PMO method, show that the following sigmatropic migrations[21] are allowed:

Answer

Neglecting the methyl substituent and assimilating the migrating hydrogen to a carbon atom, their respective transition states are isoconjugate with **A** and **B**:

The bold lines in **A** and **B** indicate the skeleton of the parent compound. The remaining atom is the migrating hydrogen. In the transition state, it overlaps with both the atom which it leaves and the atom to which it will bind. The only difference between **A** and benzene is an additional bond between two atoms of the same starred class: 1 and 5. The fourth corollary (p. 36) states that this bond does not modify the π energy, so **A** is aromatic. **B** must also be fairly aromatic, because it incorporates one six-membered aromatic ring and one five-membered non-aromatic ring. Treating **B** as the result of the union between a monoatomic radical and a seven-atom (branched) radical allows this conclusion to be justified more rigorously. **C** shows clearly that the interaction of their NBMOs is favorable, so **B** is aromatic.

3.4.3 Advantages and Disadvantages of the PMO Method

Advantages

1. In the PMO method, we need only to examine one interaction. In the FMO approach, two are sometimes necessary. The PMO method has firm theoretical foundations because the only first-order interaction is retained; all the neglected interactions are second order in P_{AB}.
2. The method is very simple to use. We simply optimize the stabilizing NBMO interaction.

[21] McLean S., Haynes P., *Tetrahedron Lett.*, 1964, 2385; Egger K. W., *J. Am. Chem. Soc.*, 1967, **89**, 3688; 1968, **90**, 1.

3. It requires little effort. The second corollary (p. 36) gives the NBMO's coefficients directly, so we do not have to calculate the other MOs. We can also take advantage of the many remarkable properties of alternants.

Whenever it is applicable, PMO is fast and elegant. Obviously, such simplicity has a price.

Disadvantages

1. PMO methods can be easily applied only if the system can be modeled as a hydrocarbon, and can be disconnected into two alternant radicals. It cannot be used to determine the site of electrophilic attack on azulene, for example. Such a problem can easily be solved using the frontier orbital approximation; see p. 119.
2. The PMO method makes enormous approximations. For example, carbons are used to model heteroatoms.[22] Substituents have to be either removed altogether, or modeled using (doubly filled or empty) p orbitals. Such a simplified treatment has obvious limitations in chemical terms. Although it is *possible* to employ an allyl anion as a model for an enolate, it is not particularly desirable: we lose all understanding of the difference between the reactivity at C and O.
3. A system having an odd number of atoms cannot be formulated as a combination of two radicals. Nor can some even-numbered systems. For instance, Bertran's elegant method for treating the regioselectivity of Diels–Alder reactions[23] fails if the diene and the dienophile are both disubstituted.

3.5 To Dig Deeper

1. Dewar M. J. S., Dougherty R. C., *The PMO Theory of Organic Chemistry*, Plenum Press, New York, 1975.
   An excellent introduction to applied quantum chemistry.
2. Albright T. A., Burdett J. K., Whangbo M. H., *Orbital Interactions in Chemistry*, John Wiley & Sons, Inc., New York, 1985.
   How to apply perturbation theory to organometallic chemistry.
3. Jorgensen W. L., Salem L., *The Organic Chemist's Book's of Orbitals*, Academic Press, New York, 1973.
   This book gives three dimensional perspective drawings of the MOs of 104 simple molecules. The first 50 pages explain how to build the MOs of complicated molecules from simpler fragment orbitals.
4. Jean Y., Volatron F., *An Introduction to Molecular Orbitals*, Oxford University Press, Oxford, 1993.
   Even more detailed than Jorgensen and Salem. It might seem masochistic to calculate approximate MOs by hand when precise computer-generated MOs can be obtained in

---

[22] To take the heteroatoms explicitly into account, correction terms have to be calculated and PMO method is then rather unwieldy.
[23] Bertran J., Carbo R., Moret T., *Ann. Quim.*, 1971, **67**, 489. See p. 88.

seconds (for Hückel) or a few minutes (*ab initio*), but there are two good reasons for making the effort. First, it allows computer output to be checked. Second, knowledge of how fragments combine provides a framework for understanding how two reagents will interact in the transition state.

5. Hoffmann R., *Acc. Chem. Res.*, 1971, **4**, 1.

The seminal paper on through-space and through-bond interactions.

4 Absolute and Relative Reactivities

Rule 1 *A reaction having zero frontier orbital overlap – i.e. the HO(1)–LU(2) and HO(2)–LU(1) interactions both equal zero – is forbidden.*

Rule 2 *A reaction becomes easier as the energy gap between the interacting frontier orbitals decreases.*

Forbidden reactions

Don't be misled by the name. When we state that the thermal disrotatory ring opening of cyclobutene is forbidden, we only imply that the transition state energy is 12–15 kcal mol^{-1} higher for the disrotatory than for the conrotatory process.[1] If this difference is reduced (by stabilizing the disrotatory or destabilizing the conrotatory transition state) or if large amounts of energy are made available to the molecule, then the selection rules may be violated. Forbidden reactions can occur, just as forbidden transitions can be observed in spectroscopy.[1]

Relative reactivity

For an ionic reaction, rule 2 becomes: a nucleophile (electrophile) reacts preferentially with the reaction partner having the lowest-lying LUMO (highest-lying HOMO). The validity of this rule is examined on p. 75

---

[1]Brauman J. I., Golden D. M., *J. Am. Chem. Soc.*, 1968, **90**, 1920; Lupton E. C. Jr, *Tetrahedron Lett.*, 1968, 4209; Doorakian G. A., Freedman H. H., *J. Am. Chem. Soc.*, 1968, **90**, 5310, 6896; Dahmen A., Huisgen R., *Tetrahedron Lett.*, 1969, 1465; Brauman J. I., Archie W. C. Jr, *J. Am. Chem. Soc.*, 1972, **94**, 4262.

Frontier Orbitals Nguyên Trong Anh
© 2007 John Wiley & Sons, Ltd

4.1 Absolute Reactivity

A chemical system must overcome an energetic barrier before it can react. Energy is required to compensate for the mutual repulsion of the electron clouds in the approaching reagents and to cleave the bonds which are broken in any reaction. Stabilizing interactions will diminish this energy barrier and promote the reaction. Such stabilizing interactions are negligible when the frontier overlap is zero, so the reaction will then be disfavored. This is the physical basis of rule 1. It is very easy to apply this rule if we limit ourselves to *thermal reactions involving closed-shell molecules.* The difficulties which may be encountered during photochemical or radical reactions are discussed on p. 110 and in Chapter 8.

4.1.1 Bimolecular Reactions

Cycloadditions

Consider the dimerization of ethylene. The frontier orbitals having different symmetries (**1**), their overlap is zero and the reaction is forbidden. In the Diels–Alder reaction, the FOs have the same symmetry, their overlap is positive (**2** and **3**) and the reaction becomes allowed.

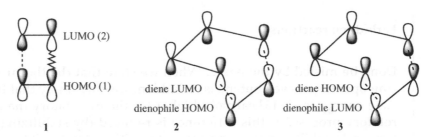

Consider now a $p + q$ cycloaddition. The *supra–supra* reaction will be thermally allowed if the HOMO (LUMO) of the p component has the same symmetry as the LUMO (HOMO) of the q component. According to Coulson's equations (p. 18), if p has $4k + 2$ electrons, its HOMO will be the $(2k + 1)$th MO, odd-numbered, therefore symmetrical. For the LUMO of the q component to be also symmetrical, its must be odd-numbered, which requires q to have $4k'$ electrons [Figures 4.1 (**1**) and 4.2a]. The total electron number is then $p + q = (4k + 2) + 4k' = 4n + 2$ electrons. If the HOMO (p) and the LUMO (q) are both antisymmetric (Figure 4.2b), then we must have $p = 4k$ and $q = 4k' + 2$. Again, $p + q = 4n + 2$.

Obviously, an *antara–antara* reaction is allowed whenever the corresponding *supra–supra* reaction is, i.e. when the FO have the same symmetry (compare Figures 4.2a and 4.2d). When the FO of the two partners have different symmetry, the *supra–supra* reaction is forbidden (Figure 4.2c) but the *supra–antara* reaction is allowed (Figure 4.2e). An analogous reasoning shows that the *supra–antara* cycloaddition is allowed when the total number of electrons is $p + q = 4n$ [Figure 4.1 (**2**)].

To summarize: *thermal supra–supra or antara–antara cycloadditions are allowed for systems having 4n + 2 electrons and supra–antara or antara–supra cycloadditions for systems having 4n electrons.*

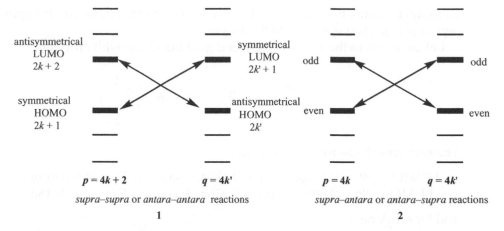

Figure 4.1 Electron number, orbital symmetry and cycloaddition stereochemistry.

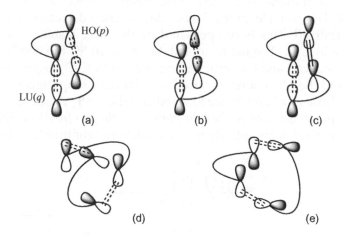

Figure 4.2 Frontier orbital interactions in cycloadditions.

Justification of the Frontier Orbital Approximation

The two-electron stabilization for non-degenerate orbitals being $2P_{ij}^2/(E_i - E_j)$, the smaller the $E_i - E_j$ gap, the greater is the gain in energy, provided that the numerator $2P_{ij}^2$ does not decrease by a larger quantity. If this condition is satisfied, the HOMO–LUMO interactions are the largest. Even then, Fukui's approximation seems mathematically indefensible. It takes into account only the two most important interactions and neglects all the others, even though they are of the *same order of magnitude*. Each of them is individually smaller than the frontier interactions, but how can we be sure that their sum is not larger? Nonetheless, checked repeatedly since 1965, frontier orbital theory has been found to be astonishingly reliable. This apparent contradiction has been explained many years ago by Fukui: *the FO approximation applies to transition states, not to the reagents.*[2]

Let us return to the Diels–Alder reaction of butadiene with ethylene:

The frontier orbitals for butadiene are:

HOMO: $\Psi_2 = 0.60(\varphi_1 - \varphi_4) + 0.37(\varphi_2 - \varphi_3)$ $E_2 = \alpha + 0.618\beta$
LUMO: $\Psi_3 = 0.60(\varphi_1 + \varphi_4) - 0.37(\varphi_2 + \varphi_3)$ $E_3 = \alpha - 0.618\beta$

and for ethylene:

HOMO: $\pi = 0.707(\varphi_5 + \varphi_6)$ $E_\pi = \alpha + \beta$
LUMO: $\pi* = 0.707(\varphi_5 - \varphi_6)$ $E_{\pi*} = \alpha - \beta$

As the reaction progresses, the C_1C_2, C_3C_4 and C_5C_6 bonds lengthen whereas the C_2C_3 bond shortens. These geometric changes induce variations of the π energies (Figure 4.3). For example, in Ψ_1, the bonding overlap decreases in C_1C_2 and C_3C_4: the MO is destabilized. This is compensated by the large stabilization due to the increased bonding in C_2C_3. The net result will be small. In Ψ_2, the bonding overlap in C_1C_2 and C_3C_4 is reduced and the antibonding overlap in C_2C_3 augmented. The effects are additive and the MO energy is raised. Similar reasoning suggests that the energy of π is raised, those of $\pi*$ and Ψ_3 are lowered and that of Ψ_4 varies very little. Therefore, *as the reaction progresses exercise the transition state,*[3] *the four frontier orbitals, and only them, tend to become quasi-degenerate. Ab initio calculations*[4] confirm this frontier gap narrowing.

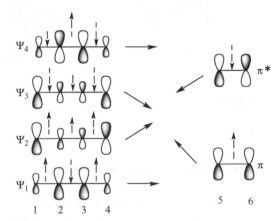

Figure 4.3 The evolution of the orbital energies during the Diels–Alder reaction. Dashed arrows indicate the partial energy variations resulting from the geometric changes as the reaction progresses. The full arrows show the changes in the total energy of each MO.

[2]Fukui K., Fujimoto H., *Bull. Chem. Soc. Jpn.*, 1969, **42**, 3392; Fukui K., *Theory of Orientation and Stereoselection*, Springer, Berlin, 1975, 25.
[3]The same argument can be made for the cycloreversion reaction.
[4]Townshend R. E., Ramunni G., Segal G., Hehre W. J., Salem L., *J. Am. Chem. Soc.*, 1976, **98**, 2190.

We have justified the FO approximation for the Diels–Alder reaction. It is clear, however, that the proof is general. The HO–LU interaction gives rise to two combinations; only the lower one (HO + λLU) is occupied:

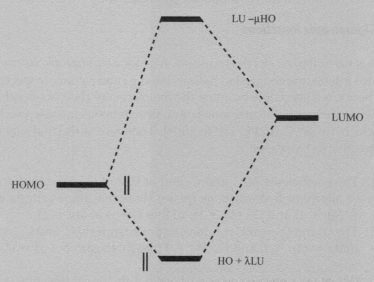

Physically, this means that electrons have been withdrawn from the HOMO to populate the LUMO. Therefore, the bonds which were initially bonding in the HOMO are weakened and those antibonding strengthened. These two effects are both destabilizing and the HOMO energy is thus raised. Conversely, the bonds which were bonding in the LUMO are strengthened and those antibonding weakened (pairing theorem, p. 36). The LUMO is then lowered and consequently the FOs approach nearer to each other.

Moreover, it can be shown[5] that the frontier coefficients increase at the reaction centers, thus enforcing the role of frontier interactions.

Hence Fukui's method is justified *in the transition state*, the FOs then being practically degenerate, with increased frontier coefficients at the reaction centers. All the other interactions are an order smaller.[6] Unfortunately, transition state orbitals cannot be obtained easily,[7] so we generally substitute the MOs of the reagents instead. This supplementary approximation,[8] introduced for practical reasons, occasionally gives results which appear to violate the frontier orbital theory. We will examine such cases in more detail later (pp. 62, 76–78).

[5]Fukui K., Koga N., Fujimoto H., *J. Am. Chem. Soc.*, 1981, **103**, 196.
[6]This may be not true for aromatics and other highly conjugated systems. The reason is that these molecules have many MOs, which are so close in energy that some of them, in particular the nearest neighbours of the FO, do not move away sufficiently and will give rise, in the transition state, to interactions of the same order of magnitude as the frontier interactions. Examples are given on pp. 113 and 120–122. The counter-examples given as 'evidence' of Fukui theory's 'inanity' generally belong to this category. See Dewar M. J. S., *Theochem*, 1989, **200**, 301.
[7]It would be pointless anyhow to make approximate frontier orbital calculations, when the transition state is already known.
[8]Which justifies the approximation $P_{ii} = 0$ (p. 26).

4.1.2   Unimolecular Reactions

For these processes, we formally decompose the molecule into two fragments and treat their recombination as a bimolecular reaction.

Sigmatropic Reactions

It is natural to model the transition state for a sigmatropic rearrangement of order (i,j) as two interacting conjugated radicals having i and j atoms, respectively. According to PMO theory, we need only examine the interaction of their nonbonding orbitals (Figure 4.4). To generate a stabilizing interaction, simultaneous inphase overlaps between C_1 and C_1' and between C_i and C_j are required. Therefore, a thermal *supra–supra* reaction will be possible in two cases:

1. The coefficients of C_i and C_1 and of C_j and C_1' have the same sign. The properties of alternant hydrocarbons (p. 36) show that these criteria will be satisfied when $i = (4p + 1)$ and $j = (4q + 1)$, so that $(i + j) = (4n + 2)$.
2. The C_i and C_j coefficients have opposite signs to C_1 and C_1', respectively. This implies that $i = (4p - 1)$ and $j = (4q - 1)$, so that again $(i + j) = (4n + 2)$.

The *antara-antara* reaction is allowed if the upper lobe of C_i has the same sign as the lower lobe of C_j, i.e. whenever the corresponding *supra–supra* reaction is permitted. A similar analysis shows that the thermal *supra–antara* (or *antara–supra*) reaction will be allowed when one of the pairs (C_1, C_i) and (C_1', C_j) has coefficients of the same sign and the other has coefficients which are opposed. This occurs when $(i + j) = 4n$. To summarize: *thermal supra–supra or antara–antara sigmatropic reactions are allowed for systems having 4n+2 electrons and supra–antara or antara–supra reactions for systems having 4n electrons.*

Electrocyclic Reactions

Consider the electrocyclic ring-opening reaction of cyclobutene. The molecule is formally divided into two fragments: the double bond and the single σ bond which is cleaved.[9] The frontier orbital interactions (σ, π^*) and (σ^*, π) relevant to the conrotatory and disrotatory reactions are given in diagrams **4**, **5**, **6** and **7**, respectively. The net overlap is positive for **4** and **5**, but zero for **6** and **7**. The conrotatory process is therefore allowed, and the disrotatory process forbidden.

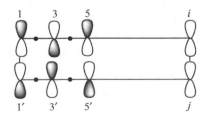

Figure 4.4   *Supra–supra* interaction of the nonbonding orbitals in two radicals i and j.

---

[9]Salem L., *Chem. Bri.*, 1969, **5**, 449.

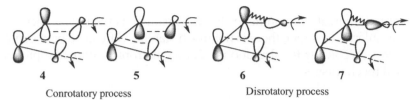

4　　　　5	6　　　　7
Conrotatory process	Disrotatory process

We have seen from Coulson's equations (p. 18) that for conjugated polyenes, the odd-numbered MOs are symmetrical and the even-numbered MOs antisymmetrical. Hence the symmetry properties of the frontier orbitals of the π fragment alternate as the polyene gains or loses a double bond. This allows us to generalize the analysis above into the selection rule: *thermal disrotatory ring openings are forbidden for 4n electron and allowed for 4n + 2 electron systems. The opposite is true for conrotatory opening.*

These selection rules can also be obtained from an analysis of polyene cyclizations. If the interaction between the terminal atoms C_1 and C_n is bonding (antibonding), it will favor (disfavor) the cyclization. Figure 4.5 shows how the contribution of any given MO Ψ_p changes as a function of the reaction stereochemistry. When p is odd (even) the conrotatory process is disfavored (favored) and the disrotatory process is favored (disfavored). Obviously, the preferred pathway can be deduced by summing the contributions of all of the occupied MOs, up to and including the HOMO:

	Conrotatory	*Disrotatory*
$p = 1$	Unfavorable	Favorable
$p = 2$	Favorable	Unfavorable
$p = 3$	Unfavorable	Favorable
⋮		

Thus, the cyclization energy is given by a sum of alternating terms which increase in magnitude as they approach the HOMO.[10] Therefore, the outcome of the reaction can be predicted by just looking at the HOMO contribution.

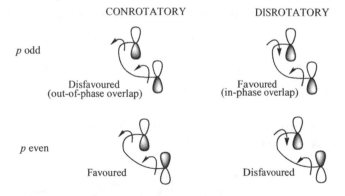

Figure 4.5　The contribution of Ψ_p to the cyclization reaction.

Limitations of the foregoing selection rules

An allowed pericyclic reaction has an aromatic transition state whereas a forbidden reaction has an antiaromatic transition state (p. 40). However, the aromaticity or

---

[10]The magnitude of the contribution of Ψ_p depends on the overlap of C_1 and C_n, which in turn depends of the coefficients of these atoms. Now, Coulson's equations show that the coefficients at the terminal atoms rise steadily, reaching a maximum in the HOMO and the LUMO, and then decline (p. 18).

antiaromaticity of annulenes tends to vanish as their sizes increase. For 22-electron annulenes or larger, there is not much stability difference between $4n + 2$ and $4n$ systems. According to calculations,[11] 5,5-sigmatropic shifts occur preferentially by diradical mechanisms.

Exercise 1 (D)   *Question 3 shows some errors to be avoided when making perturbation calculations. Relevant MOs can be found on p. 247*

The Woodward–Hoffmann rules cannot always be applied directly. For example, the intracyclic double bond complicates the analysis of Reaction (4.1), which could be an eight-electron (forbidden) or a six-electron (allowed) process.

$$\text{(reaction scheme)} \tag{4.1}$$

The problem can be resolved by using PMO or frontier orbital methods. In the PMO approach, we compare Reaction (4.1) with three other reactions, Reactions (4.2)–(4.4).

$$\text{(reaction scheme)} \qquad \text{allowed} \tag{4.2}$$

$$\text{(reaction scheme)} \qquad \text{allowed} \tag{4.3}$$

$$\text{(reaction scheme)} \qquad \text{forbidden} \tag{4.4}$$

(1) Model the transition states of Reactions (4.1)–(4.4) by the conjugated molecules **8**, **9**, **10** and **11**, and use the PMO method to calculate their resonance energies (defined as the difference between the energy of the cyclic structure and its open chain polyene counterpart). Hence, deduce the feasibility of Reaction (4.1).

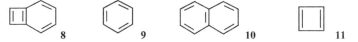

(2) Use the MOs in the Appendix and FO theory to determine the feasibility of Reaction (4.1).
(3) Use perturbation theory to calculate the frontier orbitals of **12**. Three disconnections **A**, **B** and **C** are possible:

$$\text{12} \qquad \text{A} \qquad \text{B} \qquad \text{C}$$

(a) Why is **C** a particularly bad choice?
(b) Do we need to calculate the frontier orbital energies and coefficients with high accuracy to determine the feasibility of Reaction (4.1)?
(c) Which scheme, **A** or **B**, have you chosen? Justify your choice.

[11]Nandel M., Goldfuss B., Beno B., Houk K. N., Hafner K., Lindner H.-J., *Pure Appl. Chem.*, 1999, **71**, 221; Beno B., Fennen J., Houk K. N., Lindner H.-J., Hafner K., *J. Am. Chem. Soc.*, 1998, **120**, 10490.

Answer

(1) Structure **13** below represents **8** as combination of a one-carbon radical and a heptatrienyl radical. The nonbonding orbital coefficients in the heptatrienyl radical are ± 0.5, so the resonance energy of **8**, $R(\mathbf{8})$, is given by

$$R(\mathbf{8}) = \text{Formation energy of } \mathbf{8} - \text{formation energy of octatraene}$$
$$= (2 \times 0.5\beta \times 1) - (2 \times 0.5\beta \times 1) = 0$$

The numerical coefficient of 2 appears because one electron is present in each of the interacting nonbonding orbitals. An analogous approach (structures **14–16**) shows that the resonance energies of **9, 10** and **11** are

$$R(\mathbf{9}) = (2 \times 2 \times 0.58\beta \times 1) - (2 \times 0.58\beta \times 1) = 1.16\beta$$
$$R(\mathbf{10}) = (2 \times 3 \times 0.45\beta \times 1) - (2 \times 0.45\beta \times 1) = 1.79\beta$$
$$R(\mathbf{11}) = 0 - (2 \times 0.71\beta \times 1) = -1.42\beta$$

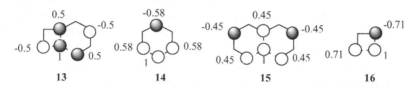

13	**14**	**15**	**16**

Hence **9** and **10** are aromatic, **11** is antiaromatic and **8** is nonaromatic. Consequently, Reaction (4.1) will not occur easily because its (nonaromatic) transition state is not stabilized. We can consider it to be 'less forbidden' than Reaction (4.4), but forbidden nonetheless. This shows clearly that there is no distinct boundary between 'forbidden' and 'allowed' reactions; they are merely the two limits of a continuum.

(2) The HOMO (LUMO) of **12** having a different symmetry than the LUMO (HOMO) of ethylene, Reaction (4.1) is forbidden. In this particular case, the FO method is simpler than the PMO approach.

(3) (a) **C** should be avoided because it does not respect the symmetry of the molecule. Therefore, it entails longer and more difficult calculations than **A** or **B**.

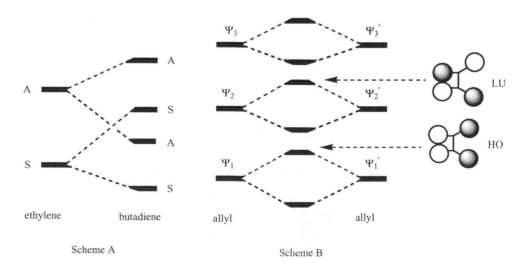

Scheme A

Scheme B

(b) For Reaction (4.1) to be allowed, the FO of **12** must have the correct symmetry to interact with those of ethylene. To a first approximation, these orbitals are the in-phase and out-of-phase combinations of the nonbonding MOs Ψ_2 and Ψ_2' of the two allyl fragments, as shown in Scheme B. *Even without any calculations*, we can conclude, from Scheme B, that Reaction (4.1) is probably forbidden. To be absolutely sure, we must check that the HOMO of **12** is $0.707(\Psi_2 + \Psi_2')$ and not $0.707(\Psi_1 - \Psi_1')$. This is indeed the case:

$$E(\psi_1 - \psi_1') = \alpha + \sqrt{2}\beta - P_{11'} = \alpha + \sqrt{2}\beta - (0.5^2 + 0.707^2)\beta = \alpha + 0.66\beta$$

correct value: $\alpha + 0.80\beta$

$$E(\psi_2 - \psi_2') = \alpha + P_{22'} = \alpha + 0.5\beta \qquad \text{\textit{correct value:} } \alpha + 0.56\beta$$

(c) **A** and **B** are both valid, so our choice is governed by the information that we wish to extract. **B** is good for determining the feasibility of Reaction (4.1) and is to be preferred also for estimating the MO energies, as it requires only first-order perturbations. It is much less appealing when calculating the coefficients of the frontier orbitals because it makes use of allyl fragments which are neither symmetrical nor antisymmetrical with respect to the symmetry plane of **12**. All the allyl orbitals then intermix and the calculation becomes long and difficult. **A** is preferable, but it is not immediately obvious whether the HOMO(**12**) derives from the butadiene Φ_2 or from the ethylene π. **B** indicates that the HOMO is symmetrical, so it must be the ethylene π orbital. With this knowledge, we can proceed to a calculation:

$$E(\text{HOMO}) = E_\pi + \frac{P_{\pi,\Phi_1}^2}{E_\pi - E_{\Phi_1}} + \frac{P_{\pi,\Phi 3}^2}{E_\pi - E_{\Phi_3}}$$

with $P_{\pi,\Phi_1} = 2 \times 0.707 \times 0.6\beta = 0.848\beta$ $\quad$ *and* $\quad$ $P_{\pi,\Phi_3} = 2 \times 0.707 \times 0.37\beta = 0.523\beta$.

Now the energy difference between the interacting orbitals π and Φ_1 (0.618β) is smaller than their resonance integral (0.848β): the rule governing the use of the second-order formulae (p. 27) is thus violated. Consequently, π and Φ_1 must be treated as quasi-degenerate and first-order equations must be used. The HOMO energy is then

$$E(\text{HOMO}) = \alpha + \beta - 0.848\beta + 0.169\beta = \alpha + 0.323\beta \ \textit{(correct value} = \alpha + 0.555\beta)$$

and its LCAO expansion is

$$\text{HOMO} = N\left(0.707\pi - 0.707\Phi_1 + \frac{P_{\pi,\Phi_3}}{E_\pi - E_{\Phi_3}}\Phi_3\right) = N[0.707(\pi - \Phi_1) + 0.323\Phi_3]$$

These calculations are represented pictorially below.

| 0.70 π | $-0.707\Phi_1$ | $-0.707\Phi_3$ | | correct values |

Note the choice of phase: the coefficients in the atoms which become bonded (the ethylene and the central atoms in the butadiene) must have the same sign in the orbitals π, Φ_1 and Φ_3. In the diagrams above, the lobes are given in white on these atoms. The negative mixing coefficient in Φ_1 causes a change in sign to gray.

Structure **12** is alternant, so we can apply the pairing theorem to evaluate the LUMO. The two FO are symmetrical about the energy α. The starred atoms have the same coefficients in both orbitals and the non-starred coefficients are equal but opposite in sign.

4.2 Relative Reactivity

4.2.1 Electrophilic Reactions

Rule 2 states that the higher the energy of the HOMO in any given compound, the better it will react with an electrophile. Thus, propene or enols are more electrophilic than ethylene, because they have HOMO energies which are higher than the ethylene value of $\alpha + \beta$ (pp. 29–33).

Exercise 2 (E)   *MOs are given on p. 249*

Why does the HOMO lie higher in 1,3-pentadiene than in isoprene?

Answer

The pentadiene (isoprene) HOMO is a butadiene HOMO perturbed by the presence of a substituent at C_1 (or C_2, respectively). The butadiene HOMO has a larger coefficient at C_1 than at C_2, so the HOMO will be more perturbed in pentadiene than in isoprene. Hence it lies at higher energy.

Geometry of 'Ate' Complexes

This problem illustrates some of the limitations of rule 2 and the frontier orbital approximation. We begin by describing the carbonyl group in terms of an interaction between an sp^2-hybridized C atom and an sp oxygen. We will place the molecule in the xy plane and orient the CO bond along the x-axis:

$$\begin{matrix} A \\\\ \\diagdown \\\\ B \end{matrix} C=O \qquad \begin{matrix} z \\\\ \\nearrow y \\\\ \\rightarrow x \end{matrix}$$

Two of the carbon sp^2 orbitals are used in bonding to A and B. The third combines with the oxygen sp hybrid to form two orbitals which we term σ_{CO} and $\sigma_{CO}{}^*$ (Figure 4.6). The last carbon valence orbital $p_z(C)$ interacts with p_z (O) to generate the π_{CO} and $\pi_{CO}{}^*$ orbitals. This lateral overlap is less efficient than the axial overlap, so the π and π^* orbitals lie at energies between σ and σ^*.[12] The oxygen p_y orbital is unaffected because it cannot

---

[12]The bonding to antibonding orbital separation is a function of the resonance integral, which, in turn, is proportional to the overlap of the two interacting orbitals (Mulliken's approximation, p. 13).

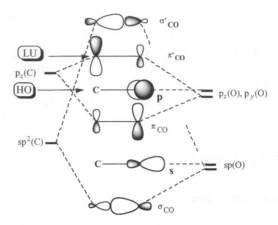

Figure 4.6   The orbitals of the carbonyl group.

interact with the carbon orbitals. The oxygen hybrid orbital whose major lobe is oriented away from the carbon atom is also essentially unchanged. These orbitals, which represent the lone pairs of the ketone, are denoted p and s, respectively. The relative energies of s and π_{CO} are difficult to determine, but this has little chemical consequence.

CO has 10 valence electrons (four provided by C and six by O), of which two are used to bind to A and B. The eight remaining electrons will occupy σ_{CO}, s, π_{CO} and p in the ground state. Hence the frontier orbitals are p (the HOMO) and π_{CO}* (the LUMO).

> This description, which uses a mixture of MO and valence bond concepts and attributes specific energies to hybrid orbitals, may appall the theoretician. It has the advantage, however, of reducing the problem to a series of two-orbital interactions, thus allowing us to find the CO orbitals very quickly.

A cation (or other electrophile) will interact principally with the HOMO of the carbonyl, i.e. the oxygen p lone pair. Hence, if all other interactions are negligible, the optimum geometry for the 'ate' complex should be **17**. However, *ab initio* calculations[13] predict structures **18** (if $M^+ = H^+$ or Me^+) and **19** (if $M^+ = Li^+$ or BH_2^+), where the cation complexes the two lone pairs simultaneously. The bent structure **18** is favored if the Lewis acid has a σ acceptor orbital, whereas the linear complex **19** is preferred if an additional π-type acceptor orbital is available. The HOMO–LUMO interaction is slightly diminished, but this is compensated by the interaction with the s lone pair.

17	**18**	**19**	**20**	**21**

[13]Raber D. J., Raber N. K., Chandrasekhar J., Schleyer P. v. R., *Inorg. Chem.*, 1984, **23**, 4076. See also: Nelson D. J., *J. Org. Chem.*, 1986, **51**, 3185; Gung B. W., Peat A. J., Snook B. M., Smith D. T., *Tetrahedron Lett.*, 1991, **32**, 453 ; Gung B. W., *Tetrahedron Lett.*, 1991, **32**, 2867.

According to MNDO calculations,[14] the *anti* complex **20** is more stable than the *syn* complex **21** by $1.8\,kcal\,mol^{-1}$. In the *syn* complex, steric repulsion increases the B–O=C angle and decreases the Me–C–H angle, leading to a stronger interaction of BF_3 with the s lone pair. The B–O bond then lengthens and the complex becomes less stable. This means that even if the HOMO–LUMO interaction does not alone determine the geometry of the complex, it remains the most important.

Exercise 3 (D) *Deserves your attention*

(1) Could we predict the failure of FO theory in determining the ate complex geometry?
(2) In a carbonyl compound, it is the π orbital, not the s lone pair, which lies just below the HOMO (Figure 4.6). So why is geometry **22** (where M^+ complexes the two highest occupied MOs) less favorable than **18**?

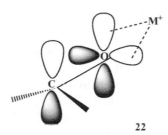

22

Answer

(1) *Frontier orbital theory must always be used with the utmost caution in structural problems, for two reasons.* The first, and main reason, is that frontier interactions are only dominant *in the transition state* and in structural problems the structures under study are stable ones. Furthermore, if a transition state may be considered with good reasons to be the perturbed initial system, as the bonds undergoing transformation are not completely formed or completely broken, this is no longer true in structural problems. The formal recombination of two fragments to give the final product corresponds to the complete formation of one or several new bonds, an important modification which may not always be treated as a perturbation. That is the second reason.

(2) Complexing the π orbital spreads the π bonding electrons in **22** over three centers (C=O–M) instead of two. The O–M bonding interaction is offset by a weakening of the CO bond. Hence the overall energy gain is low. In **18**, the two nonbonding electrons in the s lone pair become bonding, giving a net stabilization. However, this analysis is only valid when M is a pure σ acceptor; π back-donation may favor complexation to the π_{CO} orbital if M is a transition metal.

Our interpretation is validated by the work of Corcoran and Ma,[15] who studied the complexation of naphthalenones N_a and N_e by $TiCl_4$. In both cases, a 1:1 complex is formed. Despite the fact that the N_a structure makes π complexation of the

[14]Reetz M. T., Hullman M., Mossa W., Berger S., Rademacher P., Heymanns P., *J. Am. Chem. Soc.*, 1986, **108**, 2406.
[15]Corcoran R. C., Ma J., *J. Am. Chem. Soc.*, 1992, **114**, 4536.

carbonyl group possible, this event does not occur. Instead, the molecule undergoes a conformational change of the cyclohexanone portion from a chair to a twist-boat form to allow an in-plane chelation. No such conformation change is observed in the N_e case.

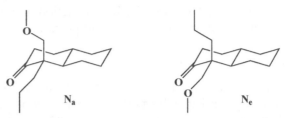

4.2.2 Nucleophilic Reactions

The Nucleophilicity of Halides

The nucleophilicity of halides in a protic solution increases in the order F^-, Cl^-, Br^-, I^-. This sequence is reversed in dipolar aprotic solution.[16]

FO theory assumes that within a series of nucleophiles the reactivity reflects their relative HOMO energies (rule 2). Due to neglect of interelectronic repulsions, particularly important in anions, the Hückel HOMO energies follow the order of the coulombic integrals (i.e. the HOMO of F^- is lowest in energy, in contradiction with the experimental ionization potential). Hartree–Fock calculations, however reproduce the intrinsic reactivity of the ions adequately. For the naked anion, the SCF HOMO energy rises as the ion becomes smaller. When the halides are solvated by formaldehyde, the HOMOs become closer to each other in energy but retain the same ordering. Solvation by three or four molecules of water reverses the reactivity order.[17]

These solvent effects can be explained in classical terms. The intrinsic (gas-phase) reactivity increases as the as the ionic radius falls. The reasons are simple: as the anion becomes smaller, the surplus electron generates greater interelectronic repulsions and the reactivity rises. Dipolar aprotic solvents interact only weakly with the halides, so they do not appreciably affect this reactivity order. In protic solvents where hydrogen bonds play an important role, the smaller the ion, the more stabilized it becomes and the reactivity order is inverted.

It is important to show, however, that frontier orbital theory is equally capable of rationalizing results which can be understood using older theories in addition to phenomena which were previously considered to be inexplicable.

Electrophilic Assistance

The reactivity of a carbonyl compound is increased by coordination to M^+, by a process known as *electrophilic assistance*. The metal cation withdraws electron density from the

[16]Tchoubar B., *Bull. Soc. Chim. Fr.*, 1964, 2069; Parker A. J., *Chem. Rev.*, 1969, **69**, 1; March J., *Advances in Organic Chemistry*, 4th edn., John Wiley & Sons, Inc., New York, 1992, p. 349 and references therein.
[17]Minot C., Anh N. T., *Tetrahedron Lett.*, 1975, 3905.

oxygen atom, thus increasing its effective electronegativity and lowering the energy of its orbitals. The energy gap between $p_z(C)$ and $p_z(O \cdots M^+)$ is increased, so they mix less efficiently. As the scheme below shows, the LUMO then moves to lower energy and becomes less symmetrical (the coefficient at C is larger). Both of these changes enhance its reactivity toward nucleophiles.

Pierre and Handel[18] have shown that if the metal ions are trapped by [2.1.1] or [2.2.1] cryptands, reductions employing $LiAlH_4$ and $NaBH_4$ in aprotic solvents are usually blocked. Loupy *et al.*, pointed out that the main effect of cationic complexation is a lowering of the substrate's LUMO. Therefore, if the carbonyl compound has a sufficiently low-lying LUMO, it can be reduced without electrophilic assistance. Indeed, benzaldehyde does react with $LiAlH_4$ even in the presence of [2.1.1] cryptand. The rate is lowered but remains acceptable, giving 45% conversion to benzyl alcohol after 1 min.[19]

An Example of Chemoselectivity: Relative Reactivities of Carbonyl Compounds with Respect to Nucleophiles

This reactivity order appears never to have been satisfactorily explained. According to one interpretation, during attack of the nucleophile, the carbon hybridization changes from sp^2 to sp^3 and the valence angles are reduced. This creates greater steric interactions for ketones ($R \leftrightarrow R'$) than for aldehydes ($R \leftrightarrow H$) and explains why the latter are more reactive:

Now, the van der Waals radii are: hydrogen 1.2, oxygen 1.4, nitrogen 1.5, chlorine 1.8 and methyl 2 Å.[20] If the above theory is correct, ketones would be the least reactive of all carbonyl compounds and acid chlorides would be less reactive than esters or even amides! Another hypothesis suggests that the resonance hybrid

[18]Pierre J. L., Handel H., *Tetrahedron Lett.*, 1974, 2317.
[19]Loupy A., Seyden-Penne J., Tchoubar B., *Tetrahedron Lett.*, 1976, 1677.
[20]Pauling L., *The Chemical Bond*, Cornell University Press, Ithaca, NY, 1967, p. 152.

lowers the positive charge at C and deactivates the ester with respect to ketones. However, this would also make acid chlorides and anhydrides less reactive than ketones. Explanations based on FO theory are more satisfactory. The MO Catalog in the Appendix gives the MOs of HCHO and MeCHO, along with those for related ketones, anhydrides, esters, amides and acid chlorides. Their LUMO energies (which should reflect their reactivity: rule 2) increase in the order aldehyde = anhydride < acid chloride < ketone < ester < amide, which is in good agreement with the experimentally observed rates. The only inaccuracy is the acid chloride being slightly less reactive than the corresponding aldehyde. This disappears if we employ transition state LUMOs rather than those of the reagents (see below).

As an exercise, let us evaluate the reagents' LUMO energies by a perturbation approach (Figure 4.7). Any R–CO–X can be viewed as the union of an R–CO fragment with an X moiety (X = H, R', Cl, NH_2, OR' or O–CO–R). The LUMO of R–CO–X will then be the π_{CO}^* orbital of the R–CO moiety, perturbed by X. A hydrogen atom cannot conjugate with a π system, so the LUMO of the aldehyde has the same energy as the π_{CO}^* orbital in the parent RCO fragment. In Hückel calculations, O, Me and Cl are all represented by a doubly occupied orbital of energy $\alpha + 2\beta$ and are differentiated by the value of the resonance integral (0.8β for oxygen, 0.7β for the methyl group and 0.4β for chlorine). A large resonance integral will provoke a large perturbation, so the LUMO lies higher in a ketone than in an acid chloride. MeO is a better donor than O, so the ester has the highest LUMO of the three functions. The amide LUMO is the highest of the series, because the nitrogen lone pair ($\alpha + 1.5\beta$) is close in energy to π_{CO}^* and the resonance integral is large (0.8β). Finally, symmetry arguments, explained in Exercise 4, prove that the LUMOs of an anhydride and an aldehyde have exactly the same energy. These results are shown in Figure 4.7. Note that *the relative positions of the LUMOs have been deduced without any calculations; we only need to employ the table of Hückel parameters given on p. 22.*

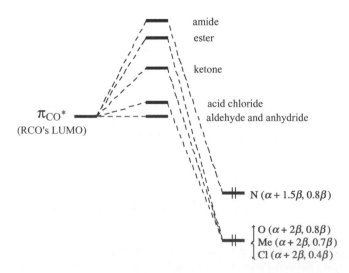

Figure 4.7   Perturbation theory estimates of some carbonyl LUMO energies.

Exercise 4 (D) *MOs are given on p. 249.*

Prove that an aldehyde RCHO and an anhydride (RCO)$_2$O have the same LUMO energy.

Answer

Breaking the anhydride down into RCO and O–CO–R fragments is clumsy because (a) the orbitals of O–CO–R are difficult to calculate by hand and (b) the symmetry properties of the molecule are not respected.

A fragmentation into O and (R–CO + CO–R) is more convenient. The fragment orbitals then comprise an oxygen lone pair ($\alpha + 2\beta$, 0.8β) two π_{CO} and two π_{CO}^* coming from the two identical R–CO subfragments. The π_{CO} pair gives rise to a symmetrical combination ($\pi_{CO} + \pi_{CO}$) and an antisymmetrical combination ($\pi_{CO} - \pi_{CO}$) having *the same energy* as π_{CO}. This is because the energy gap between bonding and antibonding levels is a function of the resonance integral [Equation (3.8)]. Here, where there is no direct bond between the two R–CO, the resonance integral is zero and both combinations are degenerate with the starting orbitals. Similarly, the π_{CO}^* pair gives rise to a symmetrical combination ($\pi_{CO}^* + \pi_{CO}^*$) and an antisymmetrical combination ($\pi_{CO}^* - \pi_{CO}^*$) having the same energy as π_{CO}^*.

antibonding combination

bonding combination

zeroing β

antisymmetric

symmetrical

Note: The MOs of an ML$_n$ complex can be found by interacting the metal orbitals with the *symmetry orbitals* of the ligands.[21] The foregoing method, where we calculate the Hückel MOs and then zero the resonance integral, permits symmetry orbitals to be obtained without recourse to group theory.

The antisymmetrical combination ($\pi_{CO}^* - \pi_{CO}^*$) has no net overlap with the oxygen lone pair, so its energy is unchanged after formation of the molecule. It becomes the anhydride LUMO. The interaction of the oxygen with the symmetric combination gives a bonding and an antibonding combination:

($\pi_{CO}^* \pm \pi_{CO}^*$) S
(R–CO····CO–R) A

($\pi_{CO}^* + \pi_{CO}^*$) – λO

O

O + λ($\pi_{CO}^* + \pi_{CO}^*$)

[21]Albright T. A., Burdett J. K., Whangbo M. H., *Orbital Interactions in Chemistry*, John Wiley & Sons, Inc., New York, 1985.

Check these deductions using the MOs from the Appendix. Note in particular that (a) the LUMO energy is the same in RCHO and (RCO)$_2$O, (b) the anhydride LUMO is an antisymmetrical combination of π_{CO}^* levels which has a coefficient of zero at the central oxygen and (c) the MO lying just above the LUMO is a symmetrical π_{CO}^* combination, mixed out-of-phase with the oxygen orbital.

Exercise 5 (E)   *Deserves your attention. Relevant MOs are given on p. 250*

(1) Without using calculations, show that the LUMO in the acid R–COOH lies at higher energy than in the ketone R–CO–R but lower than in the ester R–CO–OR.

(2) With this knowledge, can we conclude from rule 2 that an acid will be more reactive than an ester with respect to a given nucleophile?

Answer

(1) The parameters on p. 22 show that O is a better donor than Me, so it raises the energy of π_{CO}^* to a greater degree. Since MeO is an even better donor, we can deduce the order LUMO(ketone) < LUMO(acid) < LUMO(ester). This is confirmed by numerical calculations:

$$\text{LUMO(acetone)} = \alpha - 0.878\beta$$
$$\text{LUMO(acetic acid)} = \alpha - 0.917\beta$$
$$\text{LUMO(methyle acetate)} = \alpha - 0.925\beta$$

(2) No. Any nucleophile is more or less basic, so the first step of a reaction with RCO$_2$H is usually a deprotonation:

$$\text{Nu}^- + \text{R–CO}_2\text{H} \rightleftharpoons \text{NuH} + \text{R–CO}_2^-$$

The O$^-$ lone pair lies at very high energy, well above an uncharged nitrogen. Consequently, the LUMO in R–COO$^-$ is even higher than an amide's, which means that the carboxylate anion is particularly unreactive. A comparison of the MOs in the acid and the ester has no meaning because the acid is *not* the reactive species.

Exercise 6 (E)   *MOs are given on p. 251*

Class the following compounds according to their sensitivity toward hydrolysis:

| 23 | 24 | 25 | 26 |

Answer

Dividing **27** into (**28** + X), where X = H, Me, OH and NH$_2$, and calculating the LUMOs by perturbations as for the carbonyls above, we find that the rates of hydrolysis are **23** > **24** > **25** > **26**.

27 → 28 + X (H, Me, OH, NH$_2$)

Exercise 7 (D)   *Deserves your attention*

It seems not unreasonable to consider that in **23–26**, the weaker the C=N bond, the more reactive is the compound. Let us break **27** down into (**28** + X), with X = H, Me, OH and NH$_2$, and look how the C=N π bond strength is modified by X. Interaction of the X lone pair with the vacant $\pi_{CN}{}^*$ gives rise to two new orbitals, of which only the bonding combination X(lone pair) + $\lambda\pi_{CN}{}^*$ is occupied:

Physically, this means that the C=N bond is weakened because the lone pair is now partially transferred to the antibonding $\pi_{CN}{}^*$ orbital. The mixing coefficient λ – and the weakening – increase with the interaction. The reactivity order predicted on the basis of the CN bond strength is then **26** > **25** > **24** > **23**, exactly opposite to the order obtained in Exercise 6! Explain this contradiction.

Answer

Bond strengths give a measure of the thermodynamic stability and LUMO levels a measure of the kinetic stability (with respect to nucleophilic attacks). The parallelism between thermodynamic and kinetic stabilities is frequent, but not compulsory. Thus 1,3-pentadiene, which is stabilized by conjugation, is more reactive than 1,4-pentadiene.

Another example is given by the reaction between hydrogen and oxygen, which is explosive ($\Delta H = -63\,\mathrm{kcal\,mol^{-1}}$). Yet, in the absence of a flame or a catalyst, a stoichiometric mixture remains indefinitely stable at room temperature. As a final example, let us examine the series Me$^+$, Et$^+$, iPr$^+$ and tBu$^+$. tBu$^+$ has the largest positive charge on the sp^2 carbon and is at the same time the least reactive of these cations. The reason is that kinetic criteria suggest that a methyl group is electron releasing, whereas thermodynamic criteria are in favor of an electron-withdrawing effect.[22]

Because thermodynamic and kinetic criteria are not usually equivalent, linear free energy relationships correlating reaction rates with thermodynamic variables should be used cautiously.[23]

[22]Minot C., Eisenstein O. Hiberty P. C., Anh N. T., *Bull. Soc. Chim. Fr II*, 1980, 119. See also p. 22.
[23]Dewar M. J. S., Dougherty R. C., *The PMO Theory of Organic Chemistry*, Plenum Press, New York, 1975, pp. 212–220.

Exercise 8 (E)   *Deserves your attention*

The cycloaddition between butadiene and ethylene is used to illustrate the Diels–Alder reaction in almost every organic textbook. Show that cyclohexene is *not* the main product of the reaction.

Answer

The reaction mixture contains *two* dienophiles: ethylene and butadiene. The HOMO–LUMO gap is 2.236β when butadiene is the dienophile and 3.236β for ethylene. Thus, rule 2 states that the dimerization of butadiene will occur preferentially and the major product will be vinylcyclohexene.[24]

Exercise 9 (E)   *MOs are given on p. 252*

Compare the reactivity of a given nucleophile toward a carbamate NH_2CO_2R, an ester $R'CO_2R$ and an amide $RCONH_2$.

Answer

Fragmentation into $(NH_2 + CO_2R)$ and $(R' + CO_2R)$ indicates that the carbamate is less reactive than the ester, for the same reasons that an amide is less reactive than a ketone $(RCO + R')$. Similarly, fragmentation into $(NH_2CO + OR)$ and $(NH_2CO + R)$ shows that the carbamate is less reactive than the amide for the same reasons that an ester is less reactive than a ketone. We have seen that amides are less reactive than esters (Figure 4.7).

4.2.3   Cycloadditions

Alder's rule[25]

This rule states that *cycloaddition rates rise as a diene becomes more electron rich and the dienophile more electron poor.* Cycloadditions are also accelerated in cases of 'inverse electron demand', when an electron-poor diene reacts with an electron-rich dienophile.

The explanation is straightforward. It follows from the fact that substituting any given compound with donor groups causes a general rise in the energies of its MOs. Such a compound can donate electrons readily as its occupied orbitals are high in energy. Conversely, it accepts incoming electrons only with difficulty because the Pauli principle requires that they must be placed in its (high-lying) vacant orbitals. This MO description agrees well with the organic chemist's picture of an electron-rich compound. The reverse situation is found in compounds substituted with acceptor groups, whose MOs are lowered.

Now, when the reagents are unsubstituted, the four FOs are symmetrically positioned with respect to the non-bonding level (Figure 4.8a). When substituents are present, the FOs of the electron-poor compound are lowered and those of the electron-rich compound are raised (Figure 4.8b). Thus, one pair of FOs becomes nearer in energy,

---

[24]Houk K. N., Lin Y.-T., Brown F. K., *J. Am. Chem. Soc.*, 1986, **108**, 554.
[25]Alder K., *Experientia Suppl. II*, 1955, 86; Sauer J., *Angew. Chem. Int. Ed. Engl.*, 1967, **6**, 16 and ref. therein.

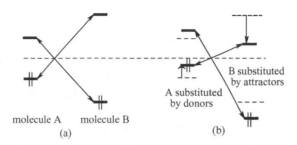

Figure 4.8 A molecular orbital interpretation of Alder's rule.

whereas the other becomes more separated. In Exercise 11, we will show that this process is not symmetrical; the reduction in the first gap is larger than the increase in the second. Hence rule 2 predicts that substituted systems will react more easily than their unsubstituted parent compounds.[26] Figure 4.8 applies both to the normal Alder rule and the inverse demand case because A could equally be the diene or the dienophile.[27]

Exercise 10 (E)

During the 1960s, organic chemists were surprised to find that the reactivity of a dipolarophile is *always* increased by substitution, whether the substituent is a donor or an acceptor.[28] This result is difficult to explain by classical effects. Can you do better using frontier orbitals?

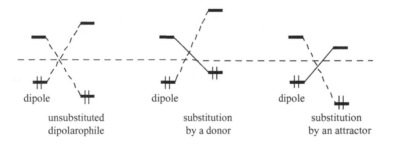

Answer

As shown in the following exercise, a donor substituent raises the frontier orbital energies whereas acceptor lowers them. Consequently, introducing a substituent on the dipolarophile causes two of the frontier orbitals to become closer in energy and two to separate (see the scheme above). These frontier orbital changes mirror those discussed in relation to the Alder rule. Again, the presence of the substituent induces a faster reaction.[29]

---

[26]Using a Taylor expansion, it can be shown that this remains true even when the gap narrowing for the first pair of FOs is exactly equal to the gap increase of the second pair.[27a]

[27](a) Eisenstein O., Anh N. T., *Bull. Soc. Chim. Fr.*, 1973, 2721, 2723; (b) Sustmann R., *Tetrahedron Lett.*, 1971, 2717, 2721; (c) Sustmann R., Trill H., *Angew. Chem. Int. Ed. Engl.*, 1972, **11**, 838; (d) Sustmann R., Schubert R., *Angew. Chem. Int. Ed. Engl.*, 1972, **11**, 840.

[28]Huisgen R., *J. Org. Chem.*, 1968, **33**, 2291.

[29]Sustmann R., *Tetrahedron Lett.*, 1971, 2717; Houk K. N., *J. Am. Chem. Soc.*, 1972, **94**, 8953.

Exercise 11 (M)

Show that a donor substituent provokes a large increase in energy of the HOMO but raises the LUMO energy to a lesser extent, whereas an acceptor lowers the LUMO more than the HOMO. Can this rule be broken?

Method: Donor substituents usually contain heteroatoms whose lone pairs can be modeled in Hückel calculations by doubly occupied orbitals. To a first approximation, acceptor substituents (polar conjugating groups such as C=O, C≡N, NO_2) can be represented by a low-lying vacant orbital (π_{CO}^* for C=O, π_{CN}^* for C≡N, etc.). In cases where a more realistic model is required, the occupied (π_{CO}, π_{CN}, etc.) orbitals are added to the calculation.

Answer

When two orbitals interact, the lower is stabilized and the higher is destabilized. The perturbation increases as the gap between the interacting orbitals decreases (p. 30). Let HO and LU be the frontier orbitals of the substituted system, HO° and LU° those of the unsubstituted parent, **D** the orbital representing a donor and **A** and **A*** those representing an acceptor.

A donor substituent is represented by an orbital lying at lower energy than the FOs of the parent hydrocarbon. **D** will raise the energy of both FOs, but the HOMO will be most strongly affected because it is closer in energy (scheme **a**). We saw a similar effect when we discussed the MOs of an enol (p. 29).

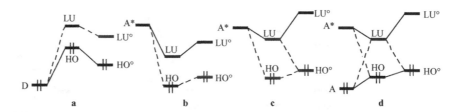

An acceptor substituent can give rise to two different cases:

1. **A*** lies at higher energy than LU° (scheme **b**, the mirror image of scheme **a**).
2. More frequently, **A*** has a lower energy than LU° (scheme **c**). The parent system being generally alternant, its FOs will be symmetric with respect to the nonbonding energy and **A*** is closer to LU° than to HO°. LU is now essentially **A***, raised by an interaction with HO° but more strongly lowered by an interaction with LU°. Scheme **d** corresponds to a more sophisticated model where the acceptor substituent is represented by the two orbitals **A** and **A***. HO is essentially HO°, lowered by an interaction with **A*** but raised by **A**. The outcome is usually a modest variation in energy. Thus, Hückel calculations using the parameters on p. 22 give the same energy for HOMOs in acrolein and ethylene. Extended Hückel calculations place the acrolein HOMO lower (−13.31 eV versus −13.22 eV for ethylene) and STO-3G calculations reverse the results (−0.3159 versus −0.3226 a.u.). Just remember that the energy of the HOMO in a system substituted by an acceptor substituent may be slightly raised or lowered. The result may depend on the type of calculation employed. The LUMO is invariably lowered.

Can a donor substituent modify the energy of the LUMO more than the HOMO, and could an acceptor do the reverse? The answer is yes, but only for bridging substituents, such as the oxygen atom in furan or the CO in cycloheptatrienone. In each case, the substituent (O or CO) can only interact with the symmetrical MOs of the hydrocarbon portion (butadiene or hexatriene). Hence the HOMO in furan has the same energy as in butadiene, but the LUMO is higher. Likewise, the LUMO in tropone lies at the same energy as in hexatriene, but the HOMO is lower.

Exercise 12 (E)

Why are Diels–Alder reactions frequently catalysed by Lewis acids?

Answer

Very often, one of the reaction partners contains a carbonyl or cyano function which can form a complex with the Lewis acid. This complexation lowers the energy of the LUMO (cf. the discussion of electrophilic assistance, p. 60), and facilitates the reaction. It is the same phenomenon as that governing the Alder rule. [30]

Exercise 13 (E)   *MOs are given on p. 253*

An alternating acetate–fumarate–acetate–fumarate polymer is produced when a mixture of vinyl acetate and dimethyl fumarate undergoes radical-induced polymerization. Why should this be?
Method. Assume that: (1) the reaction is controlled by the interaction between the radical SOMO (singly occupied MO) and the frontier orbitals of the substrates; (2) the radical adds preferentially to the unsubstituted carbon of the vinyl acetate.

Answer

The initiator R• adds to a molecule of fumarate to give a radical **A** having electron-withdrawing substituents. We saw in Exercise 11 that such a species will have a low-lying SOMO; hence it prefers to react with a substrate having a high-lying HOMO and selects a vinyl acetate molecule. This reaction generates a radical **B** having a donor substituent, a high-lying SOMO, and an affinity for the substrate having the lowest-lying LUMO. It chooses a fumarate, and the cycle continues. [31]

[30]Houk K. N., Strozier R. W., *J. Am. Chem. Soc.*, 1973, **95**, 4094; Anh N. T., Seyden-Penne J., *Tetrahedron*, 1974, **29**, 3259; Alston P. V., Ottenbrite R. M., Cohen T., *J. Org. Chem.*, 1978, **43**, 1864; Branchadell V., Oliva A., Bertran J., *Theochem*, 1986, **138**, 117; Birney D. M., Houk K. N., *J. Am. Chem. Soc.*, 1990, **112**, 4127.
[31]Fleming I., *Frontier Orbitals and Organic Chemical Reactions*, John Wiley & Sons, Inc., New York, 1976, p. 183.

Exercise 14 (E)

When a reaction mixture contains a ketone and an aldehyde, both enolizable, the only condensation product results from the attack of the ketone enolate on the aldehyde.[32]Why?

Answer

Several factors could influence this reaction; fortunately, they all favor the same outcome.

- The concentration of the ketone enolate is higher than that of the aldehyde enolate. This is true under thermodynamic control as the stability of an enolate increases with its degree of substitution. It is also true under kinetic control: since enolization is an acid–base equilibrium, the increased enolate concentration reflects the higher acidity of the ketone protons.
- The ketone enolate is more substituted and therefore has the higher-lying HOMO (although the difference is relatively small).
- The decisive factor is probably the energy of the carbonyl group LUMO. The aldehyde is a much better electrophile than the ketone.

Exercise 15 (D) *Deserves your attention. MOs are given on p. 253*

The LUMO energies in benzophenone ($\alpha - 0.294\beta$) and acetophenone ($\alpha - 0.458\beta$) are lower than in acetone ($\alpha - 0.847\beta$). So why does $NaBH_4$ reduce them seven times more *slowly* than acetone?[33]

Answer

Frontier orbital analysis can only be applied to *closely related* reagents undergoing the *same* transformation (cf. p. 114). Acetone is a saturated ketone. Acetophenone and benzophenone are conjugated ketones whose reactivity is lowered by a loss of conjugation. Since frontier interactions do not dominate, it is not even valid to compare the reactivities of acetophenone and benzophenone on the basis of their LUMO energies alone. Frontier control will be greater if the reduction is performed with the more reactive $LiAlH_4$, because the transition state will be earlier and the loss of conjugation less important. This effect also probably explains the more rapid reduction of benzaldehyde than acetone by $NaBH_4$: the transition state is early, owing to the high reactivity of the aldehyde.

Exercise 16 (E)

Why does the treatment of ethyl benzoylacetylacetonate, $MeCOCH(COPh)CO_2Et$, with aqueous ammonia (42 °C) give acetamide and ethyl benzoylacetate?

---

[32]March J., *Advanced Organic Chemistry*, 4th edn, John Wiley & Sons, Inc., New York, 1992, p. 940.
[33]Carey F. A., Sundberg R. J., *Advanced Organic Chemistry*, 1st edn, Plenum Press, New York, 1977, Vol. A, p. 335.

Answer

The attack of ammonia at the ester is rather improbable, given the presence of the two ketone functions. Reaction at the PhCO group will be disfavored by loss of conjugation. Hence we see attack at the acetyl group, with subsequent elimination of the strongest conjugate base.

Exercise 17 (E)   *Deserves your attention. MOs are given on p. 254*

When dione **29** is treated with ethylene glycol in the presence of TsOH, what will be the major product, **30** or **31**? *Hint*: See Exercise 15.

Answer

The reaction product is ketal **30**. Occasionally, the diketal is also obtained.[34] On the one hand, attack on the saturated ketone is favored by the angular methyl group, which is antiperiplanar to the nucleophile (cf. p. 152) coming from the less hindered α face. On the other hand, despite the low-lying LUMO, attack on the conjugated ketone is disfavored by loss of conjugation. The transition state is probably late, ethylene glycol being a not very reactive nucleophile. Even with a more reactive nucleophile, attack on saturated ketones may remain preferred. For example, β-mercaptoethanol in the presence of zinc chloride or boron fluoride etherate condenses with saturated ketones, but not with α,β-unsaturated ketones.

It is interesting that with the rather mild catalyst pyridine hydrochloride, benzenethiol does *not* react with saturated ketones, but reacts with α,β-unsaturated ketones to give thioenol ethers (with increased conjugation) or 1,4-additions.[35]

Exercise 18 (E)

According to Kürti and Czakó,[36] there is an important interaction between the HOMO of the ene component $CH_2=CH-C-H$ and the LUMO of the enophile. In agreement with this interpretation, the reaction is favored by electron-withdrawing substituents on the enophile. However, the rate is also enhanced if electrons are withdrawn from the ene component. For example, according to the same authors, the mechanism of the Lewis acid promoted ene reaction is believed to involve both a concerted and a cationic pathway.[37] Whether the mechanism is concerted or stepwise, a partial or full positive charge is developed at the ene component in Lewis acid promoted ene reaction.[36] Explain this contradiction.

[34]Corey E. J., Ohno M., Mitra R. B., Vatakencherry P. A., *J. Am Chem. Soc.*, 1964, **86**, 478.
[35]Fieser L. F., Fieser M., *Steroids*, Reinhold, New York, 1959, p. 308.
[36]Kürti L., Czakó B., *Strategic Applications of Name Reactions in Organic Syntheses*, Elsevier Academic Press, San Diego, London, 2005, p. 6.
[37]Snider B. B., Ron E., *J. Am. Chem. Soc.*, 1985, **107**, 8160.

Answer

There is no contradiction. When electrons are withdrawn from a molecule, its LUMO energy is lowered and this FO becomes nearer to the partner's HOMO. According to rule 2, the reaction is then facilitated. It is exactly the same problem as with the Alder's rule and the reactions with inverse electron demand.

4.3 Limitations of Rules 1 and 2

4.3.1 Some Difficulties Encountered with Rule 1

In applying rule 1, some difficulties may arise if:

- Bond rotations occur during the reaction, causing MOs with zero overlap in the initial geometry to interact in the transition state.
- An atom uses two valence AOs in the reaction. It may then be necessary to consider, in addition to the frontier orbitals, one or several more MOs.
- The reaction occurs between three or more molecules.

These difficulties are illustrated in the following examples.

Cheletropic Reactions

Cheletropic reactions are cyclizations – or the reverse fragmentations – of conjugated systems in which the two newly made σ bonds terminate on the same atom. However, a cheletropic reaction is neither a cycloaddition nor a cycloreversion. The reason is that the chelating atom uses two AOs whereas in cycloadditions, each atom uses one and only one AO. Therefore, Dewar–Zimmerman rules cannot apply to cheletropic reactions. Selection rules must be derived using either FO theory or correlation diagrams:[38] *The conjugated fragment*[39] *of 4n + 2 electron systems reacts in a disrotarory (conrotarory) mode in linear (nonlinear) reactions. In 4n electron systems, it reacts in a disrotarory (conrotarory) mode in nonlinear (linear) reactions.*

The addition of a carbene to a double bond is easily dealt with by FO theory: only the non-linear approach allows a good frontier interaction (Figure 4.9). This result is in agreement with the selection rules quoted earlier: the system having four electrons (two coming from the carbene and two from the ethylene) and its π component

---

[38]Woodward R. B., Hoffmann R, *The Conservation of Orbital Symmetry*, Verlag Chemie, Weinheim, 1970, p. 152 and references therein.

[39]Two components intervene in a cheletropic reaction: a conjugated fragment and the chelating fragment X. In a linear reaction, the two partners come together following a least motion approach. In a nonlinear reaction, the final position of X, relative to the conjugated system, is different from that taken during its approach.

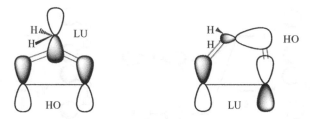

Figure 4.9 Frontier interactions between a carbene and a double bond.

reacting in a suprafacial (i.e. disrotatory) manner, the reaction should be nonlinear. In general, however, application of FO theory is less straightforward. Consider the following cheletropic reaction:[40]

As FO theory applies normally to bimolecular reactions, it is easier to study the reverse reaction: butadiene + $N_2 \rightarrow$ **32**. The MOs of butadiene are shown p. 50 and those of N_2 in Figure 4.10. *A priori*, four stereochemistries are possible: the butadiene component can react in a conrotatory or disrotatory mode and N_2 in a linear or nonlinear mode. Depending on the stereochemistry, the butadiene FOs Ψ_2 and Ψ_3 can overlap with the σ_z, π_x, π_x^*, π_y or π_y^* orbitals. Therefore, to treat all these cases, a set of *seven* 'frontier orbitals' must be taken into account.

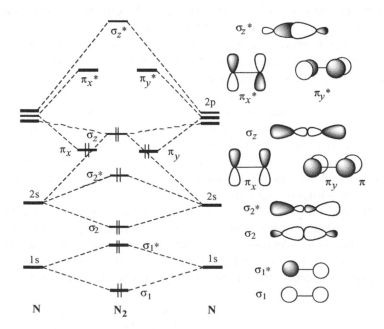

Figure 4.10 Molecular orbitals of the nitrogen molecule.

---

[40]Woodward R. B., Hoffmann R., *Angew. Chem. Int. Ed. Engl.* 1969, **8**, 781.

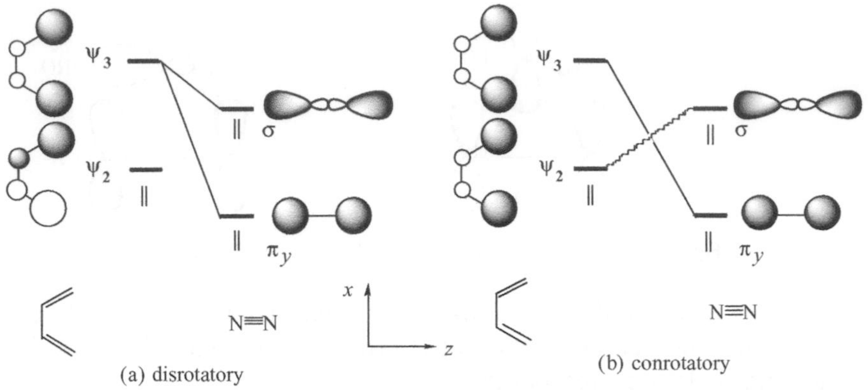

Figure 4.11 Major orbital interactions in the cheletropic reaction of diazene **32**.

Let us compare the linear reactions. In the disrotatory mode, the butadiene vacant orbital Ψ_3 interacts favorably with the nitrogen σ_z and π_y occupied orbitals. In the conrotatory mode, only the interaction between Ψ_3 and π_x is favorable and the interaction between the occupied orbitals Ψ_2 and σ_z is repulsive (Figure 4.11). Hence the disrotatory mode is expected to be favored. Note that if the six heavy atoms are coplanar, the initial overlaps of Ψ_3 with σ_z (Figure 4.11a), of Ψ_2 with σ_z and of Ψ_3 with π_x (Figure 4.11b) are zero. However, as soon as the termini of butadiene start to rotate, these overlaps become positive.

Electron count in cheletropic reactions

In a pericyclic reaction, electron counting can be effected in several ways, all equivalent. For example, in the Diels–Alder reaction, one can count the number of conjugated atoms in butadiene and in ethylene, or the number of bonds made (two σ and one π bonds) or broken (three π bonds) in the process. In all cases, a total of six intervening electrons are obtained.

This is no longer true for cheletropic reactions. *When applying the selection rules, one must **always** consider that the chelating fragment X contributes two electrons.* Erroneous conclusions can be made otherwise. Thus, in the cyclopropanation reaction, which should be considered a four-electron cheletropic reaction, only one π bond is broken. In the formation of diazene **32**, a six-electron cheletropic reaction, butadiene uses two double bonds and N_2 one lone pair and one π bond; the two components thus employ a total of eight electrons!

If X always contributes two electrons, its chemical nature should be unimportant. This is contradicted by experimental results. Whereas fragmentations of diazenes give good yields and are stereospecific,[41] heating of nitrosopyrroline **33** gives, in addition to polymers, only traces of butadiene and N_2O.[42] It can be then be expected that the validity of the selection rules is better for pericyclic than for cheletropic reactions.

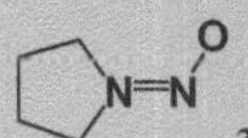

33

---

[41]Lemal D. M., McGregor S. D., *J. Am. Chem. Soc.*, 1966, **88**, 1335.
[42]Lemal D. M., McGregor S. D., *J. Am. Chem. Soc.*, 1966, **88**, 2858.

Systems Having More Than Two Components

Frontier orbital treatment of $(p + q + r)$ cycloadditions is fairly simple. We first combine p and q into one system s. The reaction between s and r is then a bimolecular reaction. To find the symmetry of the FOs of s, no calculation is required. If three-orbital interactions are neglected, it is clear that:

$$HOMO(s) = HOMO(p) - HOMO(q)$$
$$LUMO(s) = LUMO(p) + LUMO(q)$$

It is much simpler, however, to use Dewar–Zimmerman rules or PMO to treat these systems.

Exercise 19 (D)

Use the PMO method to predict the feasibility of the following reaction:

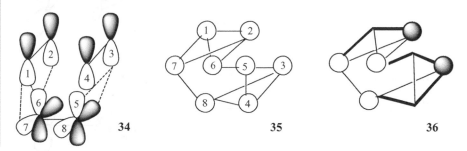

Answer

This is in fact a photochemical reaction.[43]

Let us start by examining the connectivity of the AOs intervening in the reaction. They are numbered as indicated in **34**. In **35**, overlapping AOs are represented as linked circles. As shown in **36**, this system may be considered as resulting from the union of two alternant radicals (drawn in bold lines). Union of their nonbonding MO gives two null, one favorable and three unfavorable interactions. The thermal reaction is thus not favorable.

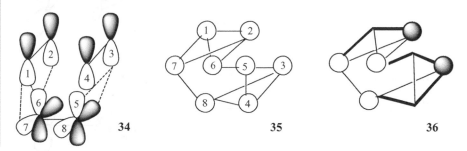

4.3.2 Problems with Rule 2

Two major problems are encountered when using rule 2. The first is practical. FO theory performs least well in the area of relative reactivity. Study of apparent violations of rule 2 shows that difficulties appear when the ordering of the FO levels differs in the starting

---

[43]Takahashi M., Kitahara Y., Murata I., Nitta T., Woods M. C., *Tetrahedron Lett.*, 1968, 3387.

materials and the transition state. It is therefore important to be able to predict when this may occur. The second problem is of theoretical interest. Is it reasonable that rule 2 takes into account only the HOMO–LUMO gap and ignores completely the overlap?

Apparent Violations of Rule 2

The relative reactivities of alkenes and alkynes

As a C≡C triple bond is shorter and stronger than a C=C double bond, the alkene π and π* should lie between the alkyne π and π* energy levels. Applying rule 2, we then draw the erroneous conclusion that alkynes are always less reactive than alkenes.[44] Where is the mistake?

The problem arises because we have employed the frontier orbitals of the starting material rather than those of the transition state. Additions to CC multiple bonds generally involve two steps. In nucleophilic reactions, the first step (addition of the nucleophile to an electron-rich species) occurs more slowly than the second (neutralization of an anionic intermediate by a counterion) and is therefore rate determining. Calculations have shown[44] that an approaching nucleophile induces bending of an alkyne and pyramidalization of an alkene. These effects diminish the resonance integrals and reduce the HOMO–LUMO gap. However, because the alkyne distorts much more easily,[45] its LUMO energy in the transition state falls below that of the alkene (Figure 4.12). 6–31G* calculations put the acetylene LUMO at 0.2224 a.u. in the starting material and 0.1574 a.u. in the transition state. The alkene values are 0.1839 and 0.1604 a.u., respectively.

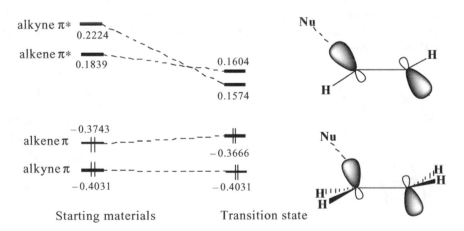

Figure 4.12   Variation in FO energies as a function of alkene and alkyne deformation.

---

[44]Alkynes are generally more reactive toward nucleophiles and less reactive toward electrophiles than their alkene counterparts. Strozier R. W., Caramella P., Houk K. N., *J. Am. Chem. Soc.*, 1979, **101**, 1340 and references therein.
[45]The CH bending force constants are 0.011 and 0.024 kcal mol⁻¹deg⁻¹ for acetylene and ethylene, respectively.

Electrophilic additions are more difficult to explain.[46] The first step is exothermic because it generates a cyclic 'onium salt', whose charge is delocalized over three centers (unless the electrophile is H[+]). However, the subsequent neutralization of the intermediate 'onium salt' by the counterion is even more favorable, so the first step is again rate-determining. When the electrophile is fairly large (e.g. Br[+]), it can interact with both ends of the unsaturated π system simultaneously, even at a substantial distance. Hence the geometric deformation of these atoms is relatively small and the transition state FOs are not very different from those in the isolated reagents. In these cases, ethylene will be more reactive than acetylene. As the radius of the electrophile decreases, the transition state becomes 'tighter', the reagents undergo greater geometric change and the reactivity of the alkyne increases with respect to the alkene. Experimental studies confirm that the rate constant ratio $k_{\text{alkene}}/k_{\text{alkyne}}$ falls in the order bromination, chlorination and protonation.[47]

The relative reactivities of carbonyl compounds

Some carbonyl compounds appear to violate rule 2. For example, acid chlorides have higher-lying LUMOs than aldehydes, but are more reactive. Yamataka *et al.*[48] calculated that NH_3 reacts more readily with F–CO–F, which has a higher LUMO (E_{LU}=0.185 a.u. at the 6–31G* level), than with H–CO–H (E_{LU} = 0.146 a.u.). However, when transition state geometries are employed, the LUMO ordering is inverted and the correct trend appears (0.032 a.u. for F–CO–F and 0.076 a.u. for H–CO–H).[46]

Staudinger reaction

A perfunctory study of the Staudinger reaction ('rich' alkene + ketene → cyclobutanone)[49] can lead us to believe in a failure of FO theory. The alkene, being rich in electrons, will react preferentially by its HOMO and the ketene by its LUMO. Also, as π_{CO}^* is lower than π_{CC}^*, the cycloaddition should give a methylene oxetane and not a cyclobutanone.

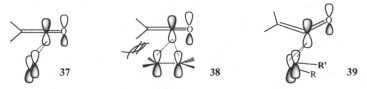

We must here anticipate a little and use some concepts developed in the following chapters. If the reaction is considered as a nucleophilic addition of the alkene to the ketene, the alkene should attack the central atom of the ketene, making an obtuse angle with C=O (p. 144). If it is considered as an electrophilic addition of ketene on the alkene, the central atom of the ketene should attack the double bond on its center

[46]Maurel F., Thesis, University of Paris VI, 1995.
[47]Yates K., Schmid G. H., Regulski T. W., Garratt D. G., Leung H.-W., McDonald R., *J. Am. Chem. Soc.*, 1973, **95**, 160.
[48]Yamataka H., Nagase S., Ando T., Hanafusa T., *J. Am. Chem. Soc.*, 1986, **108**, 601.
[49]Hyatt J. A., Raynolds P. W., *Org. React.*, 1994, **45**, 159.

(p. 172). Combining both results leads to **37** for the initial approach of the two partners. Approach **38**, which derives from **37** by rotation around a vertical axis, is disfavored by steric hindrance. However, **37** is only a first approximation. In fact, as soon as the reactants are sufficiently close, electrons are transferred from the alkene to the ketene and both molecules become distorted. The olefinic carbons are pyramidalized and the ketene is bent, as shown in **39**. In fact, the transition structure calculated by Wang and Houk[50] is very close to **39**. With two partial CC bonds, **39** clearly is an unstable structure. Rotation of the alkene around a vertical axis will bring one of its termini closer to the oxygen atom or to the terminal carbon of the ketene. Overlap is better with carbon, which has more diffuse orbitals than oxygen, and cyclobutanone formation is favored.

An imine would not react with C=O by its π orbital but by the nitrogen lone pair. The intermediate would not then be a bridged species like **39** but a zwitterion.[51] The mechanism remains unchanged, however: nucleophilic attack on C=O giving an intermediate which closes on the terminal carbon of ketene.

Exceptions and how to predict them

As we have seen, violations of rule 2 appear when changes occur in frontier orbital ordering as the reaction evolves from the starting material to the transition state. This can only arise when strong transition state interactions seriously deform the molecular geometry. Consequently, *rule 2 will be respected when the frontier orbital interactions are weak*, which is the situation in reactions with *early transition states.*

However, strong FO interactions do not necessarily lead to violation of rule 2. An inversion of frontier orbital ordering requires that the FO interactions are strong ***and*** *that the reactive sites are easily deformable.* Fortunately, this second criterion can be predicted in a number of ways. Infrared spectrometry could be used to show that acetylene is more deformable than ethylene, by looking at the CH bending frequencies. RCOCl and F–CO–F are very electron rich (having four π electrons for three atoms and six for four atoms, respectively). Therefore, the approach of a nucleophile, bringing in still more electrons, is expected to induce significant structural distortions, which are the source of the high reactivity of these compounds.

Exercises 15 and 17 show that rule 2 should be applied cautiously in nucleophilic reactions of unsaturated ketones. This is true also for electrophilic reactions; for example, osmium tetraoxide selectively oxidizes **40** at the isolated double bond (yield ~90%),[52] despite the fact that its HOMO is lower in energy.

40

----

[50]Wang X. B., Houk K. N., *J. Am. Chem. Soc.*, 1990, **112**, 1754.
[51]Sordo J. A., González J., Sordo T. L., *J. Am. Chem. Soc.*, 1992, **114**, 6249.
[52]Woodward R. B., Sondheimer F., Taub D., Heusler K., McLamore W. M., *J. Am. Chem. Soc.*, 1952, **74**, 4223.

Why Overlap is Ignored

From a theoretical standpoint, it seems incorrect that rule 2 takes into account only the denominator term $(E_{HO} - E_{LU})$ while neglecting the numerator $2P_{HO-LU}^2$. Nonetheless, it correctly predicts that the reactivity of series of dienophiles $CH_2=CHCN$, $CNCH=CHCN$, $CNCH=C(CN)_2$ and $(CN)_2C=C(CN)_2$ toward cyclopentadiene increases with increasing conjugation, whereas studies incorporateing numerator terms[53] indicated the reverse!

Calculations by Houk *et al.*[54] have shown that an increase in reactivity loosens the transition state of a Diels–Alder reaction. This agrees with the Hammond postulate, assuming that an early transition state is rather loose and a late transition state rather tight. Now, an increase in reactivity usually means that the HOMO(1)–LUMO(2) energy gap becomes smaller *and* the frontier coefficients larger, at least for a couple of reactive sites (see p. 49). In a loose transition state, the intermolecular distance is large and, consequently, the frontier overlap is small. Conversely, the frontier overlap is larger in a late transition state. Hence, whenever the frontier coefficients become larger, their overlap becomes smaller. Their product P_{ij} remains almost constant and can therefore be neglected.

'Formal' Frontier Orbitals and 'Chemical' Frontier Orbitals

The FO approximation was introduced in 1952 when all calculations were done using the Hückel method. In these conditions, the π frontier orbitals were also the chemically reactive MOs. Semi-empirical and *ab initio* calculations, which are frequently used now, take into account both σ and π orbitals. With these methods, the orbitals to be considered in FO theory are not necessarily the 'formal' frontier orbitals, i.e. the highest occupied and lowest unoccupied MOs! Consider, for example, the reduction of acetone by $LiAlH_4$. The 'chemically' important MO is of course the π^*_{CO} of the 'ate' complex, which is in fact the LUMO + 1. The formal LUMO is the empty s orbital of the lithium cation.

Exercise 20 (E)   *MOs are given on p. 255*

Is a thionoester more reactive than a dithioester, or vice versa?

R—C(=S)—OR' thionoester   R—C(=S)—SR' dithioester

---

[53]In these Hückel calculations,[27a] the intramolecular distance is presumed to be the same for each transition state. Therefore, the atomic overlaps are identical and the P_{ij} values are proportional to the frontier orbital coefficients.

[54]Houk K. N., Loncharich R. J., Blake J. F., Jorgensen W. L., *J. Am. Chem. Soc.*, 1989, **111**, 9172; Jorgensen W. L., Lim D., Blake J. F., *J. Am. Chem. Soc.*, 1993, **115**, 1936.

Answer[55]

When R = R′ = Me, the thionoester LUMO lies at $-0.56\,$eV and that of the dithioester at $-0.94\,$eV, according to AM1 calculations. Its LUMO being lower in energy, the dithioester is the more reactive.

Is this conclusion predictable? Yes. The thionoester is RC=S perturbed by OR′ whereas the dithioester is RC=S perturbed by SR′. The CO bond being stronger than the CS bond (85.5 versus $65.5\,$kcal$\,$mol^{-1}), the π^*_{CS} orbital is raised more by OR′ than by SR′. Moreover, reactions with esters being addition–elimination reactions, the weakness of the CS bond facilitates the elimination stage. *Always verify that your calculated results make sense chemically.*

Note that in the thionoester, as in the dithioester, the HOMO *stricto sensu* is a sulfur lone pair, the highest occupied π orbital being the HOMO − 1.

Exercise 21 (M)   *MOs are given on p. 255*

Generalizing the Alder rule. After considering Alder's rule, we can naturally ask the following questions:

(a) Will a cycloaddition be favored by the presence of a donor on the diene and an acceptor on the dienophile, or the reverse?
(b) Can we apply our concepts to 6 + 4, 8 + 2, … cycloadditions, etc.?

These problems can be treated in two steps by showing:

(1) That the LUMO energies in 'electron-poor' compounds do not vary significantly. Hence, we can ignore the influence of acceptor substituents, at least to a large degree.
(2) That it is desirable to place a donor substituent on the compound having the higher-lying HOMO.

Answer[56]

(1) Our rationale focuses on the Diels–Alder reaction, but it can also be applied to other thermal cycloadditions. The acceptor is modeled as a low-energy vacant orbital A whose energy reflects the nature of the substituent. Hückel values for common acceptors are C≡N (A $= \alpha - 0.781\beta$), C=O (A $= \alpha - 0.618\beta$) and NO$_2$ (A is almost non-bonding). As π^*_{CN} lies between π^*_{CC} and Ψ_3, the cyanobutadiene LUMO ($\alpha - 0.325\beta$) is Ψ_3 lowered in energy by interaction with A. On the other hand, the acrylonitrile LUMO ($\alpha - 0.460\beta$) is A lowered in energy by interaction with π^*_{CC}, which explains why it lies relatively close to the cyanobutadiene LUMO (scheme **a** below). For more powerful attractors such as C=O or NO$_2$, A lies at the same

[55]Hartke K., *Phosphorus, Sulfur, and Silicon*, 1991, **58**, 224.
[56]Anh N. T., Canadell E., Eisenstein O., *Tetrahedron*, 1978, **34**, 2283. For a completely different approach, see: Fujimoto H., Inagaki S., Fukui K., *J. Am. Chem. Soc.*, 1976, **98**, 2670; Fujimoto H., Sugiyama T., *J. Am. Chem. Soc.*, 1977, **99**, 15.

level as or lower than Ψ_3. A is then lowered more strongly by Ψ_3 than by π^*, but also raised more powerfully by its interaction with Ψ_2. Hence, despite its low-lying LUMO, the diene, once substituted, is not much more favored than the dienophile (scheme **b**). Thus, the HOMO of pentadienal lies at $\alpha - 0.241\beta$, not far removed from that of acrolein ($\alpha - 0.347\beta$).

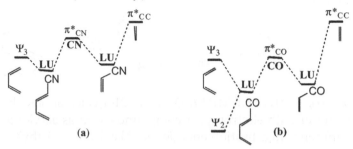

(a) (b)

(2) A donor substituent may be represented by a doubly occupied orbital D, at ($\alpha + 2\beta$) for O and at ($\alpha + 1.5\beta$) for N. Hence D *always* lies lower than the HOMOs of the diene and the dienophile. Scheme **c** illustrates the tricky case involving a first-order D–π interaction and a second-order D–Ψ_2 term. One may wonder if the HOMO of the substituted dienophile cannot be higher than that of the diene. In fact, we just need to take a double bond for D to see that the HOMO of hexatriene lies higher than the butadiene HOMO: when π is raised to the level of Ψ_2, the latter rises further, so that the diene always has the higher energy.

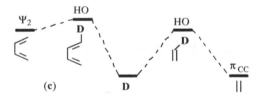

(c) D ‖

Question 1 shows that an acceptor has only a minor influence on the LUMO energy, so donors will play the dominant role. Question 2 indicates that *the HOMO–LUMO gap will be minimized by the presence of a donor on the compound having the higher-lying HOMO.*

Comments

1. The rule above does not simply imply that a donor should invariably be positioned on the electron-rich component whereas the attractor should be placed on the acceptor: butadiene is also a better acceptor than ethylene.
2. Cyclic compounds whose molecular symmetry prevents the substituent from interacting with certain orbitals are exceptions. See the article by Anh *et al.* for details.
3. If the donor is a sulfur atom, its lone pair is practically nonbonding (the electronegativities of sulfur and carbon are similar), above Ψ_2 and scheme **c** is no longer valid. If the sulfur is modeled by a carbanion, the sulfur-substituted diene and the dienophile are represented by the pentadienyl and the allyl anions, respectively. Their HOMOs will both be nonbonding. Hence FO theory predicts that placing the sulfur on the dienophile and the attractor on the diene may be slightly more favorable than the opposite substitution pattern.

Exercise 22 (E)   *MOs are given on p. 257*

2-Azabutadiene, $CH_2=N-CH=CH_2$, is an 'electron-poor' compound: its orbitals are lower lying than those of butadiene. Experiments show that electron-poor dienophiles are completely unreactive toward **A**, but those which are 'electron-rich' give Diels–Alder adducts:[57]

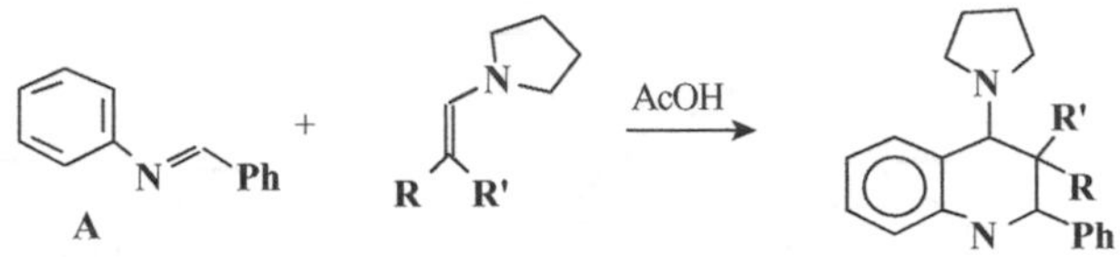

Even so, 3–21G and MP2/6–31G*//3–21G calculations show that 2-azabutadiene reacts better with electron-*poor* compounds such as acrylonitrile than with ethylene. The difference in activation energies is ~2 kcal mol^{-1}.[58] Why?

Answer

Acrylonitrile has a lower-lying LUMO and a higher-lying HOMO than ethylene. The results are a simple expression of rule 2.

Exercise 23 (E)   *MOs are given on p. 258*

Use perturbation theory to show that 2-azabutadiene is more reactive than 1-azabutadiene.

Answer

Both compounds can be viewed as a combination of ethylene and imine. Their MOs are given below:

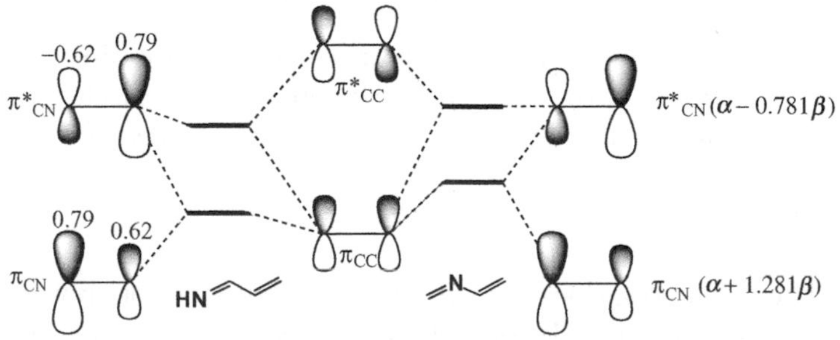

The calculation is classical, provided that we remember to treat the quasi-degenerate combinations of π_{CC} with π_{CN} and π^*_{CC} with π^*_{CN} using first-order perturbation equations (p. 30). For example, the 1-azabutadiene HOMO is π_{CC} perturbed by π_{CN} and π^*_{CN}. Therefore, it lies at

[57]Nomura Y., Kimura M., Takeuchi Y., Tomodo S., *Chem. Lett.*, 1978, 267; Shono T., Matsumura Y., Inoue K., Ohmizu H., Kashimura S., *J. Am. Chem. Soc.*, 1982, **104**, 5753; Povarov L. S., *Russ. Chem. Rev.*, 1967, **36**, 656.
[58]Gonzalez J., Houk K. N., *J. Org. Chem.*, 1992, **57**, 3031.

$$(\alpha + \beta) + (0.707 \times 0.62\beta) + \frac{(0.707 \times 0.79\beta)^2}{(\alpha + \beta) - (\alpha - 0.781\beta)} = \alpha + 0.737\beta$$

The first term $(\alpha + \beta)$ is the energy of the π_{CC} orbital. The second term $(0.707 \times 0.62\beta)$ is the correction due to first-order interaction with π_{CN} and the third term is the second-order correction due to π^*_{CN}. In the same manner, the 1-azabutadiene LUMO energy is given by

$$(\alpha - 0.781\beta) + (0.79 \times 0.707\beta) + \frac{(0.79 \times 0.707\beta)^2}{(\alpha - 0.781\beta) - (\alpha + \beta)} = \alpha - 0.373\beta$$

Analogous calculations give $\alpha + 0.549\beta$ and $\alpha - 0.494\beta$ for 2-azabutadiene. Since the HOMO–LUMO gap is smaller in the latter case, 2-azabutadiene is more reactive. The reactivity difference will be accentuated in reactions with electron-poor dienophiles.

The reaction with 1-azabutadiene is also unfavorable for other reasons. To begin with, the 4 + 2 cyclization is inherently less exothermic than with 2-azabutadiene (see Exercise 20, p. 123). Furthermore, the enamine product will rapidly undergo side-reactions with adventitious electrophiles. An imine–enamine tautomerism which transforms R–N=CH–CH=CH–CH$_3$ into RNH–CH=CH–CH=CH$_2$ also contributes to lowering the yield. Finally, electron-poor dienophiles may undergo a competing reaction with the nitrogen lone pair.

Exercise 24 (E)   *MOs are given on p. 259*

Trione *et al.*[59] have studied the interactions of three dienophiles:

with three differently substituted 2-cyano-1-azabutadienes (having conjugating, electron- donating and electron-withdrawing substituents at nitrogen):

A shows broadly the same reactivity toward each dienophile (110 °C, 7 days). Predictably, **B** reacts faster, particularly with ethyl vinyl ether (25 °C, 1 day). **C** shows no reactivity toward any of these dienophiles, even under the conditions employed for **A**. Explain these results.

Hint: see Exercises 23 and 25.

---

[59]Trione C., Toledo L. M., Kuduk S. D., Fowler F. W., Grierson D. S., *J. Org. Chem.*, 1993, **58**, 2075.

Answer

The FO gap in **A** is reduced by conjugation with the phenyl groups. In **B**, the ester reduces the FO gap and lowers the absolute orbital energies, which makes the compound particularly reactive toward electron-rich dienophiles. **C** is relatively unreactive because the effects of the CN and OMe groups cancel to a large degree. Substituents in the 2-position usually have minor effects (see Exercise 2, p. 57), but this is not necessarily the case here. The presence of the azadiene nitrogen increases the LUMO coefficient on the 2-position, so the CN substituent can lower the LUMO energy without significantly affecting the HOMO. As a result, the diene would normally show enhanced reactivity toward electron-rich dienophiles. The methoxy group has a detrimental effect: it diminishes reactivity toward electron-rich dienophiles but is insufficiently electron donating to confer any nucleophilic character upon the diene. From the next exercise, we can see that conjugated oximes are inert and hydrazones only show any useful activity at around 100 °C. Since the MeO group has a donor power somewhere between those of OH and NMe_2, it will not be able to activate the diene sufficiently to give useful reactivity toward electron-poor dienophiles.

Using Streitwieser's parameters (p. 22) with $\beta_{OMe} = 0.56\beta$ and $\beta_{ON} = 0.8\beta$, we can calculate FO energies of $\alpha + 0.484\beta$ and $\alpha - 0.171\beta$ for **A**, $\alpha + 0.679\beta$ and $\alpha - 0.056\beta$ for **B** and finally $\alpha + 0.479\beta$ and $\alpha - 0.298\beta$ for **C**. Since the HOMO–LUMO gap is largest for **C**, it is less reactive than the others.

Exercise 25 (E)   *MOs are given on p. 263*

The reactivity of azadienes has been thoroughly studied because their Diels–Alder reaction products can be used in alkaloid syntheses. The HOMO–LUMO gaps for 1-azabutadiene, 2-azabutadiene and butadiene are 1.272β, 1.230β and 1.236β, respectively. The reactivity of the latter pair is comparable, but the 2-azabutadiene reacts slightly less well with electron-poor dienophiles because its frontier orbitals are found at lower energy ($\alpha + 0.683\beta$ and $\alpha - 0.547\beta$ as against $\alpha \pm 0.618\beta$ for butadiene). This is unfortunate because most dienophiles fall into this category. However, in this exercise we will deal with the problem of increasing the reactivity of 1-azabutadienes, which normally give disappointing results in Diels–Alder reactions (see also the previous exercise and ref.[60])

(1) Ghosez's group made an early attempt to resolve this problem.[61] They showed that the conjugate hydrazone **A** gives regioselective Diels–Alder reactions with many dienophiles (100 °C, between 100 and 200 h) but that oxime **B** does not react, even after prolonged refluxing in acetonitrile. Why? Given these results, deduce whether they used electron-rich or electron-poor dienophiles.

A   B

---

[60]Boger D. L., *Tetrahedron*, 1983, **39**, 2869 ; Boger D. L., Weinreb S. N., *Hetero Diels–Alder Methodology in Organic Synthesis*, Academic Press, San Diego, 1987.
[61]Serckx-Poncin B., Hesbain-Frisque A. M., Ghosez L., *Tetrahedron Lett.*, 1982, 3261.

(2) Teng and Fowler[62] showed that *N*-acyl-2-cyano-1-azadienes react with electron-rich (e.g. CH_2=CHOEt) and electron-poor dienophiles (e.g. methyl acrylate) under relatively mild conditions. Stirring the reaction below at room temperature for 22 h gave a 61% yield for the intramolecular cyclization:

Boger and co-workers[63] developed the chemistry of conjugated *N*-sulfonylimines, which behave in a similar fashion Some of them are shown below.

Why do the Fowler and Boger reagents react so much more readily than Ghosez's compounds? To which type of dienophile are they best suited?

Answer

(1) Ghosez and co-workers used standard electron-poor dienophiles (quinone, acrylonitrile, methyl acrylate, maleic anhydride) for their experiments, hence the choice of donor substituents to increase the electron density of the azadiene (Alder's rule). However, the intrinsically electron-deficient diene can only be made sufficiently nucleophilic by the presence of exceptionally good donors. The oxygen lone pair is relatively low-lying ($\alpha + 2\beta$), so it does not confer sufficient reactivity for the oxime to react. AM1 calculations validate this qualitative reasoning: the oxime **B** HOMO lies at -9.47 eV versus -8.56 eV for **A**'s HOMO.

(2) The analysis of Alder's rule showed that any substituent, donor or acceptor, will enhance the reactivity of a diene. Fowler's and Boger's dienes are more reactive than Ghosez's reagents because it is easier to increase the electrophilicity of an azadiene than to transform it into a nucleophile.

Furthermore, the presence of two or even three good acceptors will obviously outweigh the influence of a single donor. The electron-withdrawing groups also have a practical advantage: they stabilize the enamine functionality in the product, making it less likely to decompose, especially as the reactions occur at lower temperatures.

Alder's rule suggests that the Boger and Fowler dienes should react more readily with electron-rich than electron-poor dienophiles. However, the global conclusion of rule 2 is that the reaction rate will increase as the frontier orbital gap between the reaction partners decreases. The most reactive dienes have very small HOMO–LUMO

[62]Teng M., Fowler F. W., *J. Org. Chem.*, 1990, **55**, 5646 and references therein
[63]Boger D. L., Corbett W. L., *J. Org. Chem.*, 1993, **58**, 2068 and references cited therein.

gaps, which allow them to react efficiently with both electron-rich and electron-poor dienophiles. AM1 calculations on the sulfonimine suggest that the LUMO lies 1.5 eV and the HOMO 1 eV lower than in 1-azabutadiene.[64]

Note, however (see Appendix), that the HO(**A**)–LU(acraldehyde) energy gap is slightly smaller than the HO(propene)–LU(sulfonimine) energy gap. Hence rule 2 should always be applied with caution.

---

[64]Boger D.L; Corbett W.L; Curran T.T; Kasper A.M; *J Am.Chem. Soc;* 1991, 113, 1713.

5 Regioselectivity

5.1 Cycloadditions

In principle, any cycloaddition involving two dissymmetric compounds can give head-to-head or head-to-tail products. These two compounds are usually obtained in nonequal proportions whose ratio cannot be adequately explained by steric or coulombic factors.[1] For example, Reaction (5.1) favors the more hindered product. In Reaction (5.2), the product is formed by creating bonds between atoms having the same charge in the starting material.

$$(5.1)$$

$$(5.2)$$

---

[1]Sauer J., *Angew. Chem. Int. Ed. Engl.*, 1967, **6**, 16 and references cited therein.

Frontier Orbitals   Nguyên Trong Anh
© 2007 John Wiley & Sons, Ltd

Bertran *et al.*[2] developed an elegant rationalization of these phenomena. It consists in modeling a donor substituent by a filled 2p orbital of energy α and an acceptor by an empty orbital of the same energy. One of the partners will usually be more electron rich than the other, so the reaction will occur preferentially between the HOMO of the electron-rich and the LUMO of the electron-poor component. Monosubstituted reagents can then be modeled as alternant ions, so the FOs are nonbonding orbitals. The interactions produced by orientations **A** and **B** are stabilizing whereas those by **C** and **D** are zero. Hence formation of the latter compounds is disfavored.

Bertran *et al.* method can only be applied when one reagent is monosubstituted by a donor group and the other by a single acceptor. A more general treatment recognizes that two (nonsymmetrical) reagents will interact at the sites where their FO lobes are largest.[3] This implies that the transition state in such a reaction is dissymmetric, one bond forming more rapidly than the other:

Obviously, the orientation of the cycloaddition is fixed by the formation of the first bond. In the example below, the initial interaction between **A** and **C** dictates that the second must occur between **B** and **D**:

Hence, the problem lies in determining which bond will form most easily. Rule 3 states that it will link the sites having the largest frontier orbital interaction.

Example. We will look at the reaction between 2-methoxybutadiene and methyl acrylate. The frontier orbital energies and coefficients at the reactive sites are

[2]Bertran J., Carbo R., Moret T., *Ann. Quim.*, 1971, **67**, 489.
[3]Eisenstein O., Lefour J. M., Anh N. T., *Chem. Commun.*, 1971, 969; Eisenstein O., Lefour J. M., Anh N. T., Hudson R. F., *Tetrahedron*, 1977, **33**, 523.

We only need to look for the most favorable bond. The stabilization gained by linking atoms 1 and 1' is given by

$$\frac{(0.6 \times 0.67 \times \beta)^2}{(\alpha + 0.545\beta) - (\alpha - 0.436\beta)} + \frac{(0.55 \times 0.58 \times \beta)^2}{(\alpha + \beta) - (\alpha - 0.652\beta)} = 0.226\beta$$

Similar calculations give

0.100β for a 1–2' bond

0.202β for a 4–1' bond

0.101β for a 4–2' bond

Hence the 1–1' bond forms most readily. This pathway leads to the compound below, which is the major experimentally observed product:

> **Comment**. To model the transition states leading to the different regioisomers, we have used intermediates where one bond is fully formed and the other is non-existent. It is equally possible to imagine a model where both of the σ bonds are formed to the same degree.[4] The first model exaggerates the asynchronicity of the process, whereas the second neglects it entirely. The first gives better results because neglecting asynchronicity means neglecting the most important factor in the area ofr egioselectivity.[5]

Almost invariably, one HOMO–LUMO pair has a smaller energy gap than the other. *The bond which forms most easily will link the atoms having the largest coefficients in the closest-lying FOs.*[6] Hence no numerical calculations are necessary. In the foregoing reaction, the nearest pair of FOs comprises the methoxybutadiene HOMO and the methylacrylate LUMO. Their largest coefficients are found at 1 (0.67) and 1' (0.67), respectively, so the 1–1' bond is the first to be formed.

No mechanistic assumptions are made in this approach, which is therefore perfectly general.[7] It has been used successfully to predict the outcome of Diels–Alder reactions,[3] 1,3 dipolar cycloadditions[8] and 2 + 2 cycloadditions.[5] Its validity has been confirmed

[4]Feuer J., Herndon W. C., Hall L. H., *Tetrahedron*, 1968, **24**, 2575.

[5]Minot C., Anh N. T., *Tetrahedron*, 1977, **33**, 533.

[6]This rule should be applied cautiously if the competitive sites belong to different rows of the periodic table (e.g. C versus S. See Exercise 2, p. 92, and Exercise 3, p. 93).

[7]However, the application of FO criteria to photochemical reactions is much more difficult because it is often necessary to superimpose different electronic configurations to obtain a realistic excited state model. See also p. 233.

[8]Bastide J., Henri-Rousseau O., *Bull. Soc. Chim. Fr.*, 1973, 2290 ; Houk K. N., Sims J., Watts C. R., Luskus L. J., *J. Am. Chem. Soc.*, 1973, **95**, 7301 ; Houk K. N., Sims J., Duke R. E. Jr, Strozier R. W., George J. K., *J. Am. Chem. Soc.*, 1973, **95**, 7287; Minato T., Yamabe S., Inagaki S., Fujimoto H., Fukui K., *Bull. Chem. Soc. Jpn*; 1974, **47**, 1619.

by *ab initio*[9] calculations, which show that cycloadditions are indeed asynchronous and that rule 3 accurately predicts the shortest bond in the transition state.

Exercise 1 (E)

Deserves your attention. MOs are given on p. 264
   Show, without any numerical calculations, that:

(1) The nonsubstituted carbon atom has the highest HOMO coefficent in donor-substituted dienophiles and highest LUMO coefficient in acceptor-substituted dienophiles
(2) The highest HOMO coefficient in a diene will be localized on position 4 if a donor substituent is present at C_1, but on position 1 if the donor appears at C_2.
(3) The highest LUMO coefficient in a diene will also be localized on position 4 if an acceptor substituent is present at C_1, but will be on position 1 for an acceptor at C_2.

Hint: Use a low-energy vacant orbital to model the acceptor substituent (for example a π^*_{CO} at $\alpha - 0.6\beta$) and a doubly occupied high-energy orbital to represent the donor (e.g. a nitrogen lone pair at $\alpha + 1.5\beta$). These representative values allow us to see how the energies of the substituent orbitals compare with those in butadiene and ethylene.

Answer

(1) The perturbation of ethylene by a donor substituent was given on p. 29. For an acceptor, we define the energy of the unperturbed ethylene π bonding orbital as $\alpha + \beta$, the unperturbed π^* antibonding orbital as $\alpha - \beta$ and the energy of the acceptor substituent **A** as $\alpha - 0.6\beta$. The perturbation diagram (a) below shows that the LUMO of the system comprises **A** mixed in-phase with π^* and out-of-phase with π. Note that in this diagram, **A** and the substituted carbon orbital in π and π^* *must* have the same sign (Cf. remark on p. 29). Scheme (b) outlines the three different components (**A**, $\lambda\pi^*$ and $-\mu\pi$) and how they contribute to the LUMO.

   Check that the rule (the nonsubstituted atom has the highest HOMO coefficient in electron-rich and the highest LUMO coefficient in electron-poor dienophiles) is general, using examples in the MO Catalog in the Appendix (enol, propene, acrolein, styrene, acrylonitrile).

(2) The butadiene orbitals are denoted Ψ_i and the orbital representing the donor substituent D. The MO diagram (c) below shows that the HOMO of the 1-substituted

[9]E.g. (a) Burke L. A., *J. Org. Chem.*, 1985, **50**, 3149 ; (b) Houk K. N., Loncharich R. J., Blake J. F., Jorgensen W. L., *J. Am. Chem. Soc.*, 1989, **111**, 9172 ; (c) Loncharich R. J., Brown F. K., Houk K. N., *J. Org. Chem.*, 1989, **54**, 1129 ; (d) Valenti E., Pericàs M. A., Moyano A., *J. Org. Chem.*, 1990, **55**, 3582 (AM1 calculations) ; (e) Birney D. M., Houk K. N., *J. Am. Chem. Soc.*, 1990, **112**, 4127 ; (f) Gonzalez J., Houk K. N., *J. Org. Chem.*, 1992, **57**, 3031 ; (g) Singleton D. A., Leung S. W., *J. Org. Chem.*, 1992, **57**, 4796 ; (h) Brown F. K., Singh U. C., Kollman P. A., Raimondi L., Houk K. N., Bock C. W., *J. Org. Chem.*, 1992, **57**, 4862 ; (i) Bachrach S. M., Liu M., *J. Org. Chem.*, 1992, **57**, 6736 ; (j) Dâu M. E. T. H., Flament J. P., Lefour J. M., Riche C., Grierson D. S., *Tetrahedron Lett.*, 1992, **33**, 2343 ; (k) Cioslowski J., Sauer J., Hetzenegger J., Karcher T., Hierstetter T., *J. Am. Chem. Soc.*, 1993, **115**, 1353 ; (l) Jorgensen W. L., Lim D., Blake J. F., *J. Am. Chem. Soc.*, 1993, **115**, 2936 ; (m) McCarrick M. A., Wu Y. -D., Houk K. N., *J. Org. Chem.*, 1993, **58**, 3330.

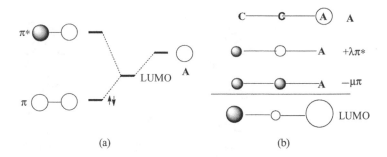

(a)　　　　　　　　　　　(b)

butadiene is the Ψ_2 orbital perturbed by an out-of-phase combination with D. Ψ_1 and Ψ_3 become mixed into the HOMO because of their interactions with D. Their contributions determine the relative values of the HOMO coefficients in the 1- and 4- positions. Mixing with Ψ_4 can be ignored because the $(E_2 - E_4)$ energy gap is large and the Ψ_4 to D resonance integral $(P_{4,D})$ is small (because the coefficient of C_1 in Ψ_4 is small). The mixing coefficient of Ψ_2 and Ψ_3 is negative:

$$\frac{P_{2,D}P_{3,D}}{(E_2 - E_D)(E_2 - E_3)} = \frac{(-)(-)}{(+)(-)} < 0$$

whereas the mixing coefficient of Ψ_2 with Ψ_1 is positive:

$$\frac{P_{2,D}P_{1,D}}{(E_2 - E_D)(E_2 - E_1)} = \frac{(-)(-)}{(+)(+)} > 0$$

The contribution made by each of these components to the HOMO is shown in (d). Related catalog examples are 1,3-pentadiene and 1-methoxy- and 1-aminobutadiene.

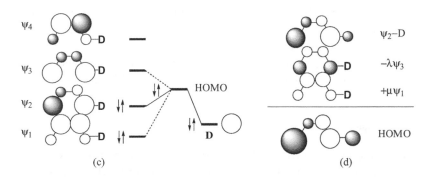

(c)　　　　　　　　　　　(d)

$P_{1,D}$ and $P_{4,D}$ have non-negligible values if a donor is present in the 2-position, because Ψ_1 and Ψ_4 have significant coefficients at this site. As a consequence, every Ψ_i orbital contributes to the HOMO (see scheme e below). The coefficients at the central carbons are difficult to evaluate because some of the contributions cancel, but there is no doubt that the coefficient at C_1 is much greater than that at C_4. Catalog examples are isoprene and 2-amino- and 2-methoxybutadiene.

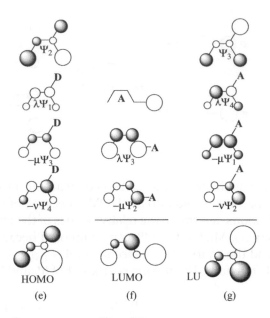

HOMO   LUMO   LU
(e)   (f)   (g)

(3) A schematic breakdown of the LUMOs is given in (f) and (g). Mixing of A with Ψ_1 and Ψ_4 can be ignored in scheme (f) (cf. question 1). Examples in the catalog are pentadienal and 2-formylbutadiene.

Exercise 2

Question 1 is relatively difficult. Questions 2 and 3 are easy
Examine the FOs of $CH_2{=}S$, $O{=}CH{-}CH{=}S$ and $Me{-}CH{=}S$ (given p. 265).

(1) In $CH_2{=}S$, the LUMO coefficient at S is smaller than that at C. In $Me{-}CH{=}S$, this difference increases slightly, but in $O{=}CH{-}CH{=}S$, the S coefficient is larger. Can you predict these changes, *without any numerical calculations*?
(2) Predict the principal products from the reactions of 2-*tert*-butyldimethylsiloxy-butadiene with $CH_2{=}S$, $O{=}CH{-}CH{=}S$ and $Me{-}CH{=}S$.
(3) Will the regioselectivity in the reactions above change if we replace the thiocarbonyls by $CH_2{=}O$, $O{=}CH{-}CH{=}O$ and $Me{-}CH{=}O$?

Answer[10]

(1) The change in the LUMO coefficients is due to the mixing of π_{CS} into π^*_{CS}, due to the perturbation induced by the substituent (Me or C=O). In $Me{-}CH{=}S$, the mixing coefficient is given by (p. 30)

$$\frac{P_{Me,LU(CS)}P_{Me,HO(CS)}}{[E_{LU(CS)} - E_{Me}][E_{LU(CS)} - E_{HO(CS)}]} = \frac{(-)(-)}{(+)(+)} = (+)$$

─────────────

[10] Vedejs E., Perry D. A., Rondan N. G., *J. Am. Chem. Soc.*, 1983, **105**, 7001.

The mixing coefficient is therefore positive. When using perturbation theory, remember that in the mixing MOs, the coefficient of the substituted atom must have the same sign as the substituent (p. 29). As the Me group is located on the C atom of C=S, if the π_{CS} is written as

$$\pi_{CS} = 0.65C + 0.76S$$

then π^*_{CS} must be written as

$$\pi^*_{CS} = 0.76C - 0.65\,S \quad \text{and not} \quad \pi^*_{CS} = -0.76C + 0.65S$$

Mixing π_{CS} into π^*_{CS} will then diminish the S coefficient. The solution is even simpler for O=CH—CH=S. As π^*_{CO} lies between π^*_{CS} and π_{CS}, the O=CH—CH=S LUMO is essentially π^*_{CO} mixed in phase with π^*_{CS} and out-of-phase with π_{CS}:

$$\text{LUMO} = N(\pi^*_{CO} + \lambda\pi^*_{CS} - \mu\pi_{CS})$$

where N is a normalization coefficient and λ and μ positive mixing coefficients. This formula shows that the perturbation by the C=O substituent will increase the S coefficient.

(2) The reagent, being an electron-rich diene, will react through C_1 where its HOMO is most heavily localized. Hence, the first bond will form between C_1 and the site of highest LUMO coefficient in the reaction partner: S for O=CH—CH=S and C in CH$_2$=S and Me–CH=S. The major products will be **1, 2** and **3**.

      1   **2**   **3**

2-Methoxybutadiene can be used to model the diene in a qualitative treatment. AM1 calculations give very similar S and C LUMO coefficients for O=CH—CH=S; however, the difference is more pronounced in STO-3G and 3–21G calculations. Note that even with AM1 calculations (C coefficient = -0.64; S coefficient = 0.65), a preference is still expected for attack at sulfur, as sulfur orbitals are much more diffuse than carbon orbitals (covalent radii: 1.02 and 0.77 Å respectively).

(3) The MOs for CH$_2$=O, O=CH—CH=O and Me–CH=O are given in the catalog. The carbon has the highest LUMO coefficient, so the favored products are

      4   **5**   **6**

Exercise 3 (E)

MOs are given on p. 267

Predict the predominant product in each of the reactions below:

Answer[11]

(a) We will use the numbering scheme employed in the Appendix. Compound **8** is produced when the 1–4 bond forms first. An initial 4–4 bond will lead to **9**. The respective stabilization energies are:

$$1-4: \quad \frac{(0.46 \times 0.39 \times \beta)^2}{E_{HO} - E_{LU}} + \frac{(0.51 \times 0.33 \times \beta)^2}{E_{HO} - E_{LU}} = \frac{0.06}{E_{HO} - E_{LU}} \beta_{CS}$$

$$4-4: \quad \frac{2 \times (0.33 \times 0.39 \times \beta)^2}{E_{HO} - E_{LU}} = \frac{0.03}{E_{HO} - E_{LU}} \beta_{CC}$$

Furthermore, as $\beta_{CS} > \beta_{CC}$, **8** is certainly the main product.

Note: The sulfur orbitals being diffuse (see the previous exercise), β_{SS} should be the largest of all resonance integrals. We therefore expect the major product to be in fact **17**. Indeed, the reaction initially gives a mixture of **8** and **17**, but NMR studies show that the latter disappears completely upon standing for 1 week at room temperature.[12]

---

[11]Reaction (a): Pradère J. P., Bouet G., Quiniou H., *Tetrahedron Lett.*, 1972, 3471. Reactions (b) and (c): Pradère J. P., N'Guessan Y. T., Quiniou H., Tonnard F., *Tetrahedron*, 1975, **31**, 3059.
[12]Guémas J. P., Quiniou H., *Sulfur Lett.*, 1984, **2**, 121.

(b) The smallest FO gap is between the HOMO of **10** and the LUMO of **11**. The largest coefficients are at the sulfur and the terminal carbon of the acrylonitrile. Compound **12** should be the major product.

(c) The highest LUMO coefficient in **14** is at C_4. Hence, the principal product is **15**.

Exercise 4 (E)

MOs are given on p. 268.

Justify the regiochemistry of Reactions (5.1) and (5.2) on p. 87.

Answer

Pentadiene can be used to model *tert*-butylbutadiene in Reaction (5.1). The carbonyl group affects the dienophile much more than the methyl, which can then be neglected. The ester function can also be replaced by an aldehyde (verify that ethyl 2-methacrylate can be simulated by either methyl 2-methacrylate, 2-methylacrolein, methyl acrylate or acrolein). In each case, the first-formed bond will link the atom having the highest HOMO coefficient in the diene to the atom having the highest LUMO coefficient in the dienophile.

To calculate the dimerization of acrolein, we need to evaluate four possible bond formations. Numbering from oxygen, show that the 4–4′ bond has an energy of 0.256β, the 1–4′ bond 0.182β, the 4–3′ bond 0.143β and the 1–3′ bond 0.07β.

Exercise 5 (E)

Retrosynthetic analysis of fumagillol **18** leads to two possible synthetic schemes, (a) and (b), both starting with a Diels–Alder reaction:[13]

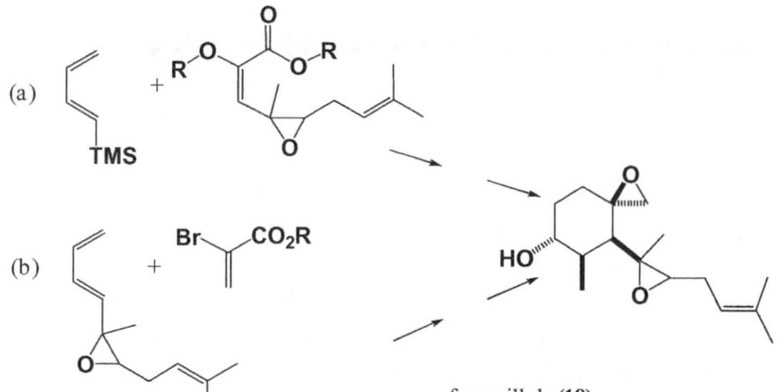

fumagillol **(18)**

Which one would you choose?

Answer[14]

Scheme (b) was used in the published synthesis. Scheme (a) may lead to the wrong substitution pattern. TMS is a weak electron donor, so the more reactive site of the

[13]Corey E. J., Cheng X.-M., *The Logic of Chemical Synthesis*, J. Wiley & Sons, N.Y., 1989, p. 19.
[14]Corey E. J., Snyder B.B., *J. Am Chem. Soc.*, 1972, **94**, 2549.

diene would be the nonsubstituted terminus. In the dienophile, OR and CO_2R have the strongest effects, so we expect the reactive site to be the monosubstituted carbon.

5.2 Electrophilic Reactions

5.2.1 Markovnikov's Rule

Two explanations are generally given for Markovnikov's rule. According to the first, the rule results from the preferential formation of the most stable carbocation in the system. This implicitly assumes that the transition state resembles the product, which implies that the addition of an electrophile to a double bond is endothermic. Unfortunately, this is not always justified.[15] The second interpretation suggests that the presence of electron-releasing methyl group makes the double bond more 'electron-rich'. The π electrons are mobile, so the electron density increase tends to be greater on the terminal carbon. Libit and Hoffmann[16] have shown, however, that deformation of the π cloud does indeed occur, but practically no electron has been transferred from the methyl group.

The FO explanation is very simple. An electrophile will attack the propene at the site of the highest HOMO coefficient, i.e. the nonsubstituted carbon atom (Exercise 1, p. 32). This is a simple application of rule 3. This rule implies that the transition state FOs are not very different from the FOs of the starting materials, and it probably works best for reactions with 'early' transition states. The main advantage of the FO approach over that based on the stability of the intermediate cations is its generality: it applies not only to Markovnikov's rule but also C-substitution of enolates.

5.2.2 Regioselectivity Involving Enols and Enolates

Even in the 1960s, organic chemists had difficulty in understanding why enolates, whose highest charge density is found at oxygen, frequently react through their carbon atoms. An interpretation, proposed in the late 1960s, is based on the theory of hard and soft acids and bases (HSAB) theory.[17] A soft (or hard) reagent will preferentially attack the carbon (or oxygen) atoms, because these are the soft (and hard) sites on the enolate, respectively. In most cases, the HSAB and FO theories are analogous. A reaction involving hard reaction partners is under charge control,[18] whereas reactions between soft reagents are under frontier control. However, the phenomenological HSAB theory does not interpret subtle differences particularly well. Why, for instance, should esters give *C*-alkylation when anhydrides give *O*-alkylation? Equally, why should RBr attack

---

[15]The protonation involves cleaving a carbon π bond (at a cost of ~50 kcalmol^{-1}) and the formation of a CH bond (which produces ~100 kcalmol^{-1}). The positive charge is transferred from a hydrogen to a carbon center. Relevant ionization potentials: H(1s) $\leq$ 13.6 eV, C(2p) $\leq$ 11.4 eV.

[16]Libit L., Hoffmann R., *J. Am. Chem. Soc.*, 1974, **96**, 1370.

[17]Pearson R. G., *J. Chem. Educ.*, 1987, **64**, 561; Ho T. L., *Tetrahedron*, 1985, **41**, 1; Ho T. L., *Hard and Soft Acids and Bases Principle in Organic Chemistry*, Academic Press, New York, 1977.

[18]Klopman G., *J. Am. Chem. Soc.*, 1968, **90**, 223.

at the carbon atom in solution and the oxygen in the gas phase? The FO approach is more useful in these cases.

Rule 3 states that the reaction of the enolate will occur at carbon if it is under frontier orbital control, i.e. if the electrophile LUMO is low enough in energy that the $(LU–\Psi_2)$ interaction in much larger than the $(LU–\Psi_1)$ interaction, Ψ_1 and Ψ_2 being the occupied π orbitals of the enolate (Figure 5.1). Consider now an electrophile whose LUMO is sufficiently high lying that there is not much difference between $(E_{LU} – E_1)$ and $(E_{LU} – E_2)$. Remember that the interactions of the electrophile with Ψ_1 and Ψ_2 are given by

$$\frac{P_{LU,\Psi_1}{}^2}{E_{LU} – E_1} \quad \text{and} \quad \frac{P_{LU,\Psi_2}{}^2}{E_{LU} – E_2}$$

The denominators not being significantly different, these two expressions are of the same order of magnitude. In other words, the FO interaction is no longer predominant and interactions of the electrophile LUMO with both Ψ_1 and Ψ_2 are to be taken into account. The regioselectivity will then depend on the C and O coefficients *in both of the occupied orbitals*. Ψ_1 is much more dissymmetric than Ψ_2 in an enolate, so the attack at oxygen is favored.

O-Substitution will therefore be associated with electrophiles having high-lying LUMO's. Electrophiles can be subdivided into two classes: (1) carbonyl compounds, which have a low-lying π^*_{CO} LUMO and normally react at C; and (2) compounds of general formula R–X, wherein the LUMO is a σ^*_{CX} orbital whose energy is variable. Figure 5.2 shows that strong CX bonds are associated with high σ^*_{CX} orbital energies. Indeed, the mixing of the X and carbon valence AOs gives rise to one σ_{CX} and one σ^*_{CX} orbital. The destabilization Δ' of the σ^*_{CX} with respect to $\varphi(C)$ is greater than the stabilization Δ of σ_{CX} with respect to $\varphi(X)$. Δ may be taken as a measure of the bond strength. Thus, the stronger is the CX bond, the larger are Δ and Δ', and the higher the energy of the corresponding antibonding orbital tends to be. Table 5.1 reproduces some bond dissociation energies given by Huheey.[19] They refer to homolytic processes, but their trends are adequate for our analysis. The CF bond is too strong to be cleaved by an enolate, so the best candidates for *O*-alkylation will be RX where X = OR'. Experiments have shown alkyl sulfates and sulfonates to be excellent *O*-alkylation reagents.[20]

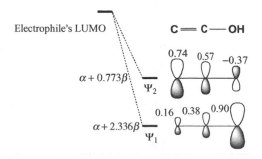

Figure 5.1 Interactions of the electrophile's LUMO with the enol occupied MOs.

[19]Huheey J. E., *Inorganic Chemistry*, 3rd Edn, Harper & Row, New York, 1983, p. A37.
[20]Carey F. A., Sundberg R. J., *Advanced Organic Chemistry*, Plenum/Rosetta, New York, 1977, Part B, p. 16.

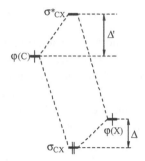

Figure 5.2   Relationship between the bond strength and antibonding orbital energy.

Silicon is less electronegative than carbon (1.8 as against 2.5 on the Pauling scale) and the Si–Cl bond strength is much higher than that of the CO π bond. Consequently, the σ^*_{SiCl} orbital lies at very high energy and an enolate will react with R_3SiCl to give an enol ether. Thermodynamic factors also favor *O*-silylation. Thus, theory reproduces the experimental studies,[21] which show that the percentage of *O*-alkylation increases in the order RI, RBr, RCl.

At first sight, it is difficult to understand why enolates should undergo *O*-methylations with diazomethane.[20] The reason is that the CN bond of diazomethane is stronger than the 73 kcalmol^{-1} shown in Table 5.1, because it has a partial double bond character (as we can see from an MO analysis or the resonance hybrids below). Consequently, σ^*_{CN} is relatively high-lying, so *O*-methylation occurs:

$$\overset{-}{C}H_2 - \overset{+}{N} \equiv N \leftrightarrow \overset{-}{C}H_2 - N = \overset{+}{N} \leftrightarrow CH_2 = \overset{+}{N} = \overset{-}{N} \leftrightarrow \overset{+}{C}H_2 - N = \overset{-}{N}$$

Why acid chlorides and anhydrides undergo *O*-acylation has not been properly explained. According to one explanation, as the reactivity of the electrophile increases,

Table 5.1   Some σ bond dissociation energies (kcal/mol^{-1})

C–I	51	Si–I	56
C–S	65	Si–S	70 (?)
C–Br	68	Si–Br	74
C–N	73		
C–Cl	78	Si–Cl	91
C–C	82.5	Si–C	76
C–O	85.5	Si–O	108
C–F	116	Si–F	135

[21]Kurts A. L., Genkina N. K., Masias A., Beletskaya I. P., Reutov O. A., *Tetrahedron*, 1971, **27**, 4777; Sarthou P., Guibé F., Bram G., *Chem. Commun.*, 1974, 377 ; LeNoble W. J., Morris H. F., *J. Org. Chem.*, 1969, **34**, 1969; LeNoble W. J., Puerta J. E., *Tetrahedron Lett.*, 1966, 1087; Brieger G., Pelletier W. M., *Tetrahedron Lett.*, 1965, 3555.

the transition state becomes earlier and resembles the enolate. The enolate oxygen atom then reacts preferentially because it has higher charge. Nonetheless, some highly reactive species (aldehydes, for example) react at carbon. The following interpretation can be suggested. Carbonyl compounds attack preferentially at the enolate carbon, as dictated by frontier orbital considerations. Aldehydes and ketones undergo single-step reactions, so they give aldols. However, acylations are more complicated and proceed in two steps (Figure 5.3). For unreactive species such as esters, the first (addition) is slower than the second (elimination) step, so *C*-acylation occurs. The elimination step will be rate limiting for highly reactive compounds such as anhydrides and acid chlorides, so the regioselectivity is determined by the relative stability of the equilibrating intermediates **1** and **2**. It is easy to show (cf. p. 210) that in **2**, the CX bond, being weakened by two oxygens, is more easily broken than in **1**, where it is weakened by only one oxygen.

We will conclude by looking at the differences which are sometimes observed between reactions in solution and in the gas phase. STO-3G calculations indicate that the FO coefficient at carbon is larger than at oxygen (-0.76 versus 0.65). However, 3–21G, 3–21+G and AM1 analyses[22] of the reaction between $CH_2{=}CHO^-$ and MeF have shown that the lowest energy transition state gives *O*-methylation, but that *C*-methylation is more exothermic. When the systems are solvated using Onsager's counterfield model,[23] the energy gap between the *O*- and *C*-methylation transition states is reduced, to a greater extent for MeCl than MeF.[24]

Experimentally, MeBr gives *O*-methylation in the gas-phase reactions[25] but RBr favors *C*-alkylation in solution. In general, charge control dominates gas-phase reactions because strong Coulombic effects[26] are not attenuated by counterions or solvents. Since solution reactions are usually under frontier orbital control, the reactivity

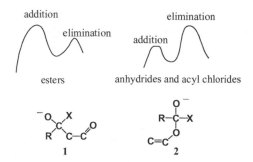

Figure 5.3 Regioselectivity in enolate acylation reactions.

[22]Houk K. N., Paddon-Row M. N., *J. Am. Chem. Soc.*, 1986, **108**, 2659; Sanchez Marcos E., Maraver J., Anguiano J., Bertran J., *J. Chim. Phys.*, 1987, **84**, 765.

[23]Onsager L., *J. Am. Chem. Soc.*, 1936, **56**, 1486.

[24]Maurel F., unpublished results.

[25]Jones M. E., Kass S. R., Filley J., Barkley R. M., Ellison G. B., *J. Am. Chem. Soc.*, 1985, **107**, 109.

[26]More than 30 orders of magnitude greater than the force of gravity. Feynman calculates that two grains of sand of 1 mm diameter separated by 30 m would exert a force of 3 million tons on each other if they were completely ionized.

differences can be spectacular.[27] For instance, the gas-phase reaction of CD_3O^- with $CH_3CO_2CH_3$ follows an S_N2 pathway, which gives CD_3OCH_3 and $CH_3CO_2^-$ rather than transesterification products.[24b] An even more surprising series has been described by Bartmess and co-workers.[27e] They found that the preferred reaction of HO^- with acetonitrile in the total absence of solvent is deprotonation:

$$HO^- + CH_3CN \quad \rightarrow \quad H_2O + {}^-CH_2CN$$

but that an S_N2 reaction predominates when the anion is monosolvated:

$$(H_2O)HO^- + CH_3CN \quad \rightarrow \quad CH_3HO + CN^-(H_2O)$$

and hydrolysis is observed in solution:

$$(H_2O)_n HO^- + CH_3CN \quad \rightarrow \quad CH_3CO_2^-(H_2O)_n + NH_4OH$$

Regioselectivity in Sulfur Compounds

The chemistry of sulfur compounds is replete with beautiful examples of regioselectivity inversions,[28] but unfortunately not all of them can be explained by FO theory.

Regioselectivity of Enethiolates

It seems that carbophilic attacks have never been observed in reactions involving enethiolates **A** derived from thioketones.[29] Let us model **A** by replacing all alkyl groups with methyls. According to PM3, MNDO and STO-3G calculations, the HOMO of **A** has its largest coefficient at sulfur. Rule 3 then predicts thiophilic attacks. It is interesting to note that the subjacent π MO, which is the HO-2 MO, has also the larger coefficient at sulfur. This suggests that even with high-lying LUMO electrophiles, thiophilic attacks are preferred, in good agreement with experimental results.

[27]E.g. (a) Faigle J. F. G., Isolani P. C., Riveros J. M., *J. Am. Chem. Soc.*, 1976, **98**, 2049; (b) Comisarow M., *Can. J. Chem.*, 1977, **55**, 171; (c) Fukuda E. K., McIver R. T. Jr, *J. Am. Chem. Soc.*, 1979, **101**, 2498; (d) Bohme D. K., Mackay G. I., Tanner S. D., *J. Am. Chem. Soc.*, 1980, **102**, 407; (e) Caldwell G., Rozeboom M. D., Kiplinger J. F., Bartmess J. E., *J. Am. Chem. Soc.*, 1984, **106**, 809 ; (f) Johlman C. L., Wilkins C. L., *J. Am. Chem. Soc.*, 1985, **107**, 327.
[28]Metzner P., Thuillier A., *Sulfur Reagents in Organic Synthesis*, Academic Press, San Diego, 1994.
[29]Schaumann E., in *The Chemistry of Double-bonded Functional Groups*, Patai S. (Ed.), John Wiley & Sons. Ltd, Chichester, 1989, p. 1324; Duus F., in *Comprehensive Organic Reactions*, Barton D. H. R., Ollis W. D., Neville Jones D. (Eds.), Pergamon Press, Oxford, 1979, Vol. 3, p. 395.

However, consider now the enethiolates **B** derived from dithioesters. Experimentally, they are attacked at C by carbonyl compounds[30] (which have low-lying LUMOs) and at S by alkyl halides[31] (which have high-lying LUMOs). By analogy with the enolate case, we may expect that in the highest occupied π MO of **B**, the largest coefficient would be at C, and in a subjacent π MO, it would be at S. Unfortunately these expectations are not always fulfilled. PM3 and 3–21G calculations give the largest coefficient at sulfur in all occupied π MOs. Only with STO-3G calculations can we find a larger coefficient at C in a low-lying π MO (HOMO-3) and the difference is rather small (0.33 for C versus 0.30 for S). In the other two occupied π orbitals (HOMO and HOMO-2), the largest coefficient is at S and the differences are more marked (0.76 versus 0.41 and 0.27 versus 0.21, respectively). Thus, for the enethiolates, FO theory is unable to predict carbophilic attacks. The probable reason is that, compared with thiophilic transition states, carbophilic transition states are rather late: C orbitals are much more contracted than S orbitals and can interact with the electrophile orbitals only at relatively short distances.[32] It follows that molecular distortions and, by way of consequence, frontier orbital changes, are more important in carbophilic reactions: the transition state FOs can no longer be approximated by the enethiolate FOs.

Enethiolates add preferentially 1,4 to conjugated carbonyls.[33] The aluminum enolate derived from Et(C=S)SMe reacts through S to give 1,4-addition with 3-penten-2-one,[34] which is coherent with a frontier-controlled reaction (p. 96), assuming that this enethiolate HOMO has its larger coefficient at the S atom. On the other hand, the titanium enolate of the same compound gives a 1,2-addition at carbon, in line with a charge-controlled reaction, provided that the subjacent π MO has its larger coefficient at carbon. PM3 calculations clearly suggest the possibility of regioselectivity inversion. Indeed, the HOMO of CH_2=C(SMe)SLi has the larger coefficient at sulfur (-0.60 versus 0.44), but the subjacent HO-4 orbital has the larger coefficient at carbon (0.38 versus 0.32 at sulfur).

Regioselectivity of Thiocarbonyls

When a thiocarbonyl undergoes a nucleophilic attack at carbon, a negative charge develops at S, which can accommodates it easily (remember that a sulfur atom can stabilize an adjacent negative charge[35]). On the other hand, thiophilic attacks give

[30]Meyers A. I., Tait A., Comins D. L., *Tetrahedron Lett.*, 1978, 4657; Beslin P., Vallée Y., *Tetrahedron*, 1985, **41**, 2691.

[31]Schuijl P. J. W., Brandsma L., Arens J. F., *Recl. Trav. Chim. Pays-Bas*, 1966, **85**, 1263; Beiner J. M., Thuillier A., *C. R. Acad. Sci. Paris*, Sér. C, 1972, **274**, 642.

[32]In 'normal' transition states, the incipient bond has ~1.4 as its equilibrium value (Anh N. T., Maurel F., Lefour J. M., *New J. Chem.*, 1995, **19**, 353 and references cited therein). We can then take as first 'guesstimates' 2.55 Å for the (Electrophile···S) distance in thiophilic transition states and 2.16 Å for the (E···C) distance in carbophilic transition states. The latter transition states are thus tighter and later than the former transition states.

[33]Berrada S., Metzner P., Rakotonirina R., *Bull. Soc. Chim. Fr.*, 1985, 881.

[34]Metzner P., in *Review on Heteroatom Chemistry*, Oae S. (Ed.), MYU, Tokyo, 1989, Vol. 2, p. 152.

[35]A well-known *Umpolung* reaction uses anion of 1,3-dithiane as an acyl synthon (Corey E. J., Seebach D., *Angew. Chem.*, 1965, **77**, 1134). According to Schleyer *et al.* (Schleyer P. v. R., Clark T., Kos A. J., Spitznagel G. W., Rohde C., Arad D., Houk K. N., Rondan N. G., *J. Am. Chem. Soc.*, 1984, **106**, 6467), the stabilizing effect is not due to sulfur d orbitals, but to the polarizability of the sulfur atom and/or electron transfer from the carbanion to the adjacent low-lying σ_{SC}^*.

rise to a negative charge at carbon, which is less favorable. Hence thiophilic attacks occur only if the thiocarbonyl is substituted by electron-attracting groups (CF_3, C=S, S, etc.). For example, alkyl Grignard reagents react with dithioesters R'–CS–SR_2 by thiophilic additions.[36] A thiophilic addition has been observed for the bis-thionoester[37] MeO–CS–CS–OMe. An electron-donating substituent destabilizes a carbanion and disfavors thiophilic attacks. Hence only nucleophilic attacks at carbon have been reported for thionoesters[38] R–CS–OR' and for thioamides[39] R'–CS–NR_2.

On the whole, FO theory accounts for thiocarbonyl regioselectivity in a fairly satis-factory manner. AM1, PM3, MNDO, 3–21G and STO-3G calculations on thioacetone all give the largest LUMO coefficient at carbon, thus suggesting that carbophilic attack is the rule. The sulfur coefficient is larger in the LUMO of CF_3–CS–CF_3, although the AM1 difference is negligible. For Ph–CS–Me, STO-3G calculations give the largest coefficient at S, which is probably incorrect (see Exercise 19, p. 123).

AM1, 3–21G and STO-3G calculations all give the largest LUMO coefficient on C for Me–CS–SMe. Hence FO theory cannot predict thiophilic attacks on dithioesters.

5.2.3 FO Theory and Ionic Reactions

FO theory can be used fruitfully for preliminary explorations by computational chemists. However, its main value lies in its capacity to provide chemists with *rapid* qualitative solutions to their problems. So do not waste your time refining the frontier orbitals. For conjugated molecules, Hückel MOs may be sufficient. Using the Hückel method for ionic reactions is, however, questionable and SCF calculations raise some other problems.

[36]Léger L., Saquet M., *Bull. Soc Chim. Fr.*, 1975, 657; Meyers A. I., Tait T. A., Comins D. L., *Tetrahedron Lett.*, 1978, 4657.
[37]Hartke K., Gillmann T., *Liebigs Ann. Chem.*, 1986, 1718.
[38]Narashiman L., Sanitra R., Ramachandran J., Sastry V. V. S. K., *Chem. Commun.*, 1978, 719.
[39]Beak P., Yamamoto J., Upton C. J., *J. Org. Chem.*, 1975, **40**, 3052; Tominaga Y., Kohra S., Hosomi A., *Tetrahedron Lett.*, 1987, **28**, 1529.

FO Study of Ionic Reactions Using Hückel Calculations

The parameters given p. 22 are for neutral species, so it may be necessary to change them when dealing with charged compounds. The *effective* electronegativity of an atom increases (decreases) when it has a positive (negative) charge, and the Coulomb integral can be modified accordingly. For example, a carbocation is more electron attracting than an oxygen, so we can take, say,

$$\alpha(C^+) = \alpha + 1.2\beta$$

The precise value chosen is of little importance; the essence of the process is to represent a difference in electronegativity.[5] This appears very simple. Things become slightly more complicated in practice, as the following examples show. In Exercise 3 (p. 20), we saw that standard parameters correctly localize the charge of the allyl cation on C_1 and C_3. However, the calculations also suggested that the lowest-lying vacant orbital Ψ_2 is nonbonding, which implies that the cation is no more electrophilic than the corresponding radical. To correct this inaccuracy without destroying the symmetry of the system, α must be modified simultaneously for both C_1 and C_3. These atoms bear half a positive charge each, whereas the value given above is for a whole charge. Hence we need an intermediate value, such as $\alpha + 0.6\beta$. In the pentadienyl cation, where the charge is spread over three atoms, a value of $\alpha + 0.4\beta$ is probably appropriate for $\alpha(C^+)$. The benzyl cation, $PhCH_2^+$, is even more complex because charge is more heavily localized on the CH_2 carbon than the *ortho* and *para* positions of the ring. Three different values of α become necessary: one for the benzylic atom, one for the *ortho* and *para* carbons and one for the uncharged *meta* and *ipso* carbons. This kind of correction can go on *ad infinitum*, even without taking solvent and ion pairing effects into account.

Let us try now to compare an enol and an enolate. As an enol is a neutral molecule, the parameters of its oxygen atom can be taken as ($\alpha + 2\beta$ and 0.8β). Now, the enol oxygen has a smaller negative charge than the enolate oxygen. Since the effective electronegativity of an atom decreases as its negative charge increases, the enolate oxygen must have a Coulomb integral which is less negative, say $\alpha + 1.5\beta$. These parameters give the enolate MO energies at higher levels than the corresponding enol MO energies, thus correctly reproducing the greater reactivity of the enolate. However, the oxygen net charge is then *smaller* in the enolate than in the enol! Due to the neglect of electronic repulsion in Hückel calculations, it is unfortunately impossible to obtain both the net charges and the MO energies in the correct order. It is usually more important to have the correct MO energies than the net charges. A simple solution consists in calculating charged species using the standard parameters, while bearing the following rule in mind:

Rule *A positive charge lowers the energy of every MO and a negative charge raises every MO. The energy change in any given MO depends on the LCAO coefficient at the charged atom: the greater the coefficient, the greater is the change.*

Sulfur compounds also raise a problem. A sulfur atom can stabilize an adjacent positive charge by its lone pairs. However in contrast with oxygen, it can also stabilize an adjacent negative charge by its vacant 3d orbitals. No single set of parameters can account for both of these properties simultaneously. See, however, ref. 35.

FO Study of Ionic Reactions Using SCF Methods

Given that incompletely corrected calculations will often yield poor results, it is essential to keep in mind two points when doing MO calculations:

1. You can *never* calculate a reaction, only a model for it. Let us suppose that 1s is necessary to enter into a computer the coordinates of one molecule. Just the input for 1mmol will then require 6.02×10^{20} s, i.e. about 19 billion years, or, if you prefer, the same order of magnitude as the number of seconds elapsed since the Big Bang! And if your model is chemically absurd, then no matter how carefully your calculations are done, the results will be ludicrous (see p. 243 for some illustrative examples). Remember the adage: *garbage in, garbage out!*
2. Exactly as expensive reagents do not guarantee better yields, 'realistic' physical models and refined methods do not always secure better results (Cf. pp. 242–244). Let us just examine here the enolate regioselectivity problem and convince ourselves that an isolated enolate may be a good model for solution chemistry, but *not* for gas-phase chemistry!

The FO treatment of enolate regioselectivity presented in Section 5.2.2 is based on the MOs of enol. Let us first check that the conclusions remain unchanged if the MOs of the naked enolate or of the enolate accompanied by its counterion are used instead. Figure 5.4 shows the C and O net charges in $CH_2{=}CHO^-$, $CH_2{=}CHOLi$ and $CH_2{=}CHONa$, according to PM3, STO-3G and 3–21G calculations.[40] Below each drawing are shown the energies (in eV) of the two occupied π orbitals and the ratios of the C and O coefficients in each MO.

The results are similar for the nine calculations. In the HOMO, the coefficient is larger at C than at O. Hence, according to rule 3, electrophiles having low-lying LUMOs

PM3	STO-3G	3-21G
-0.65 **O**$^-$	-0.50 **O**$^-$	-0.80 **O**$^-$
$-0.72={=}/$	$-0.36={=}/$	$-0.67={=}/$
$E_2 = -1.87$ C/O = 1.55	$E_2 = 3.15$ C/O = 1.17	$E_2 = -1.13$ C/O = 1.38
$E_1 = -6.80$ O/C = 2.39	$E_1 = -3.44$ O/C = 1.50	$E_1 = -6.86$ O/C = 2.56
-0.52 **OLi**	-0.29 **OLi**	-0.83 **OLi**
$-0.51={=}/$	$-0.19={=}/$	$-0.56={=}/$
$E_2 = -7.69$ C/O = 1.44	$E_2 = -6.44$ C/O = 1.09	$E_2 = -7.38$ C/O = 1.25
$E_1 = -12.32$ O/C = 2.48	$E_1 = -11.70$ O/C = 2.32	$E_1 = -12.90$ O/C = 4.09
-0.60 **ONa**	-0.49 **ONa**	-0.79 **ONa**
$-0.63={=}/$	$-0.30={=}/$	$-0.58={=}/$
$E_2 = -6.79$ C/O = 1.51	$E_2 = -3.67$ C/O = 1.01	$E_2 = -6.70$ C/O = 1.24
$E_1 = -11.61$ O/C = 2.42	$E_1 = -9.60$ O/C = 1.62	$E_1 = -12.14$ O/C = 3.56

Figure 5.4 π MO energies (in eV), coefficient ratios and net charges for some enolates.

---

[40]There are no AM1 parameters for Li and Na.

will attack preferentially at C. In the subjacent π orbital, however, the larger coefficient is at O and the O/C ratio in Ψ_1 is higher than the C/O ratio in Ψ_2. This means that electrophiles with LUMOs of high enough energy will give O-substitution. Thus, for qualitative interpretation, any of the nine models is suitable. Note, however, that due to electronic repulsion, the orbital energies of $CH_2{=}CHO^-$ are rather high and frontier control is somewhat exaggerated. It follows that this model will incorrectly predict that in the gas phase, C-substitution is favored! Note also that, contrary to *ab initio* calculations, PM3 (and also AM1) calculations give a larger net charge on C than on O in enolates.

FO Theory and Gas Phase Reactions

As mentioned on p. 99, gas-phase reactions are under charge control and, therefore, almost by definition, FO theory is inappropriate for their study. Such a conclusion would be precipitous. Note to begin with that only the anion behaves in an unusual manner, the comportment of its partner being 'normal'. An FO study of gas-phase S_N2 reactions $(X^- + RY \rightarrow XR + Y^-)$ is therefore perfectly possible. We can also study the competition of electrophilic sites. On the other hand, FO theory will give questionable conclusions for the regioselectivity of anions (e.g. O-alkylation versus C-alkylation of enolates). For these problems, a more thorough study, requiring in particular transition states determination, is necessary.

Can FO theory confirm that the percentage of C-alkylation of enolates increases steadily in the series RI < RBr < RCl? Table 5.2 shows the LUMO energies of MeBr, EtBr and PrBr, according to AM1, PM3, STO-3G and 3–21G calculations.

The marginal differences suggest that for most purposes, alkyl groups can be replaced by Et, or even by Me (to avoid steric and conformational complications). As negative energies imply bonding orbitals, PM3 should be avoided when LUMOs are being considered. The LUMO energy should decrease as the size of R increases. Therefore, AM1 energies are not always reliable.

From Table 5.3, it can be seen that for the problem of C- versus O-alkylation by alkyl halides, AM1, 3–21G and even PM3 calculations give better results than STO-3G calculations, which put the LUMO of bromides at higher energies than the LUMO of chlorides: *ab initio* calculations are *not* necessarily better than semi-empirical calculations, from the point of view of FO theory.

Table 5.2 LUMO energies of RBr (in eV)

	MeBr	EtBr	PrBr
AM1	0.90	0.81	0.83
PM3	-0.14	-0.15	-0.15
STO-3G	10.18	10.03	10.03
3–21G	4.79	4.83	4.81

Table 5.3   LUMO energies of MeX and EtX (in eV)

	MeI	MeBr	MeCl	EtI	EtBr	EtCl
AM1	0.52	0.90	1.60	0.45	0.80	1.50
PM3	−0.43	−0.14	1.33	−0.46	−0.15	1.24
STO-3G	8.50	10.18	9.81	8.35	10.03	9.72
3–21G	3.15	4.79	5.36	3.20	4.83	5.40

5.3 Nucleophilic Reactions

5.3.1 Additions to Conjugated Carbonyl Compounds

The Michael reaction was discovered in 1887 and the first organometallic conjugate addition was observed in 1904.[41] Nonetheless, the factors which influence the competition between 1,2- and 1,4- additions were poorly understood until 1970. Kohler[42] found that ketones and esters gave more 1,4- addition product than aldehydes. Later, Gilman and Kirby[43] suggested that 1,2- addition increases as the reactivity of the organometallic rises. A consideration of the cyclodimerization of acrolein (p. 87) suggests another interpretation, based on HSAB theory. The first bond forms between the terminal acrolein carbons, so the frontier orbitals must favor a conjugate addition *even when the reagent has a positive charge.* This must be all the more true if the reagent is negatively charged. Hence the rule[44] that a hard nucleophile gives 1,2- addition and a soft one gives 1,4-addition. This rule encompasses and expands those of Kohler[42] and of Gilman and Kirby.[43]

FO analysis allows us to go a little deeper into these reactions. According to STO-3G calculations, the largest LUMO coefficient is at C_4 in the 'free' (i.e. noncoordinated) conjugated carbonyl and at C_2 in the corresponding 'ate' complex. Therefore, Loupy and Seyden predicted that even a 'hard' reagent such as $LiAlH_4$ should be able to give

---

[41]Duval D., Géribaldi S., in *The Chemistry of Enones*, Patai S., Rappoport Z. (Eds), John Wiley & Sons, Ltd, Chichester, 1989, Part I, p. 355; Perlmutter P., *Conjugate Addition Reactions in Organic Synthesis*, Pergamon Press, Oxford, 1992.

[42]Kohler E. P., *Am. Chem. J.*, 1907, **38**, 511.

[43]Gilman H., Kirby R. H., *J. Am. Chem. Soc.*, 1941, **63**, 2046.

[44]Eisenstein O., Lefour J. M., Minot C., Anh N. T., Soussan G., *C. R. Acad. Sci. Paris, Série C*, 1972, **274**, 1310; Bottin J., Eisenstein O., Minot C., Anh N. T., *Tetrahedron Lett.*, 1972, 3015; Deschamps B., Anh N. T., Seyden-Penne J., *Tetrahedron Lett.*, 1973, 527; Durand J., Anh N. T., Huet J., *Tetrahedron Lett.*, 1974, 2397; Barbot F., Chan C. H., Miginiac P., *Tetrahedron Lett.*, 1976, 2309; Cossentini M., Deschamps B., Anh N. T., Seyden-Penne J., *Tetrahedron*, 1977, **33**, 409; Priesta W., West R., *J. Am. Chem. Soc.*, 1976, **98**, 8421; Seyferth D., Murphy G. J., Mauzé B., *J. Am. Chem. Soc.*, 1977, **99**, 5317.

conjugate additions if 'ate' complexation can be avoided.[45] Experimentally, they indeed found that the 1,2- addition predominates when 2-cyclohexenone is treated with excess LiAlH$_4$ (diethyl ether, room temperature, 15 min), a product mixture containing 98% cyclohexenol and 2% cyclohexanone is obtained in an overall yield of 98%. In the presence of [2.1.1] cryptand, the regioselectivity is inverted: 77% of cyclohexanone and 23% of cyclohexenol are obtained in a chemical yield of 80%.

If the Loupy–Seyden interpretation is correct, conjugated carbonyls, when used as dienophiles, should be able to react by the C=O group *when strongly complexed*.[46] There do not appear to be any cases of esters showing such reactivity (see Exercise 6 for a possible reason), but Danishefsky and Kerwin[47] has used a Lewis acid-catalysed reaction to prepare dihydropyrans from conjugated aldehydes:

Exercise 6 (M)

Deserves your attention. MOs are given on p. 268
   Justify Kohler's empirical rule.

Answer

Conjugated aldehydes, ketones and esters may be modeled by acrolein, methylvinyl ketone and methyl acrylate, respectively. Their LUMOs are shown below:

The ratio of the coefficients at C$_4$ and C$_2$ increases from aldehyde to ketone to ester. The degree of 1,4-addition should increase in the same order, so we have Kohler's rule. Can we be sure of its generality after analyzing so few compounds? A simple perturbation calculation proves that we can. The LUMOs of conjugated carbonyls are the π^*_{CO} of HC=O, R–C=O and RO–C=O perturbed by the π^*_{CC} (and more weakly by the π_{CC}) of the C=C double bond. The higher the π^*_{CO} (see p. 62 for the ordering), the stronger it will mix with

[45]Loupy A., Seyden J., *Tetrahedron Lett.*, 1978, 2571.
[46]I thank É. Bézard for drawing my attention to this point.
[47]Danishefsky S., Kerwin J. F. Jr, *J. Org. Chem.*, 1982, **47**, 3183. For a cycloaddition with acrylo-nitrile, see Janz G. J., Duncan N. E., *J. Am. Chem. Soc.*, 1953, **75**, 5389.

π^*_{CC} and the larger the C_4 coefficient. Therefore, this coefficient will fall in the order ester, ketone, aldehyde. Steric effects, which are accentuated by the Dunitz–Bürgi non-perpendicular approach trajectory (p. 144), also favor 1,2-additions to aldehydes (see below).

This exercise is a nice demonstration of how simple qualitative methods can be used to derive general rules. As trends can only be established by comparison of large data sets and as *ab initio* methods solve one molecular structure at a time, they are much less easy to use in this fashion.

A cautionary note must be sounded here, the above explanation being slightly oversimplified. First, electrophilic assistance does not necessarily imply 1,2- addition. Michael reactions occur, even in the absence of cryptands! Equally, a rigorously purified organocadmium reagent will not react with enones; Li^+ or Mg^{2+} impurities are needed to promote the conjugate addition.

Furthermore, any cations present in the reaction mixture do not automatically confer electrophilic assistance. Deschamps[48] has shown that conjugated phenones are not complexed by Li^+ because the phenyl group diminishes the basicity of the oxygen.[49] Lefour and Loupy[50] proposed that nucleophilic additions to a carbonyl compound can proceed by two mechanisms. The first, *complexation control*, generally occurs when the nucleophile forms a weak ion pair with the counterion. The cation is then associated with the carbonyl group rather than with the nucleophile, giving a 'naked' nucleophile and an activated carbonyl.[51] Complexation-controlled reactions are therefore usually rapid. 1,2-Addition is preferred, particularly for Li^+ cations (because the strong complexation lowers the LUMO energy and increases the coefficient at C_2) and *for soft nucleophiles* (because the frontier orbital control is increased).[52]

When the cation remains coordinated to the nucleophile, the reaction is under *association control*. Association-controlled reactions are usually slow, because the substrate is not activated and the nucleophile is deactivated. This general class can be subdivided into two groups. In the first (which occurs in the C-alkylation of enolates), the metal is not directly bound to the reactive site. If the transition state is acyclic, conjugate additions will dominate because there is no electrophilic assistance. If it is cyclic, chelation favors addition to the carbonyl:

The second subgroup has the reaction center bound directly to the metal, a situation which occurs with many organometallic reagents. The reaction then involves the cleavage of a strong metal–carbon bond. Lefour and Loupy[50] believe that nucleophilic assistance is necessary for this process. The regioselectivity of the reaction depends on the nature of the metal involved. Hard metals tend to promote addition to the carbonyl

---

[48]Deschamps B., *Tetrahedron*, 1978, **34**, 2009, and references therein.
[49]Seguin J. R., Beaupere D., Bauer P., Uzan R., *Bull. Soc. Chim. Fr.*, 1974, 167.
[50]Lefour J. M., Loupy A., *Tetrahedron*, 1978, **34**, 2597.
[51]Four species are present in the reaction mixture: the nucleophile and the carbonyl, each either naked or associated with the cation. As the equilibrium will be displaced constantly, the reaction occurs between the most reactive species: the naked nucleophile and the complexed carbonyl.
[52]This prediction contradicts HSAB theory.

group and soft ones to the C=C double bond. The addition of cryptands or crown ethers to the reaction medium provides an experimental test which may be used to distinguish between the two Lefour–Loupy mechanisms. These additives slow complexation-controlled reactions but accelerate association-controlled processes. Solvent effects modify the nature of the ion pair, so they are capable of changing the reaction profile from one mechanism to the other. Hence, the competition between 1,2- and 1,4- additions is sensitive to solvent effects.[53]

Obviously, the nature of the nucleophile is also important. Charge-delocalized anions tend to favor 1,4- attack, while reagents having well-localized charges prefer 1,2-addition.[54] Bertrand *et al.*[53] found that the percentage of 1,2-addition is *inversely* proportional to the negative charge on the enolate, which is not in favor of a charge-controlled reaction but rather supports Loupy and Seyden's contention[45] that these reactions are under frontier orbital control.

Other factors may also intervene. Studies on a number of enones have shown that conjugative effects and steric hindrance have to be considered. The non-perpendicular Dunitz–Bürgi reaction trajectory (p. 144) means that an incoming nucleophile can easily be hindered by the presence of substituents:

Thus, enones having two substituents at C_4 often undergo 1,2-addition. Conversely, ketones and esters are inherently more prone to conjugate additions than aldehydes. Phenones give 1,4-additions almost exclusively.[48] They are not particularly basic and are less prone to give strong 'ate' complexes, which favor 1,2-additions.[45] Furthermore, a 1,2-addition would diminish the conjugation, which is unfavorable (cf. Exercise 15, p. 70, and Exercise 17, p. 71). The lost of conjugation intervenes mostly in late transition states and this agrees with the Gilman–Kirby empirical rule. Also, 1,2-versus 1,4-regioselectivity has been correlated with the aggregation state of the anion solution structures.[55] See also Exercise 10, p. 214.

Exercise 7 (E)

MOs are given on p. 269

Generally, cryptands and crown ethers inhibit additions to a CO group (p. 61). Why do they promote the reaction below?

[53]Bertrand J., Gorrichon L., Maroni P., Meyer R., *Tetrahedron Lett.*, 1982, **23**, 3267.
[54]Kyriakou G., Roux-Schmitt M. C., Seyden-Penne J., *Tetrahedron*, 1975, **31**, 1883; Deschamps B., Seyden-Penne J., *Tetrahedron*, 1977, **33**, 413 ; Wartski L., El-Bouz M., Seyden-Penne J., Dumont W., Krief A., *Tetrahedron Lett.*, 1979, 1543 ; Wartski L., El-Bouz M., Seyden-Penne J., *J. Organomet. Chem.*, 1979, **177**, 17 and reference cited therein See also ref. 48.
[55]Croisat D., Seyden-Penne J., Strzalko T., Wartski L., Corset J., Froment F., *J. Org. Chem.*, 1992, **57**, 6435; Strzalko T., Seyden-Penne J., Wartski L., Froment F., Corset J., *Tetrahedron Lett.*, 1994, **35**, 3935; Strzalko T., Seyden-Penne J., Wartski L., Corset J., Castella-Ventura M., Froment F., *J. Org. Chem.*, 1998, **63**, 3295; Corset J., Castella-Ventura M., Froment F., Strzalko T., Wartski L., *J. Org. Chem.*, 2003, **68**, 3902.

Ph-CHO +   (structure: Ph and OMe on a C=C with Cl and O⁻ M⁺)

Answer

The reaction is 'association controlled'. This implies that there is no electrophilic assistance and the incoming enolate simply behaves as an ion pair. The cation remains bound to the enolate because (a) its basicity is increased by the presence of the methoxy group (the HOMO energy of a simple enolate is $\alpha + 0.682\beta$, whereas its methoxy substituted derivative is $\alpha + 0.564\beta$), and (b) the presence of the phenyl group lowers the basicty of the benzaldehyde.[56] The cryptand accelerates the reaction because it activates the enolate without deactivating the carbonyl functionality.

5.4 Radical Reactions

At first sight, radical reactions seem to be ideally adapted to frontier orbital analysis. They are usually exothermic, so their transition states resemble the reagents. Therefore, the starting material frontier orbitals should provide us with a very good approximation. Furthermore, radicals are 'soft' reagents, whose reaction partners are often neutral molecules. Frontier control is dominant in such reactions.

However, it is wise to be careful. If the SOMO (singly occupied MO) lies close in energy to other orbitals, the radical must be described as a combination of several electronic configurations (by *configuration interaction*[57]). Thus, the ground state (a) may be affected by mixing with excited states, such as:

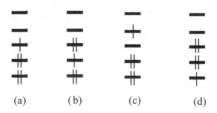

(a)   (b)   (c)   (d)

These excited states do not all contribute equally and each configuration will interact differently with any given substrate. For example, the most important interactions for configurations (a) and (b) are

[56]Seguin J. R., Beaupere D., Bauer P., Uzan R., *Bull. Soc. Chim. Fr.*, 1974, 167.
[57]Configuration interaction (CI) is the MO equivalent of valence bond theory's resonance. The difference is that we combine limiting electronic configurations rather than limiting formulae. Electrons revolving around the nucleus are often compared to men circling around a beautiful woman. In Hückel calculations, no man pays any attention to any other man (*independent electrons*), which is not very realistic. Hartree–Fock (HF) calculations take into account the average repulsion: the rivals tend to avoid one another. CI is a simple way to depict the instantaneous correlation. It underlines the fact that at each instant, every man tries to keep the greatest distance between himself and each of his rivals.

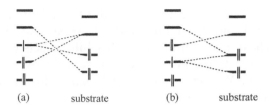

(a)   substrate   (b)   substrate

Considering the very large number of interactions to be taken into account, the FO approach loses much of its charm. Even when a radical can be described by a single configuration, scheme (a) above shows that it may be necessary to consider four orbital interactions; including the three-electron SOMO–HOMO interaction, which may pose a problem as it can be either attractive or repulsive (p. 12).

Furthermore, FO theory can only treat reactions under kinetic control. Reactions under thermodynamic control may give different results. Thus, calculations indicate that kinetic control favors 6-*endo* cyclization of 2-oxo-5-hexenyl radicals[58] whereas thermodynamic control gives 5-*exo* cyclization.[59]

Nonetheless, as we saw in Exercise 13 (p. 69), FO methods can be used to explore radical chemistry.[60] Here is another example. When $CH_3^{\bullet}$ reacts with propionic acid, the ratio of attack at C_2 and C_3 is 5:1. For $Cl^{\bullet}$, it is 1:50:

Fossey[61] rationalized these results by proposing that hyperconjugation with the electron-attracting carbonyl group lowers the $\sigma(C_2H)$ and $\sigma^*(C_2H)$ orbitals below the $\sigma(C_3H)$ and $\sigma^*(C_3H)$ orbitals. Chlorine being more electronegative than carbon, the $Cl^{\bullet}$ SOMO is at lower energy than that in $CH_3^{\bullet}$. Hence the former reacts more efficiently with $\sigma(C_3H)$ whereas the latter is better matched to $\sigma^*(C_2H)$:

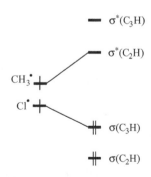

---

[58]See p. 145 for the definitions of '6-*endo*' and '5-*exo*'.

[59]Leach A. G., Wang R., Wohlhieter G. E., Khan S. I., Jung M. E., Houk K. N., *J. Am. Chem. Soc.*, 2003, **125**, 4271.

[60]Fujimoto H., Yamabe S., Minato T., Fukui K., *J. Am. Chem. Soc.*, 1972, **94**, 9205; Fleming I., *Frontier Orbitals and Organic Chemical Reactions*, John Wiley & Sons, Inc., New York, **1976**; Fossey J., Lefort D., Sorba J., *Les Radicaux Libres en Chimie Organique*, Masson, Paris, 1993.

[61]Fossey J., Thesis, Paris, 1974.

This interpretation is consistent with the nucleophilic properties which are generally associated with the methyl radical. In passing, note that Canadell's rule states that any radical, nucleophilic or electrophilic, reacts with an alkene at the site having the largest HOMO coefficient.[62] Canadell and co-workers argue that the three-electron SOMO–HOMO interaction is stabilizing, due to the energetic proximity of these orbitals. See, however, p. 12.

5.5 Periselectivity

Several reactions might be possible in certain systems. For example, the thermal reaction of fulvene with butadiene can potentially proceed through three different Woodward–Hoffmann-allowed *supra–supra* cycloadditions:

Houk *et al.* introduced the term *periselectivity*[63] to indicate discrimination between several possible pericyclic reactions. Fleming[60] suggested that a molecule undergoing a pericyclic reaction will deploy the longest possible conjugated system in the cyclization reaction and provided about 20 examples (of cycloadditions and sigmatropic, cheletropic and electrocyclic reactions) which justified his hypothesis. His theory is supported by Coulson's equations (p. 18), which state that the largest MO coefficients in a conjugated system are found at the termini.

Herndon and co-workers[4] developed a model for predicting regioselectivity which has been adapted to periselectivity problems by Paddon-Row.[64] It makes two assumptions: (a) the two reaction partners approach in parallel planes and (b) the distance between these planes is the same in all reactions. The second is a serious constraint, which explains a success rate of 14 correct predictions out of 17 cases (82%). The model is more reasonable when applied to regioselectivity, where two different orientations *of the same cycloaddition* are compared (122 correct predicttions in 133 cases, i.e. 91.7%).[5] Nonetheless, to the best of our knowledge, FO theory provides the only simple way to study periselectivity available at present.

Exercise 8 (E)

MOs are given on p. 269

(1) Which atom in fulvene will be most easily attacked by (a) electrophiles and (b) nucleophiles? Is fulvene an electrophile, nucleophile or a non-discriminating reagent?

---

[62]Poblet J. M., Canadell E., Sordo T., *Can. J. Chem.*, 1983, **61**, 2068.
[63]Houk K. N., Luskus L. J., Bhacca N. S., *J. Am. Chem. Soc.*, 1970, **92**, 6392.
[64]Paddon-Row M. N., *Aust. J. Chem.*, 1974, **27**, 299.

(2) What will be the principal product when fulvene reacts with (a) 1-aminobutadiene and (b) cyclopentadienone?

Answer[65]

(1) An electrophile will attack the sites having the largest HOMO coefficient (C_3 and C_6 using the MO catalog numbering scheme) whereas a nucleophile will attack at the site of the highest LUMO density (C_1). Note that the intermediate generated by the attack of the nucleophile at C_1 is aromatic. Fulvene is basically an electrophile because its LUMO ($\alpha - 0.254\beta$) lies closer to α than its HOMO ($\alpha + 0.618\beta$) The fulvene LUMO lies lower than π^*_{CO} in formaldehyde, whereas the HOMO has the same energy as in butadiene.

(2) In the first reaction, the dominant FO interaction occurs between the fulvene LUMO and the 1-aminobutadiene HOMO. It is easy to prove that the 6 + 4 reaction is favored; the bond which forms most easily links C_1 in the fulvene to C_4 in the diene.

Applying Paddon-Row's method to the second reaction, and taking into account all four FOs, suggests that the 6 + 4 cycloaddition is the most favorable reaction (having an interaction energy of 0.546β) followed by the 4 + 2 wherein the fulvene acts as a diene (interaction energy 0.359β) and finally 4 + 2 where the fulvene provides the dienophile component (interaction energy 0.284β). Nonetheless, experiments prove that this last compound is the main product.

5.6 Limitations of Rule 3

Rule 3 has been tested on 200 4 + 2[3] and 2 + 2[5] cycloadditions. Using a non-synchronous model (where one σ bonds forms before the other) and Hückel FOs, the regioselectivity is predicted correctly in more than 95% of cases. The result is independent of the parameters, provided that they are chemically reasonable, by which we mean that the Coulomb and resonance integrals must have their largest values for the most electronegative atom and the strongest bond, respectively. This exceptional success rate results from a happy coincidence of a number of factors, most notably: (1) frontier orbital control is important in cycloadditions, including 2 + 2; (2) coefficients are compared *within the same orbital*; (3) with few exceptions, the isomeric transition state geometries are similar.

The rate of success falls whenever these conditions are not met. For example, condition (1) is not always satisfied in *aromatic substitution* reactions. The FOs of polycyclic aromatic hydrocarbons are not well separated from the other MOs,[66] so subjacent orbital control may intervene. Particular care should be taken with non-alternant hydrocarbons because they tend to react under charge control. This is even truer in heteroaromatic systems (cf. Exercise 15, p. 119, and Exercise 16, p. 121), where Hückel

---

[65]Houk K. N., George J. K., Duke R. E. Jr, *Tetrahedron*, 1974, **30**, 523.
[66]As the number of conjugated atoms increases, the energy gaps between the MOs decrease.

calculations are not always reliable.[67] When the reactive site is the heteroatom itself, the HOMO is often a lone pair.[68] Use of Hückel π MOs, which ignore non conjugated lone pairs, then makes erroneous conclusions likely. For example, if we treat the reaction of a ketene with formaldimine as a 2 + 2 cycloaddition, straightforward application of rule 3 predicts that the major product will be azetidinone **28**. If we regard it as a nucleophilic attack by the imine followed by a cyclization, we would expect the β-lactam **29** instead.[69]

Condition (2) is only properly satisfied if the two sites under comparison involve the same chemical element. If the elements are different, their orbitals are not of the same size and their interactions with the incoming reagent will depend on overlaps in addition to on frontier orbital coefficients. FO analysis is then much more complicated. Great care must be taken if the two elements are from different rows of the periodic table, C and S, for example.

Condition (3) is often poorly satisfied for periselectivity, where the difficulties associated with rules 2 and 3 are often combined. Furthermore, whereas regioselectivity can be treated qualitatively, periselectivity requires a quantitative approach, so assumptions concerning transition state geometries must also be made.

Frontier orbital theory can give erroneous results when it is used for comparing *different reactions* (*e.g.* in periselectivity) because it only takes into account stabilizing interactions. Now, the fact that activation energies are positive indicates that *destabilizing interactions are larger than stabilizing interaction*. Therefore, FO theory is only valid when the repulsive terms, which are essentially associated with bond breaking processes, vary less than the FO terms: comparisons should be restricted to *similar compounds undergoing the same reaction*. 'Similar' means compounds having the same reacting functional groups, so propene is similar to ethylene rather than to its isomer cyclopropane. The 'same reaction' condition ensures that the processes to be compared have the same number of broken bonds. Consider, for example, the reaction of EtCl with an anion. The LUMO of the staggered conformation of EtCl is an in-phase combination of σ^*_{CCl} with a small contribution from σ^*_{CH}, the antibonding orbital of the antiperiplanar CH bond:

[67]For example, reactions which have been treated by MNDO and PM3 methods include: Cheney B. V., *J. Org. Chem.*, 1994, **59**, 773; Matsuoka T., Harano K., Hisano T., *Heterocycles*, 1994, **37**, 257. I have been unable to reproduce the experimental results using Hückel calculations.

[68]A subjacent lone pair MO may still be the *effective* HOMO. A reaction at the lone pair converts two nonbonding electrons into two bonding electrons, which is obviously highly favorable.

[69]For experimental studies, see: Pacansky J., Chang J. S., Brown D. W., Schwarz W. J., *J. Org. Chem.*, 1982, **47**, 2233. Theoretical studies: Yamabe S., Minato T., Osamura Y., *Chem. Commun.*, 1992, 26; Fang D., Fu X., *Int. J. Quantum Chem.*, 1992, **44**, 669; Sordo J. A., Gonzalez J., Sordo T. L., *J. Am. Chem. Soc.*, 1992, **114**, 6249; Cossio F. P., Ugalde J. M., Lopez X., Lecea B., Palomo C., *J. Am. Chem. Soc.*, 1993, **115**, 995; Lopez R., Sordo T. L., Sordo J. A., Gonzalez J., *J. Org. Chem.*, 1993, **58**, 7036; Assfeld X., Ruiz-Lopez M. F., Gonzalez J., Lopez R., Sordo J. A., Sordo T. L., *J. Comput. Chem.*, 1994, **15**, 479.

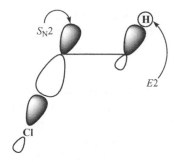

Therefore, the largest coefficient is found on the carbon bearing the chlorine. A 'naïve' interpretation of rule 3 would suggest that substitution will always be favored over elimination in this compound, irrespective of the anion employed. This line of reasoning is triply flawed. First, the transition state for the elimination reaction depend on *four* parameters: the breaking of two bonds (CCl and CH) and the making of two others (C=C and AH, A being the anion). The transition state of the substitution reaction depends on *two* parameters: the breaking of CCl and the making of AC. Rule 3 compares only the ease of formation of the AH and AC bonds, neglecting all the other parameters. Just the fact that the number of bonds broken is different for the two reactions suffices to prevent the use of FO approximation. Second, it is unreasonable, in a bimolecular reaction, to take into account only the substrate's LUMO and forget completely the reagent. Third, the transition states being different, their FOs will differ from one another and also with those of the starting materials. Therefore, no conclusion can by drawn from the analysis of only the latter. In fact, in the elimination transition state, the CH bond is already largely broken, so σ^*_{CH} appears at lower energy than σ^*_{CCl}.[70] Thus, the substrate LUMO is localized on the CH bond in the transition state.[71] The marked changes in the frontier orbitals as a reaction progresses explain why pronounced regioselectivities are sometimes observed in reactions where the coefficients at the competing sites are very similar in the starting materials. It is also at the root of the success of Fukui's theory: *in the transition state, the HOMO and the LUMO approach each other in energy and their coefficients increase at the reacting sites.*

The $E2$–S_N2 competition is reasonably easy to evaluate, because basicity (an affinity for H^+) is mainly under charge control and nucleophilicity (an affinity for C^+) mainly under frontier control.

Exercise 9 (M)

Can be solved without any calculations. MOs are given on p. 270
   One synthesis of occidentalol[72] begins as follows

[70]Remember that weakening a bond causes its HOMO to rise and its LUMO to fall (cf. p. 98). See also: Fukui K., Koga N., Fujimoto H., *J. Am. Chem. Soc.*, 1981, **103**, 196.
[71]Minato T., Yamabe S., *J. Am. Chem. Soc.*, 1988, **110**, 4586.
[72]Watt D. S., Corey E. J., *Tetrahedron Lett.*, 1972, 4651.

If 1-methylcyclohexene is used instead of 4-methyl-3-cyclohexenone, the regiochemistry inverts so the methyl group appears next to the ester in the product. Why?

Answer

The scheme begins with a Diels–Alder reaction, followed by a 4 + 2 cycloreversion which extrudes CO_2. In the Diels–Alder reaction, the diene is substituted at one terminus by two good electron-withdrawing groups (two esters) and at the other terminus by a poor donor (–O–CO). The attracting effect prevails and the diene can therefore be modeled by CH_2═CH—CH═CH—A, A being an electron-attracting substituent. Such a diene reacts preferentially by its LUMO, which has its largest coefficient at the non-substituted terminus. The double bond in 1-methylcyclohexene is substituted by three donor alkyl groups: it will react by its HOMO, whose largest coefficient is on the less substituted extremity. Rule 3 then predicts that the major product has the methyl group next to the ester. The regioselectivity inversion accompanying the introduction of the ketone function suggests that 4-methyl-3-cyclohexenone reacts in fact as a *conjugated enol*. As oxygen is a better donor than methyl, the more reactive site is then the carbon bearing the methyl group. Although the enol tautomer is not present in great quantity, it is more reactive (rule 2) and its concentration is kept significant by the displacement of the keto–enol equilibrium. Rule 3 then predicts the correct regiochemistry.

Use the MOs given in the MO Catalog in the Appendix to check these conclusions.

Exercise 10 (E)

MOs are given on p. 271

Where will alkylation tend to occur on the ethyl acetoacetate dianion?[73]

Answer

On the terminal carbon, where the largest HOMO coefficient is localized. This relatively trivial exercise serves to underline that FO theory can *also* rationalize results usually explained by more classical concepts.

Exercise 11 (M)

MOs are given on p. 272

Explain the following reaction:[74]

[73]Hauser C. R., Harris T. M., *J. Am. Chem. Soc.*, 1958, **80**, 6360; Thompson C. M., Green D. L. C., *Tetrahedron*, 1991, **47**, 4223.
[74]Vollhardt K. P. C., *Organic Chemistry*, Freeman, New York, 1987, p. 1178.

Answer

Steric hindrance prohibits any phenyl groups from being coplanar with the five-membered ring. Hence the starting material can be modeled as cyclopentadienone. The highest LUMO coefficient is localized at oxygen, so this is where the *t*BuLi attacks. This is not surprising because the intermediate is aromatic.

The HOMO in this intermediate has its highest coefficient at C_1 (i.e. the carbon bound to the oxygen). Consequently, this site is probably coordinated to the Li^+ counterion. Hence MeI should attack at C_3, where the second highest coefficient is found.

Note: When cyclopentadienone is viewed as a combination of CO with butadiene, it is apparent that two of its MOs are identical with the antisymmetric Ψ_2 and Ψ_4 butadiene MOs. Show that the large LUMO coefficient at O results from a mixing of π^*_{CO} with π_{CO}, mixing mediated by Ψ_1.

Exercise 12 (E)

MOs are given on p. 272
  The resonance scheme

  shows that every carbon in pyrrole is susceptible to electrophilic attack. So, is there any difference between the α and β positions?

Answer

The α carbons, where its HOMO coefficients are highest, are the more reactive. This result can be easily obtained by a perturbational calculation. In fact, pyrrole can be regarded as a nitrogen-bridged butadiene. The antisymmetric nature of the Ψ_2 and Ψ_4 butadiene orbitals prevents any net overlap the nitrogen, so their energies are unchanged. Ψ_1 being quasi-degenerate with the nitrogen lone pair, their first-order interaction gives two orbitals of energy $\alpha + 0.7\beta$ and $\alpha + 0.24\beta$. Hence Ψ_2 (of energy $\alpha + 0.618\beta$) is the pyrrole HOMO.

Exercise 13 (D)

MOs are given on p. 272
  The conjugated ketone $C_2H_5-CO-CH{=}CH-C_2H_5$ can give rise to enolates E_1 and E_2:

Use frontier orbital theory to determine whether MeI will react faster with E_1 or E_2. What is the regioselectivity of the reaction?

Answer

We are only interested in the π system, so the ethyl groups can be replaced by methyls. An enolate which remains bound to its counterion is usually better represented as an enol than a naked enolate. Thus enols E'_1 and E'_2 can be used to model E_1 and E_2:

The E'_1 HOMO is the higher lying, $(\alpha + 0.305\beta$ versus $\alpha + 0.346\beta)$, so rule 2 states that this compound should be more reactive. If you did not notice that the relative energies of these HOMOs can be deduced without doing calculations, re-read Exercise 2 on p. 57.

The regioselectivity at E_2 poses no great difficulties. The electrophile can only reasonably attack at O or $C_{\alpha'}$ and MeI is relatively 'soft'. Hence reaction occurs at $C_{\alpha'}$ because it has easily the largest HOMO coefficient. The situation is more complicated for E_1. Three reactive sites, O, C_α and C_γ are present. Hückel calculations show that the E'_1 HOMO favors attack at C_γ but the subjacent (HOMO − 1) favors $C_{\alpha'}$ whereas (HOMO − 2) and (HOMO − 4) prefer O. Note that the energy gap between the HOMO and (HOMO − 1) is only 1.081β (compared with 1.563β in $CH_2{=}CHOH$), so subjacent effects are stronger in a dienol than in an enol. As a consequence, the softest electrophiles will give γ attack (frontier control), slightly harder reagents will show α regioselectivity (subjacent control) and the hardest reagents will attack at oxygen (charge control). These predictions seem to be in agreement with the available experimental data. For example, the aldol addition of a prenal dienolate occurs in the γ position,[75] but the harder reagent MeI gives an α attack.[76] However, this interpretation still awaits full verification.

Exercise 14 (E)

MOs are given on p. 273

A mixture of butadiene and acrolein is heated. What relative abundances of the compounds below will probably appear in the product mixture?

---

[75]Duhamel L., Guillemont J., Poirier J. M., Chabardes P., *Tetrahedron Lett.*, 1991, **32**, 4495, 4499.
[76]The classical scheme for introducing a *gem*-dimethyl group at C_4 in di- and triterpenes involves the methylation of the corresponding 3-keto-Δ-4.

Answer

Supra–supra 2 + 2 cycloadditions being forbidden, **2** and **3** are the least likely products. The remaining compounds can be subdivided into three groups:

(1) **6**, **7**, **8** and **9** are products resulting from the reaction of acrolein and butadiene (HOMO–LUMO gap $= 0.937\beta$).
(2) **1** is the cyclodimer of butadiene (HOMO–LUMO gap $= 1.2\beta$).
(3) **4** and **5**; the acrolein cyclodimerization products (HOMO–LUMO gap $= 1.347\beta$).

Alder's rule states that the reactions having the smallest frontier orbital gaps will occur most easily. Hence the products will appear in the order group 1 > group 2 > group 3 > **2, 3**. The relative proportions of the compounds in group 1 reflect simple regioselectivity effects; the bond most likely to form will link the sites having the highest LUMO coefficient in acrolein and highest HOMO coefficient in butadiene. Therefore, **6** and **7** are more likely than **9**, which, in turn, is more probable than **8**. The choice between **6** and **7** comes down their ease of ring closure. The product of the coefficients is greater in **7**, but the CC overlap is better than the CO overlap, so these compounds will appear in the order $7 \geqslant 6 > 9 > 8 > 1$. A similar analysis for group 3 and the 2 + 2 reactions gives $1 > 4 > 5 > > 2 > 3$. This simple analysis only considers the two FOs having the smallest HOMO–LUMO gap. A more sophisticated model, taking each of the four FOs into account and using an asynchronous model where one of the σ bonds forms more rapidly than the other, gives: $6 > 7 > 8 > 9 > 1 > 4 > 5 > > 2 > 3$.

Exercise 15 (E)

Deserves your attention. MO's are given on p. 273
    Predict the site of electrophilic attack at naphthalene, azulene, indole and benzofuran.

Answer.[77]

The arrows indicate the site of the highest HOMO coefficient in each compound. They are also the experimentally preferred sites for attack by electrophiles. The agreement is noteworthy, particularly the different regioselectivities of indole and benzofuran. However, the reasons for this success are different for each case.

    Rule 3 should be reliable for naphthalene because the $\alpha(0.43)$ and $\beta(0.26)$ coefficients are dissimilar. The subjacent orbital lies only 0.382β below the HOMO and has markedly different coefficients ($\alpha = 0$, $\beta = 0.410$). However, it will not compete with the HOMO because the electrophiles are charged; this implies that their LUMOs are

---

[77]Fleming I., *Frontier Orbitals and Organic Chemical Reactions,* John Wiley & Sons, Inc., New York, 1976, p. 58 and references therein.

low lying, so subjacent interference will be minimal. Equally, the alternant nature of the hydrocarbon ensures that the carbon centers are essentially neutral and eliminates charge control influences.

For azulene, the predicted sites are reliable because all of the regiochemical influences are coherent. We can best start with a 'chemical' analysis of the molecule. The normal rules of aromaticity suggest that the five-membered ring should have a negative charge whereas the seven-membered ring should be positive. Hence any electrophilic attack must occur at positions 1, 2 or 3 (catalog numbering). Reaction at the ring junctions can be excluded because it destroys the aromaticity of the system. Attack in the 2-position would give a quinonoid type intermediate, which is not a particularly stable structure. Hence FO calculations, which give coefficients of zero at position 2 and 0.54 at position 1, are chemically reasonable. It is easy to show that subjacent considerations are not important. However, azulene is nonalternant, so charge control may not be negligible. Fortunately, the charge and frontier orbital effects reinforce each other, the charges on 1 and 2 being -0.17 and -0.05 units, respectively.

Indole and benzofuran combine two problems: they are nonalternant *and* they contain heteroatoms. The indole frontier orbital coefficients in the 8- and 9-positions[78] are very similar, 0.491 and 0.493, respectively. The subjacent orbital lies only 0.256β below the HOMO, but it has very small coefficients in these positions. The overriding factor seems to be the net charge (-0.12 at position 9, essentially neutral at position 8), which strongly favors attack at position 9. The case of benzofuran is still more complicated. There is a difference between the frontier orbital coefficients at position 8 (0.54) and 9 (0.47), but charge control still prefers the 9-position (-0.10 versus -0.03 units for the 8-position). The experimental results show that the frontier orbital terms dominate.

Would it have been possible to predict the regioselectivity differences between indole and benzofuran without doing calculations? To a degree, yes. Indole and benzofuran can reasonably be viewed as perturbed styrenes. The parameters given on p. 22 show that the oxygen atom will perturb the system less than the nitrogen. Styrene itself undergoes attack at the terminal carbon, and benzofuran resembles styrene sufficiently for attack at position 8 to be retained. However, because of the increased perturbation by the nitrogen atom, indole should resemble both styrene and an enamine. This causes the attack to switch to the 9-position.

This relatively trivial exercise has been discussed in detail to stress two points. The first is general: frontier orbital theory will give the correct answers in only 70–80% of cases. Hence it is always necessary to check that theoretical predictions are 'chemically reasonable'. If they are, all well and good: we can be reasonably confident. If not, there may be an interesting phenomenon to look into. However, we must *never* use a computer or a theory as a 'black box'. The second point refers specifically to aromatic substitution reactions. If the substrate is nonalternant, or if the conjugated system incorporates heteroatoms, it is essential to examine subjacent orbital and charge control effects in addition to frontier orbital preferences. These other factors are particularly important when the conjugated system incorporates large numbers of atoms because the FOs are then not significantly different in energy from the other MOs. Hence FO theory holds well for furan and benzofuran but fails for dibenzofuran, as we will see in the next exercise.

---

[78]This unusual numbering is due to the Hückel software!

Exercise 16 (M)

Deserves your attention. MOs are given on p. 274

Which sites at dibenzofuran are the most susceptible to attack (a) under frontier control and (b) under charge control?

Answer[79]

In diagram **A** below, the site predicted for frontier orbital attack is shown by the solid arrow and for charge-controlled reactions by the dotted arrows. Diagrams **B** and **C** show the experimental results. Only the nitration reaction occurs at position 4 (using catalog numbering); all of the remaining reactions (Friedel–Crafts alkylations and acylations, sulfonation, halogenation, etc.) prefer the 5-position. According to Keumi *et al.*, the preference for the 4-position increases with the similarity between the NO_2^+ ion and the electrophile. Nitration clearly occurs under FO control, in a reaction which probably has an early transition state. However, the remaining reactions are difficult to explain in charge control terms: this pathway should give equal proportions of attack at positions 3 and 5, but experiments show that the 3-substituted compound rarely comprises more than 5% of the product.

We can try to refine our model. The reaction occurs in the presence of strong acid, so it is possible that the first step involves a protonation at oxygen. This requires a second calculation incorporating a modified value of $\alpha + 3\beta$ for the oxygen Coulomb integral. However, the result indicates an even greater degree of position 3 attack in charge-controlled reactions and no change under frontier orbital control: the HOMO having a zero coefficient at oxygen is unaffected by its parameters.

What if we model a late transition state using a Wheland intermediate? The Hückel π energies, which are as follows:

| 18.466β | 18.407β | 18.411β | 18.486β |

suggest that the attack will occur preferentially at position 6, and then at position 3. Obviously, FO theory, even when improved upon by taking into account also charge and subjacent control, is incapable of explaining the reactivity of a highly conjugated molecule such as DBF (which is also heteroatomic and nonalternant). The reasons for this failure deserve some analysis. Because the orbitals of DBF are similar in energy (the six highest-lying occupied MOs are all found between $\alpha + 1.944\beta$ and α

---

[79]Keumi T., Tomioka N., Hamanaka K., Kakihara H., Fukushima M., Morita T., Kitajima H., *J. Org. Chem.*, 1991, **56**, 4671.

+ 0.705β) they intermix easily as the reagent approaches (this can be proved using the three-orbital perturbation equation given on p. 30). This deformation of the DBF electron clouds occurs so readily that the atomic point charges and FO coefficients in the starting material will be very different from those in the transition state. Consequently these *static indices* (i.e. based on the characteristics of an *isolated, nonperturbed* reagent) do not operate. The calculations of the Wheland complexes are also useless because they simply eliminate the reactive site from the conjugated system and neglect the properties of the incoming reagent. In problems of this kind, it is essential to use at least SCF methods and a transition state involving both reagents.

Keumi *et al.* applied MNDO calculations to the π and σ complexes to explore the early and late transition states, respectively. Their data confirm that an early transition state favors attack at position 4; the later ones give attack at position 5.

Exercise 17 (D)

(1) Why do highly enolizable ketones give more *O*-alkylation product than normal ketones?[80]
(2) 1,3-Cyclopentanedione and 1,3-cyclohexanedione are highly enolized ketones. Will they give the same proportions of *O*- and *C*-alkylations under the same reaction conditions (of solvent, cation, etc.)?

Answer

(1) Such enols are usually conjugated and *C*-alkylation breaks the conjugation.
(2) The alkylation of cyclopentanedione can occur at C or O without any significant change in the molecular structure, which remains essentially planar in both cases. This geometry allows conjugation to operate fully and favors *O*-alkylation.[81] The reactivity of cyclohexanedione reflects two antagonistic effects: conjugation favors *O*-alkylation whereas conformational criteria favor *C*-alkylation. The *C*-alkylated dione can adopt a slightly deformed chair structure, but the enol is compelled to take up an unfavorable envelope configuration. Hence the percentage of *O*-alkylation will be lower than in cyclopentanedione. Experiments (with potassium enolates and Ph–CO–CH$_2$Br in aqueous dioxane at 80°C) show that the O:C ratio is 80:20 for cyclopentanedione and 48:52 for cyclohexanedione[82]

Exercise 18 (E)

Why does alkylation of diphenylacetophenone in DMSO give large quantities of enol ether?

Answer

Aprotic solvent and conjugation both favor *O*-alkylations.[83]

---

[80]House H. O., *Modern Synthetic Reactions*, 2nd edn, Benjamin, Menlo Park, CA 1972, 523.
[81]In highly conjugated systems, the MOs become closer to one another and charge control, which favors *O*-substitution, plays a more important role.
[82]Rosenthal D., Davis K. H. Jr, *J. Chem. Soc. C*, 1966, 1973.
[83]Zook H. D., Russo T. J., Ferrand E. F., Stotz D. S., *J. Org. Chem.*, 1968, **33**, 2222.

Exercise 19 (E)

Predict the product formed in the reaction of a magnesium reagent with *t*BuCS–CO$_2$Et. The LUMO of the model substrate Me–CS–CO$_2$Me (STO-3G calculations)[84] is

Answer[85]

Rule 3 correctly predicts the experimentally observed thiophilic addition. Simple thio-aldehydes and thioketones always have a greater LUMO coefficient at C than S. The results are opposite here and this reversal reflects the experimental results nicely.

Exercise 20 (D)

MOs are given on p. 274

Oxazole **A** is often used as a diene in Diels–Alder reactions. However, isoxazole **B** appears to be inert. Their different reactivities are surprising, considering that their MOs are similar. Propose an explanation. *Bond lengths and strengths*: CC 1.54 Å and 82.6 kcalmol^{-1}; C=C 1.34 Å and 144 ± 5 kcalmol^{-1}; CN 1.47 Å and 72.8 kcalmol^{-1}; C=N 1.27 Å and 147 kcal/mol^{-1}; NO 1.40 Å and 48 kcalmol^{-1}.[86]

A   **B**

Answer

The oxazole cycloaddition implies the transformation of two C=C bonds into two C–C bonds, but the isoxazole converts one C=N and one C=C bond into a C–C and a C–N bond. The data given above show that the reaction involving **A** is more exothermic than for **B** by approximately 10 kcal/mol^{-1}. According to MP2/6–31G*//3–21G calculations, the reaction of ethylene with **A** is exothermic by 19.7 kcal/mol^{-1} and endothermic with **B** by 4 kcal/mol^{-1}.[87]

The frontier orbitals are found at $\alpha + 0.676\beta$ and $\alpha - 0.895\beta$ for **A** and $\alpha + 0.811\beta$ and $\alpha - 0.782\beta$ for **B**. Both **A** and **B** are less reactive than butadiene because their HO-MOs lie below $\alpha + 0.618\beta$ and their LUMOs above $\alpha - 0.618\beta$. However, both isomers have a similar HOMO–LUMO gap (1.571β for **A** and 1.591β for **B**), so their different reactivities are not caused by the frontier orbital energy difference. **B**'s behavior must be explained by other factors. The NO bond fragility probably limits its applications because forcing conditions cannot be employed. A second problem arises because the isoxazole FOs are almost symmetrically arranged about the nonbonding level; this

---

[84]Whereas optimization by STO-3G calculations leads to a planar structure for this model compound, both AM1 and PM3 calculations favor a *gauche* structure, the S=C–C=O dihedral angle being a 85°51′ (AM1) and 91°76′ (PM3).

[85]Metzner P., Vialle J., Vibet A., *Tetrahedron*, 1978, **34**, 2289.

[86]Huheey J. E., *Inorganic Chemistry*, 3rd edn, Harper & Row, New York, 1983, pp. A37–A38.

[87]Gonzalez J., Taylor E. C., Houk K. N., *J. Org. Chem.*, 1992, **57**, 3753.

makes **B** a relatively indiscriminating diene. **A** is more useful because it has a marked preference for electron-poor dienophiles, which are the more common. A third problem concerns the dissymmetry in the transition state. If we look at the cycloaddition with ethylene, the frontier orbital terms for atom 1 in **A** (using the numbering scheme in the Appendix) are approximately

$$\frac{0.53^2}{1.676} + \frac{0.68^2}{1.895} = 0.412\beta$$

and for atom 4:

$$\frac{0.63^2}{1.676} + \frac{0.52^2}{1.895} = 0.380\beta$$

Hence we can expect a dissymmetric transition state having the shorter bond at C_1. Note that we need to consider all four FOs: although the nucleophilic character of **A** is significant, it is not so great that the influence of the LUMO can be ignored entirely.

An analogous calculation for isoxazole **B** gives 0.389β for N_1 and 0.394β for C_4. The transition state will be more dissymmetric than indicated by these not very different values, because orbital overlap falls exponentially with distance and the AOs of nitrogen are smaller than those of carbon. The isoxazole reacts less efficiently because the formation of the second (CN) bond is difficult and this difficulty increases with increasing dissymmetry of the dienophile.

Houk's 3–21G transition states for the reaction of ethylene with **A** and **B** are given below. The interatomic distances of 2.093, 2.171, 2.068 and 2.087 Å correspond to 136, 141, 134 and 142% of their equilibrium distances, respectively. The dissymmetry of the transition state (estimated by the difference in the degree of formation of the partial bonds) is therefore 5% for **A** and 8% for **B**. This would be even more marked if we were to consider a substituted dienophile. Note also that the 'later' transition state in the case of **B** is consistent with lower reactivity.

Exercise 21 (E)

FO theory suggests that the CC bond will form more rapidly than the CO bond in a cycloaddition between butadiene and formaldehyde. 3–21G calculations[88] give, however, a CC transition state distance (2.133 Å) longer than the CO distance (1.998 Å). Explain this contradiction.

Answer

There is no contradiction: 2.133 Å is 1.385 times the CC equilibrium bond length (1.54 Å) and 1.998 Å is 1.397 times the CO equilibrium bond length (1.43 Å). The CC bond is forming slightly more rapidly than the CO bond in the transition state.

---

[88]McCarrick M. A., Wu Y.-D., *J. Org. Chem.*, 1993, **58**, 3330.

Exercise 22 (E)

Requires no calculation

What compound (**A** or **B**) is the major product in the following cycloadditions (Z = CN, CHO, Ph, CH=CH$_2$, SiMe$_3$, H and Me)?

R = H, OMe

Answer[89]

The nonsynchronous model (one σ bond is formed more readily than the other) will be used here. The diene, being substituted at positions 1 and 3 by donor groups, is electron rich and reacts preferentially via carbon 4. Hence the problem lies in determining with which atom of the dienophile, S or C, position 4 will link most easily. The thiocarbonyl group, being electron poor, will receive electrons from its partner. If C=S reacts at carbon, a negative charge will develop at the sulfur atom, which can cope with it easily. A negative charge on the carbon atom is not so stable, so reactions at sulfur occur only if Z is an electron-withdrawing group. Therefore, the major product will be **B** if Z = SiMe$_3$, H, Me and **A** if Z = CN, CHO. A problem arises when Z = Ph and Z = CH=CH$_2$: is the conjugative delocalization sufficient to stabilize the incipient carbanion? Experimentally, it is found that **B** is the major product for these two reactions.

Note that FO theory gives here fairly good results. According to 3–21G calculations (see Vedejs *et al.*[89]), the major product is **B** when the larger LUMO coefficient is at carbon and **A** when it is at sulfur. An anomalous result occurs with Z = CN: the two coefficients are equal, but the experimental yields are 4% for **B** and 70% for **A**.

Exercise 23 (E)

Protonation of pentadienyle anion gives rise to 1,4-pentadiene, instead of the expected conjugated 1,3-pentadiene. Is it really so surprising, when we look at the HOMO coefficients and the charge distribution (STO-3G calculations)?

Answer

Frontier control favors attack on C$_3$ (rule 3). Note, however, that charge control also favors this regiochemistry if the influence of neighboring charged atoms is taken into account. C$_3$ has in its proximity two charged atoms. The termini have only one such, the second charged atom being further away.

---

[89] Vedejs E., Perry D.A., Houk K. N., Rondan N. G., *J. Am. Chem. Soc.*, 1983, **105**, 6999.

Exercise 24 (E)

Do not use Hückel calculations. MOs are given on p. 275

It is known that dienolates react preferentially at position α. Also, despite increasing steric hindrance, substitution tends to boost the nucleophilicity of the reactive site. Suggest a justification.

Answer

The HOMOs of three dienolates (R′ = R″ = H; R′ = H, R″ = Me and R′ = Me, R″ = H) are shown in the MO Catalog in the Appendix. According to AM1 and STO-3G calculations, the $C_\alpha : C_\gamma$ ratios of the HOMO coefficients are 1.25 (AM1) and 1.27 (STO-3G) for the first compound, 1.32 and 1.32 for the second and 1.18 and 1.24 for the third. Rule 3 then predicts that attacks at α is always preferred and this trend is enhanced by substitution at C_α and diminished by substitution at C_γ. To the best of our knowledge, this trend has never received a theoretical interpretation.

Note that Hückel calculations give the larger HOMO coefficient at C_γ. Note also that other factors may intervene to favor this regioselectivity (see p. 109 and Exercise 13 p. 117).

Exercise 25 (E)

MOs are given on p. 275

Predict the regioselectivity in the Diels–Alder reactions of acrolein and *N*-acetyl-2-cyano-4-phenyl-1-azabutadiene with 1-hexene, styrene, ethyl vinyl ether and methyl acrylate. The FOs of the azadiene are given below; the others may be found in the Appendix.

Answer

The azadiene FOs are close in energy, so they must both be considered in every reaction, even those with electron-rich dienophiles. Hexene and the ether can be modeled as propene and vinyl methyl ether, respectively; the calculations then give the following (the figures are the values of FO interaction in the corresponding incipient bond):

Therefore, FO theory predicts that the major products will be as follows:

which is experimentally correct, except for the reaction between styrene and the azadiene.[90]

Exercise 26 (E)

To be solved without any calculations

(1) Classify, according to their energies, the LUMOs of tropone, 3- and 4-ethoxycarbonyltropones and 3- and 4-methoxytropones.
(2) Predict the structures of the major 6 + 4 adducts resulting from the reactions of 3- and 4-ethoxycarbonyltropones with isoprene and 1-acetoxybutadiene.
(3) Would the structures of the major 6 + 4 adducts of 3- and 4-methoxytropones be as easy to predict?

---

[90]Teng M., Fowler F. W., *J. Org. Chem.*, 1990, **55**, 5646.

Answer[91]

(1) If tropone is considered as resulting from the union of a hexatriene (MO given on p. 256) and a carbonyl group, then it is easy to see that its LUMO is identical with the hexatriene LUMO. The LUMO coefficient at position 4 being larger than that at position 3, a substituent at this position will have a stronger influence (Cf. Exercise 2, p. 57). As a donor substituent raises the MO energies and an attractor lowers them, the LUMOs ordering is therefore 4-ethoxycarbonyltropone < 3-ethoxycarbonyltropone < tropone < 3-methoxytropone < 4-methoxytropone.

(2) The dienes being electron rich and the substituted tropone electron poor, their main interaction would be between the HOMO(diene) and the LUMO(tropone). The regioselectivity is determined by the first-formed bond, which links the atom having the largest HOMO coefficient (C_1 for isoprene, C_4 for 1-acetoxybutadiene) to the atom having the largest LUMO coefficient of tropone. At a first approximation, we may consider that the 3-substituent affects mostly the 2–3 double bond and the 4-substituent the 4,5–6,7 diene; the reactive site is then C_2 for 3-ethoxycarbonyltropone and C_7 for 4-ethoxycarbonyltropone. The major products therefore are the following compounds:

(3) The methoxy substituent, being electron releasing, raises all the orbital energies: the HOMO(tropone)–LUMO(diene) interaction can no longer be neglected. The regiochemistry becomes harder to predict, especially as the FOs are distorted in opposite senses. When the largest HOMO coefficient is at C_2, the largest LUMO coefficient is at C_7 and vice versa. The Hückel FOs of the four tropones confirm this qualitative reasoning.

[91]Garst M. E., Roberts V. A., Houk K. N., Rondan N. G., *J. Am. Chem. Soc.*, 1984, **106**, 3882.

6 Stereoselectivity

> Competition between two trajectories of attack
>
> **Rule 4** *The best trajectory of attack follows the best frontier orbital overlap.*

6.1 Pericyclic Reactions

6.1.1 Electrocyclic Reactions

Torquoselectivity

The Woodward–Hoffmann rules state that the thermal opening of cyclobutenes occurs by a conrotatory process. This ring opening gives two different and distinguishable products if the cyclobutene has substituents about the C_3C_4 bond (cyclobutene numbering). They result from *in* or *out* opening:

It does not depend on steric factors.[1] For example, Curry and Stevens have shown that compounds such as **1** may open to give products whose bulky group is found in the 'inside' position:

---

[1](a) Curry M. J., Stevens I. D. R., *Perkin Trans.* 2, 1980, 1391; (b) Dolbier W. R., Koroniak H., Burton D. J., Bailey A. R., Shaw G. S., Hansen S. W., *J. Am. Chem. Soc.*, 1984, **106**, 1871; (c) Dolbier W. R., Koroniak H., Burton D. J., Heinze P. L., *Tetrahedron Lett.*, 1986, **27**, 4387; (d) Dolbier W. R., Koroniak H., Burton D. J., Heinze P. L., Bailey A. R., Shaw G. S., Hansen S. W., *J. Am. Chem. Soc.*, 1987, **109**, 219; (e) Piers E., Lu Y. -F., *J. Org. Chem.*, 1989, **54**, 2267; (f) Dolbier W. R., Gray T. A., Keaffaber J. J., Celewicz L., Koroniak J., *J. Am. Chem. Soc.*, 1990, **112**, 363; (g) Godt A., Schluter A.-D., *Chem. Ber.*, 1991, **124**, 149; (h) Binns F., Hayes R., Ingham S., Saengchantara S. T., Turner R. W., Wallace T. W., *Tetrahedron*, 1992, **48**, 515; (i) Piers E., Ellis K. A., *Tetrahedron Lett.*, 1993, **34**, 187.

Frontier Orbitals   Nguyên Trong Anh
© 2007 John Wiley & Sons, Ltd

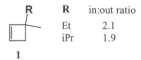

R	in:out ratio
Et	2.1
iPr	1.9

1

Torquoselectivity has been defined as the preference for inward or outward rotation of substituents and a theoretical interpretation proposed by Rondan and Houk.[2a,b] According to 3–21G calculations, the transition state C_3C_4 bond length is fairly normal at ~2.2 Å so the transition state is neither particularly early nor late. Nonetheless, bond cleavage is sufficiently advanced for the transition state FOs to be assimilated to the bond orbitals σ_{34} and σ^*_{34}. Let us formally decompose the molecule into two fragments, the ring and the substituent R, and examine the interactions in the transition state. The analysis may be simplified by using a p orbital to represent the R group (HOMO or LUMO, as appropriate). During an *out* opening, this p orbital overlaps only with the AO localized on C_3. During an *in* opening, it interacts with C_3 and C_4 simultaneously (Figure 6.1).

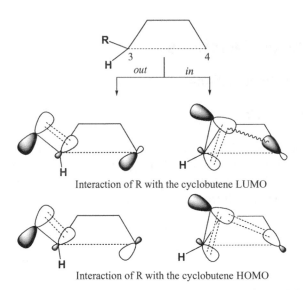

Interaction of R with the cyclobutene LUMO

Interaction of R with the cyclobutene HOMO

Figure 6.1   Frontier overlaps between an R substituent and a cyclobutene ring.

[2](a) Kirmse W., Rondan N. G., Houk K. N., *J. Am. Chem. Soc.*, 1984, **106**, 7989; (b) Rondan N. G., Houk K. N., *J. Am. Chem. Soc.*, 1985, **107**, 2099. See also: (c) Rudolf K., Spellmeyer D. C., Houk K. N., *J. Org. Chem.*, 1987, **52**, 3708; (d) Buda A. B., Wang Y., Houk K. N., *J. Org. Chem.*, 1989, **54**, 2264; (e) Kallel E. A., Houk K. N., *J. Org. Chem.*, 1989, **54**, 6006; (f) Kallel E. A.,Wang Y., Spellmeyer D. C., Houk K. N., *J Am. Chem. Soc.*, 1990, **112**, 6759; (g) Jefford C. W., Wang Y., Houk K. N., *J. Org. Chem.*, 1991, **56**, 856; (h) Niwayama S., Houk K. N., *Tetrahedron Lett.*, 1992, **33**, 883; (i) Jefford C. W., Bernardinelli G., Wang Y., Spellmeyer D.C., Buda A., Houk K. N., *J. Am. Chem. Soc.*, 1992, **114**, 1157; (j) Nakamura S., Houk K. N., *Heterocycles*, 1993, **35**, 631; (k) Niwayama S., Houk K. N., Kusumi T., *Tetrahedron Lett.*, 1994, **35**, 527; (l) Nakamura K., Houk K.N., *J. Org. Chem.*, 1995, **60**, 686; (m) Niwayama S., Kallel E. A., Sheu C., Houk K. N., *J. Org. Chem.*, 1996, **61**, 2517; (n) Niwayama S., Kallel E. A., Spellmeyer D. C., Sheu C., Houk K. N., *J. Org. Chem.*, 1996, **61**, 2813; (o) Dolbier W. R., Koroniak H., Houk K. N., Sheu C., *Acc. Chem. Res.*, 1996, **29**, 471; (p) Walker M. J., Hietbrink B. N., Thomas B. E. IV, Nakamura K., Kallel E. A., Houk K. N., *J. Am. Chem. Soc.*, 2001, **123**, 6669; (q) Lee P. S., Zhang X., Houk K. N., *J. Am. Chem. Soc.*, 2003, **125**, 5072.

We can now consider the effect of the substituent. Electron-donor R groups can be represented by a doubly occupied orbital (cf. Exercise 11, p. 68). Such an orbital will generate a stabilizing two-electron interaction with the cyclobutene LUMO. This stabilization increases with increasing orbital overlap and favors the *out* reaction.[3] The destabilizing four-electron interaction of R with the cyclobutene HOMO increases with overlap and inhibits the *in* reaction. The *out* mode being favored and the *in* mode disfavored (Figure 6.2), we are reasonably sure that *a donor substituent will move to the outside*. Electronic factors normally prevail over steric factors, as shown by the opening of 3-methoxy-3-*tert*-butylcyclobutene in which the *tert*-butyl group takes exclusively the inside position.[4]

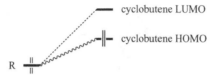

Figure 6.2 Major frontier interactions for an electron-donating substituent. Wavy line = unfavorable interaction; dotted line = favorable interaction.

Electron-accepting R groups are best represented by two low-lying orbitals, one filled and one empty. Figure 6.1 shows that the interaction of the HOMOs will inhibit the in pathway, whereas the HOMO(cyclobutene)–LUMO(R) interaction will promote it. The LUMO(cyclobutene)–HOMO(R) also disfavors the *in* pathway. These results are summarized in Figure 6.3: interaction (1) favors the *in* mode and interactions (2) and (3) impede it. This means that the energy gap between the cyclobutene HOMO and the R LUMO must be as small as possible for an *in* reaction to occur. In other words, *R must be a powerful acceptor.* However, Rondan and Houk believe that even then, interaction (3) may largely cancel interaction (1) and the torquoselectivity will not be very pronounced.

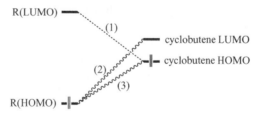

Figure 6.3 Major interactions for an electron-withdrawing substituent (*in* mode).

Extensions of the Rondan–Houk Treatment and Quantitative Analyses

Generalizations

Can we extend this approach? The analysis shown in Figure 6.1 remains valid for all conrotatory processes: donors should rotate preferentially outward and attractors

---

[3]In the *in* mode, the overlap between R and C_4 is negative.
[4]Houk K. N., Spellmeyer D. C., Jefford C. W., Rimbaud C. G., Wang Y., Miller R. D., *J. Org. Chem.*, 1988, **53**, 2125.

should rotate inward. However, other factors may also intervene. For example, neither *cis,cis*-1,3,5,7-octatetraene nor *cis,cis,cis*-1,3,5-cyclooctatriene is planar, so that the transition structure of their conrotatory interconversion takes a helical conformation. According to *ab initio* calculations,[5] steric effects then prevail over electronic effects and direct the torquoselectivity of cyclooctatetraene.

A *priori*, we would expect disrotatory reactions to show poorer torquoselectivity than conrotatory reactions for two reasons. Consider, for example, the hexatriene ↔ cyclohexadiene interconversion. On the one hand, the overlap between R and the distal carbon C_6 is similar for the *in* and *out* pathways, as in the *in* mode, the major lobe at C_6 is oriented away from R:

out mode   in mode

On the other hand, hexatriene must take a helical structure.[6] By analogy with the octatetraene case, it may be inferred that electronic effects will have a smaller influence and steric effects a larger influence than in the cyclobutene torquoselectivity. This is indeed what was found by Houk and co-workers.[7]

Quantitative Analyses

In cases where the 3-substituent of cyclobutene is saturated, e.g. alkyl or silyl[8] groups, the simple Rondan–Houk treatment often does not work. Thus, Curry and Stevens[1] have shown that an ethyl group, which is at least as good donor as a methyl group, when competing with the latter, prefers the *in* position. On the basis of his calculations, Houk[9] suggests that this anomaly is due to conformational effects: the inward rotation of an ethyl group gives rise to more favorable *gauche* interactions than that of a methyl group.

These quantitative calculations allow some interesting clarifications. For example, with the qualitative treatment, it is difficult to understand why $CO_2{}^-$, which should be a good donor, prefers the *in* position in the opening of 3-carboxylate-3-methylcyclobutene.[10] Now *ab initio* calculations show that, compared with the

[5]Thomas B. E. IV, Evanseck J. D., Houk K. N., *J. Am. Chem. Soc.*, 1993, **115**, 4165. Geometry optimizations by 3–21 G calculations, energies calculated using Møller–Plesset theory and the 6–31G* basis set.

[6]For this reason, in his vitamin B_{12} synthesis, Woodward initially predicted the wrong stereochemistry for the electrocyclization of a hexatriene. This led to the discovery of the conservation of orbital symmetry (Woodward R. B., in *Aromaticity*, Special Publication No. 21, Chemical Society, London, 1967, 217).

[7]Evanseck J. D., Thomas B. E. IV, Houk K. N., *J. Org. Chem.*, 1995, **60**, 7134.

[8]Lee P. S., Zhang X., Houk K. N., *J. Am. Chem. Soc.*, 2003, **125**, 5072.

[9]Kallel E. A., Wang Y., Spellmeyer D.C., Houk K. N., *J. Am. Chem. Soc.*, 1990, **112**, 6759.

[10]Niwayama S., Wang Y., Houk K. N., *Tetrahedron Lett.*, 1995, **35**, 6201. The apparent discrepancy with the 6–31G(d)//3–21G calculations (Table 1, ref. 8) disappears when solvation is included. Professor Houk is thanked for this personal communication.

cyclobutene case, the outward rotation of 3-methylcyclobutene is favored by 1.2 kcal mol^{-1} and that of 3-carboxylatecyclobutene by 4.4 kcal mol^{-1}. On the other hand, the inward rotation 3-methylcyclobutene is disfavored by 5.7 kcal mol^{-1} and that of 3-carboxylatecyclobutene by 0.7 kcal mol^{-1}. Hence the latter influence prevails and this explains the rather surprising result.

Exercise 1 (D)

When analyzing problems of relative reactivity or regioselectivity, we considered the interaction between the closest-lying HOMO–LUMO pair and ignored all of the others. Why is it not possible here to ignore interactions (2) and (3)?

Answer

There is no real inconsistency. The reactions so far have involved real molecules which interact weakly in the transition state: the new bonds are only very partially formed. This means that overlap is weak, because it decreases exponentially with distance and the intermolecular separations are large (2 Å or more). In perturbation equations, the numerators P_{ij}^2 are negligible, so the denominators $(E_i - E_j)$ determine the outcome.

In this example, the combining fragments interact strongly because the bonds are fully made after the 'perturbation' process. The distance between the R group and carbon C$_3$ is about 1.5 Å, so P_{ij} values are large. The cyclobutene 'fragment' is a transition state and the energy gap between its FOs is very small. Consequently, (2) and (3) are of the same order of magnitude as (1). When R is a weaker acceptor, the energy of its FOs rises: interaction (1) diminishes and (2) and (3) increase. Hence a point comes where the latter dominate.

Exercise 2 (M)

A silyl substituents on a cyclobutene tends to rotate inwards,[11] as exemplified by the following reactions:

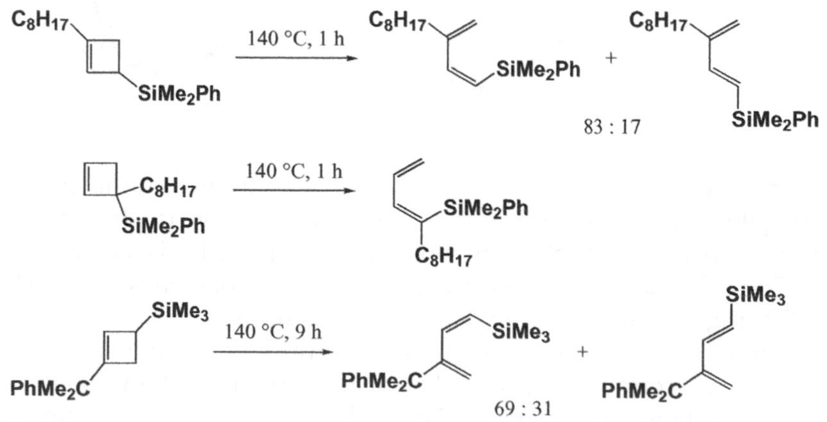

---

[11]Murakami M., Miyamoto Y., Ito Y., *Angew. Chem. Int. Ed.*, 2001, **40**, 189.

Murakami et al. justify this preference by an attracting interaction between the occupied σ_{34} orbital (corresponding to the cyclobutene C_3C_4 bond) and the low-lying vacant σ^*_{SiC} orbital of the silyl substituent.[11] Besides this interaction, Houk and co-workers[8] also take into account electrostatic interactions, which play a lesser role, however. Inagaki and co-workers[12] consider the major factor to be the electron-donating effect from the σ_{SiC} into the incipient π^* orbital:

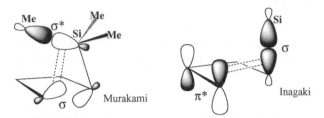

Who is right? Who is wrong?

Answer

Nobody is wrong... and nobody is completely right! The Si–C bond being weak, the energy of the occupied σ_{SiC} orbital is high and that of the vacant σ^*_{SiC} orbital is low. In other words, the Si–C bond can both donate *and* receive electrons. However, Inagaki and co-workers' argument is less correct than Murakami and co-workers' arguments, for the following reasons. Due to the positive charge on silicon, the σ_{SiC} and σ^*_{SiC} orbitals are lowered, so the electron-withdrawing effect is stronger (*cf* Exercise 1, p. 189). Also, FO theory should be applied to the transition state and the transition state LUMO is mainly localized on the C_3C_4 bond. Therefore, electrons are donated mostly into the σ^*_{34} bond:

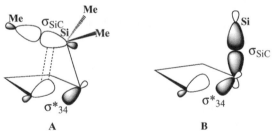

In other words, interaction **A** should be more important than that considered by Inagaki and co-workers. Interaction **B** (donation from σ_{SiC} into σ^*_{34}) is negligible, these orbitals being practically orthogonal and their overlap is small.

The calculated activation energies for 3-R-cyclobutenes (R = CH_3, CH_2F, CHF_2 and CF_3) ring openings show that, compared with the R = Me case, the inward rotation is facilitated for the mono- and difluoro derivatives, but impeded when R = CF_3. Houk and co-workers[8] attributes this last result to repulsion between the fluors and the ring π system. However, when R = NH_3^+, the inward rotation is definitely disfavored, suggesting that *in fine* electrostatic factors play only a minor role.

[12]Ikeda H., Kato T., Inagaki S., *Chem. Lett.*, 2001, 270.

Exercise 3 (E)

What compound will result from the thermal opening of bicyclo[3.2.0]hept-6-en-2-one:

A   B

Answer[13]

The cyclobutene ring is substituted by a donor alkyl chain, which tends to rotate outwards, and by an attractor carbonyl which tends to rotate inwards: the product is therefore **B**.

Exercise 4(E)

3-Acetylcyclobutene opens thermally with an *in:out* ratio of 0.5. The ratio rises to 4 in the presence of ZnI_2. Explain.

Answer[14]

The acetyl group forms an 'ate' complex with the Lewis acid which is a much stronger acceptor.

Exercise 5 (E)

The activation energy for the ring opening of *cis*-dimethyl-3,4-cyclobutene is lower than that of dimethyl-3,3-cyclobutene by 2.1 kcal mol^{-1}.[1a] However, in both cases, one methyl goes inwards and the other outwards. Why is there a difference?

Answer[9]

Both transition states have about the same energy, but in *cis*-dimethyl-3,4-cyclobutene the substituents are eclipsed, thus destabilizing the initial state and lowering the activation energy.

An Exercise in Qualitative Analysis

The Rondan–Houk theory rationalizes many experimental results and leads to several predictions which have since been confirmed. Exceptions have been explained by Houk and co-workers, using numerical calculations. In a sense, the problem of electrocyclic reaction torquoselectivity may be considered solved.

[13]Bajorek T., Werstiuk N. H., *Chem. Commun.*, 2002, 648.
[14]Niwayama S., Houk K. N., *Tetrahedron Lett.*, 1993, **34**, 1251.

However, some anomalies are not easily anticipated. We have discussed the 3-carboxylate-3-methylcyclobutene case. It would be useful to be able to predict these exceptions. The problem amounts to determining the limitations of the Rondan–Houk theory, which requires an analysis of the approximations made. Clearly, one important approximation is the assimilation of the cyclobutene FOs to the σ_{34} and σ^*_{34} orbitals, because the frontier coefficients at C_1 and C_2 are far from negligible.[15] Dropping this approximation but representing R by a carbon 2p orbital, we can then model the molecule by a pentadiene (numbered as in Figure 6.1):

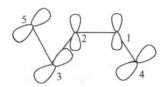

With this model, we need only apply the method already used to derive the selection rules for electrocyclic reactions (p. 53). From the Coulson equations, we can deduce that in the *in* conrotatory cyclization of pentadiene, the Ψ_1 MO generates a destabilizing C_5–C_4 secondary interaction, Ψ_2 a stabilizing and Ψ_3 a destabilizing interaction. The absolute values of these contributions rise steadily because the terminal coefficients increase from Ψ_1 to Ψ_3. Therefore, the sign of their sum is given by the HOMO contribution. If R is an attractor, the HOMO is Ψ_2 and rotation inwards is favored. If R is a donor, the HOMO is Ψ_3 and rotation inwards is disfavored. As the Coulson equations are valid only for polyenes, these conclusions are correct insofar as R can be modeled by a carbon 2p orbital. It follows that the Rondan–Houk theory works better for conjugative than for saturated substituents.

Representing R by *one* orbital was another approximation of the Rondan–Houk model. This model may fail if R can be approximated by an allyl fragment. The reason is that the HOMO of NO_2 or CO_2^- has a coefficient of zero at its central atom: the overlap between R and C_4 is then nil. Consequently, four of the five interactions shown in Figures 6.2 and 6.3 disappear and prediction is no longer possible.

Exercise 6 (D)

Houk and co-workers' calculations[16] indicate that in the electrocyclic ring openings of 1,2-dihydroazete, 1,2-dihydropyridine and 1,2-dihydroazocine, there is a marked preference for the outward rotation of the N–H bond for the four- and six-electron systems. No strong preference is observed for the eight-electron system. Suggest an explanation.

Hint. Two AOs belonging to the same atom are orthogonal. In other words, the overlaps between the nitrogen lone pair and the σ_{CN} and σ_{CN}^* orbitals are negligible. A modification of the Rondan–Houk model is therefore required.

[15]These FOs are shown in Fig. 1 of ref. 8.
[16]Walker M. J., Hietbrink B. N., Thomas B. E. IV, Nakamura K., Kallel E. A., Houk K. N., *J. Org. Chem.*, 2001, **66**, 6669.

Answer

The major interaction occurs between the nitrogen lone pair and the π^* orbital of the ring. The overlap is better for the inward rotation of the lone pair, at least for the four- and six-membered rings. Due to the larger eight-membered ring, the overlaps are practically the same for the inward and outward rotations.

Exercise 7 (E)

Explain the torquoselectivity in the following cyclobutene ring-opening reactions:

in : *out* ratio

0:100

1:11

100:0

Answer[17]

The alcohol function is a donor substituent, the ester is an allyl-type acceptor and the aldehyde is a good acceptor.

Exercise 8 (E)

Explain the following results:[18]

1 : 1

11 : 1

Answer

From Exercise 7, we know that an ester group, although electron attracting, tends to rotate outwards. In the first reaction, this preference is impeded by the β-methyl group.

---

[17]Piers E., Lu Y-F., *J. Org. Chem.*, 1989, **54**, 2267.
[18]Niwayama S., Kallel E. A., Spellmeyer D. C., Sheu C., Houk K. N., *J. Org. Chem.*, 1996, **61**, 2813.

6.1.2 Sigmatropic Rearrangements

Torquoselectivity

Houk and co-workers[2] suggest that the stereoselectivity of a sigmatropic rearrangement is influenced by the substituents on the migrating bond. Taking hydrogen shifts as examples, they showed that substituents have no effects in 'linear' reactions. For nonlinear reactions, donor substituents prefer the *out* position, whereas *very* strong acceptors prefer to be *in*:

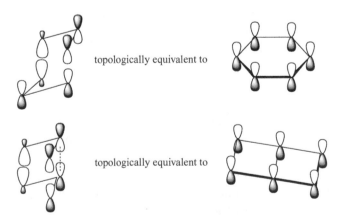

linear   nonlinear

Cope Rearrangements

In an elegant experiment, Doering and Roth have shown that the preferred transition state for a Cope rearrangement takes a chair rather than a boat form.[19] This result cannot be derived by FO theory but can be justified easily using aromaticity rules.[20] The chair transition state resembles benzene: in fact, if we use equal values for the integrals α and β, its secular determinant is identical with that of benzene. The boat transition state is different because of a secondary interaction between C_2 and C_5 (dashed line in the diagram below). This produces a structure more akin to bicyclohexatriene, a compound formed by fusing two antiaromatic cyclobutadienes. The bicyclohexatriene is less stable: its π energy is 7.657β as against 8β for benzene. Hence the Cope transition state is less stable in the boat than the chair conformation.

topologically equivalent to

topologically equivalent to

---

[19]Doering W. von E., Roth W. R., *Tetrahedron*, 1962, **18**, 67; *Angew. Chem.*, 1963, **75**, 27. In fact, the experimental results showed only that the favored transition state may take a chair *or* a twist form (Goldstein M. J., DeCamp M. R., *J. Am. Chem. Soc.*, 1974, **96**, 7356 and references cited therein).
[20]Dewar M. J. S., *Tetrahedron*, 1966, Suppl. 8, Part I, 75; *Angew. Chem. Int. Ed. Engl.*, 1971, **10**, 761.

Exercise 9 (D)

The homodienyl 1,5-sigmatropic hydrogen shift can occur by either an *endo* or an *exo* mode. Experimental results indicate that the *endo* transition state is favored by at least 12 kcal mol^{-1}.[21] Suggest an interpretation.

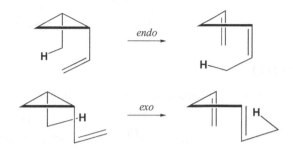

Answer[22]

The *exo* transition state resembles a *trans*-cycloheptene!

6.1.3 Cycloadditions and Their Orientations

Endo–Exo Orientation

Cycloadditions are easier to treat than unimolecular reactions; they only require an evaluation of the best FO overlap (rule 4). Let us look at the cyclodimerization of butadiene. Woodward and Hoffmann suggested that the experimentally observed *endo* compound is due to secondary interactions (shown by the double arrows above), which increase the stabilizing the FO's interaction.[23] Cisoid configurations are often adopted by the dienophile in Diels–Alder reactions,[24] as first suggested by Dewar.[20] For

---

[21]Daub J. P., Berson J. A., *Tetrahedron Lett.*, 1984, **25**, 4463.
[22]Loncharich R. J., Houk K. N., *J. Am. Chem. Soc.*, 1988, **110**, 2089.
[23]Woodward R. B., Hoffmann R., *J. Am. Chem. Soc.*, 1965, **87**, 4388; Alston P. V., Ottenbrite R. M., Cohen T., *J. Org. Chem.*, 1978, **43**, 1864.
[24]Stammen B., Berlage U., Kindermann R., Kaiser M., Günther B., Sheldrick W. S., Welzel P., Roth W. R., *J. Org. Chem.*, 1992, **57**, 6566 and references cited therein.

6 + 4 cycloadditions, *endo* reactions are disfavored by negative secondary interactions (wavy lines), which reduce the overall overlap:

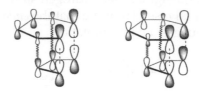

Exercise 10 (E)

Copyrolysis of **A** and maleic anhydride gives **B**. What is its stereochemistry?

A   B

Answer[25]

   The reaction begins with a conrotatory cyclobutene ring opening to give **C**. The phenyl groups open outwards to limit steric interference while retaining maximum conjugation. The second step is an 8 + 2 cycloaddition. Prove that this reaction has an *endo* preference by using the FOs of dimethylenecyclohexadiene (p. 248) and maleic anhydride (to be simulated by hexatriene, p. 256). Hence **B** has the stereochemistry given below:

C   B

Exercise 11 (M)   *MOs are given on p. 276*

Compound **A** is formed by heating tropone with cycloheptatriene. Suggest a mechanism for this reaction.

A

Answer[26]

   A can be easily visualized as a combination of the starting materials:

---

[25]Huisgen R., Seidl H., *Tetrahedron Lett.*, 1964, 3381.
[26]Itô S., Fujise Y., Woods M. C., *Tetrahedron Lett.*, 1967, 1059.

It is amusing that orbital symmetry allows a one-step reaction! Three of the four pairs of atoms which bind have coefficients of the same sign (given in bold in the diagrams) and the opposed coefficients for the last pair are smaller than the others.

Nonetheless, this pathway is unlikely because of unfavorable entropy terms. In practice, the reaction probably begins with a 6 + 4 cycloaddition:

followed by an intramolecular Diels–Alder reaction. Two points merit consideration here. First, secondary overlaps in a 6 + 4 cycloaddition disfavor *endo* orientations *and* favor *exo*. Second, tropone will act as the four electron component because the coefficients at the extremities of its LUMO are smaller (±0.39) than those in the cycloheptatriene (±0.52)

These *endo–exo* preferences are energetically small and are of the order of a kcal mol^{-1}. Consequently, factors such as dipole–dipole,[27] electrostatic,[28] steric[29] and solvent effects[27,30] can also influence the stereoselectivity. Secondary orbital interactions may not provide *all* of the answers, but no other theory can rationalize both the preferential *endo* orientation of 4 + 2 and 8 + 2 cycloadditions *and* the *exo* orientation of 6 + 4 cycloadditions so efficiently. See also Exercise 12.

Exercise 12 (E)

2,5-Dimethyl-3,4-diphenylcyclopentanedione reacts with cyclopentene and with cyclopentadiene. In one case it gives a single, clean product; in the other it gives a mixture of two isomers. Which compound gives the selective reaction, and why?

[27]Berson J. A., Hamlet Z., Mueller W. A., *J. Am. Chem. Soc.*, 1962, **84**, 297.
[28]Kahn S. D., Pau C. F., Overman L. E., Hehre W. J., *J. Am. Chem. Soc.*, 1986, **108**, 7381.
[29]Herndon W. C., Hall L. H., *Tetrahedron Lett.*, 1967, 3095; Sauer J., Sustmann R., *Angew. Chem. Int. Ed. Engl.*, 1980, **19**, 779 and references therein.; Fox M. A., Cardona R., Kiwiet N. J., *J. Org. Chem.*, 1987, **52**, 1469; Boucher J. L., Stella L., *Tetrahedron*, 1988, **44**, 3595.

Answer[31]

Cyclopentene gives an *endo-exo* mixture, but secondary overlaps with cyclopentadiene generate a pure *endo* product.

Syn–Anti Orientation

Diels–Alder reactions involving cyclic dienes give rise to bridged adducts. Any substituent X found at the bridge may adopt a *syn* or *anti* position with respect to the double bond:

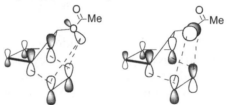

Preferential formation of a *syn* derivative could be explained by steric factors if the dienophile is bulky. However, *anti* isomers sometimes predominate (yields of up to 100% occur), which implies that the reaction can be directed by attractive interactions occurring between X and the dienophile. As with *endo–exo* preferences, it is natural to think in terms of secondary orbital overlaps.

Consider the reaction of acetoxycyclopentadiene with ethylene. In the conformation shown, the oxygen sp^2 lone pair (cf. p. 200) has the correct symmetry to interact with the ethylene π orbital in the transition state leading to the *anti* isomer. This is a four-electron combination, so it is destabilizing. However, at the same time, the oxygen p lone pair interacts favorably with the ethylene π^*. Furthermore, a rotation of the acetoxy group around the OC bond will diminish the overlap of the sp^2 lone pair with π and increases its overlap with π^*. Both lone pairs can then interact with π^*: the stabilizing interaction dominates and the *anti* isomer should be preferred. Experimentally, Winstein *et al.*[32] isolated only this isomer.

Only about 20 cases are known where the *syn* or *anti* stereochemistry of cycloadducts has been reasonably well established. FO theory correctly predicts the result in 13 out of 16 cases studied.[33] Other factors influencing the relative orientation, for example dispersion forces,[34] have been invoked.

[30]Sustmann R., Sicking W., Lamy-Schelkens H., Ghosez L., *Tetrahedron Lett.*, 1991, **32**, 1401; Sustmann R., Sicking W., *Tetrahedron*, 1992, **48**, 10293.
[31]Houk K. N., *Tetrahedron Lett.*, 1970, 2621.
[32]Winstein S., Shavatsky M., Norton C., Woodward R. B., *J. Am. Chem. Soc.*, 1955, **77**, 4183.
[33]Anh N. T., *Tetrahedron*, 1973, **29**, 3227.
[34]Williamson K. L., Hsu Y. F. L., *J. Am. Chem. Soc.*, 1970, **92**, 7385.

This FO treatment has been criticized by Inagaki *et al.*,[35] who pointed out that the acetoxycyclopentadiene HOMO should be antibonding between the lone pair and the ring π system so that when the secondary interaction between the oxygen and the dienophile is favorable, the *main* diene–dienophile frontier interaction is unfavorable (wavy lines in the diagram below):

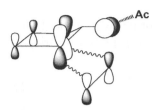

According to 3–21G calculations, the cyclopentadiene ring is fairly flat, so the acetoxy oxygen is not only far removed from C_1 and C_4, but also lies practically in the nodal plane of the π system. No through-space interaction occurs between the oxygen and the diene, either in the starting molecule or in the transition state. Hence Inagaki *et al.* objection is not confirmed, at least in this case, but neither is the FO treatment! The *anti* transition state indeed lies at lower energy than the *syn* transition state, but no significant charge transfer to the dienophile is observed. In fact, contrary to expectations, the dienophile is slightly longer in the *syn* transition state. Clearly, more work is needed to clarify this problem.

Exercise 13 (M)

Cyclooctatetraene itself does not undergo 4 + 2 cycloadditions: it is its valence isomer bicyclo[4.2.0]octatriene which reacts with dienophiles.[36] Use Dewar's PMO method to explain why the direct reaction with cyclooctatetraene is disfavored.

Answer

The transition state can be disconnected into the allyl and the heptatrienyl radicals. An analysis of the interaction of their nonbonding orbitals reveals two favorable contributions and one which is unfavorable. Consequently, the transition state is 'nonaromatic' and the reaction is not favored.

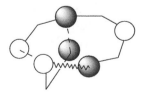

Cyclooctatetraene gives a *syn* adduct with maleic anhydride.[37]

---

[35]Inagaki S., Fujimoto H., Fukui K., *J. Am. Chem. Soc.*, 1976, **98**, 4054.
[36]Huisgen R., Mietzsch F., *Angew. Chem. Int. Ed. Engl.*, 1964, **3**, 83.
[37]Goldstein M. J., Gevirtz A. H., *Tetrahedron Lett.*, 1965, 4417; Bellus D., Helferich G., Weis C. D., *Helv. Chim. Acta*, 1971, **54**, 463.

6.2 Addition Reactions

6.2.1 Nucleophilic Additions

The Non-perpendicular Dunitz–Bürgi Attack

Traditionally, nucleophiles were thought to attack ketones perpendicularly to the carbonyl function to ensure the best overlap between the incoming nucleophile and the carbon AO[38]:

However, crystallographic studies of several aminoketones[39] whose functional groups are separated by chains of different lengths have shown that the nitrogen lone pair is always oriented toward the ketone functionality and that the NCO angle is invariably greater than 90° (107° on average). Dunitz and co-workers[39] took the view that these structures are a series of 'snapshots' describing the trajectory of a nitrogen nucleophile attacking the ketone and concluded that nucleophiles strongly prefer a nonperpendicular approach:

This effect has been reproduced in *ab initio* calculations[40] and rationalized.[41] The principle of maximum overlap states that the preferred trajectory corresponds to the best *molecular* overlap between the reaction partners (rule 4). If the nucleophile adopts a perpendicular trajectory, the *atomic* overlap with the carbon will be maximized. However, a competing out-of-phase overlap between the nucleophile HOMO and the carbonyl LUMO (shown by the wavy line) reduces the *overall* frontier orbital interaction. If the nucleophile is displaced laterally (arrow), the small diminution in the overlap with C is outweighed by the reduction in the antibonding interaction with O. This increase in the overall overlap explains the preference for attack from an obtuse angle

[38]Alder R. W., Baker R., Brown J. M., *Mechanism in Organic Chemistry*, John Wiley & Sons, Inc., New York, 1971, p. 310; Bender M. L., *Chem. Rev.*, 1960, **60**, 53.

[39]Bürgi H. B., Dunitz J. D., Shefter E., *J. Am. Chem. Soc.*, 1973, **95**, 5065; Bürgi H. B., Dunitz J. D., Lehn J. M., Wipff G., *Tetrahedron*, 1974, **30**, 1563.

[40]Bürgi H. B., Lehn J. M., Wipff G., *J. Am. Chem. Soc.*, 1974, **96**, 1956; Bürgi H. B., *Angew. Chem. Int. Ed. Engl.*, 1975, **14**, 460.

[41](a) Anh N. T., Eisenstein O., *Nouv. J. Chim.*, 1977, **1**, 61; (b) Anh N. T., *Top. Curr. Chem.*, 1980, **88**, 145.

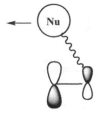

Exercise 14 (E)

Nucleophilic additions to the carbonyl group of cyclohexenones occur with significantly higher axial selectivity (average axial/equatorial ratio: 20:1) than the additions to the corresponding cyclohexanones (average ratio: 5:1). Baldwin[42] proposed that the approach vector of the nucleophiles on enones is tilted toward the side of the saturated carbon chain, so that the nucleophiles experience larger steric hindrance from the axial substituent at C_6 upon equatorial attack.

However, calculations[43] indicate that the nucleophile is slightly moved toward the side of the double bond, opposite to Baldwin's prediction. Explain.

Answer

The signs of the LUMO coefficients of cyclohexenone (which can be modeled by acrolein) are shown below. The secondary interaction with C_2 is responsible for the tilting toward the double bond.

Baldwin's Cyclization Rules

Baldwin's rules provide one of the most interesting applications of nonperpendicular attack.[44] Let us begin by defining the nomenclature. Three qualifiers characterize each reaction. The first denotes the number of atoms in the ring to be formed. The second signifies whether (*endo*) or not (*exo*) the remote atom of the bond under attack is incorporated into the ring. The third (*tet*, *trig* or *dig*) indicates the hybridization of the atom at the reactive site. The examples below should clarify these definitions:

[42]Baldwin J. E., *Chem. Commun.*, 1976, 738.
[43]Wu Y.-D., Houk K. N., Florez J., Trost B. M., *J. Org. Chem.*, 1991, **56**, 3656.
[44]Baldwin J. E., *Chem. Commun.*, 1976, 734; Baldwin J. E., Cutting J., Dupont W., Kruse L., Silberman L., Thomas R. C., *Chem. Commun.*, 1976, 736.

4-*exo-tet*

5-*exo-dig*

6-*endo-trig*

Rule 1. For tetrahedral reaction sites (e.g. for S_N2 reactions), 3- to 7-*exo-tet* reactions are favored whereas 5- and 6-*endo-tet* reactions are disfavored.

Rule 2. For trigonal reaction sites (e.g. nucleophilic attacks at double bonds), 3- to 7-*exo-trig* and 6- to 7-*endo-trig* reactions are favored whereas 3- to 5-*endo-trig* reactions are disfavored.

Rule 3. For digonal reaction sites (e.g. nucleophilic attacks at triple bonds), 5- to 7-*exo-dig* and 3- to 7-*endo-dig* are favored whereas 3- to 4-*exo-dig* reactions are disfavored.

These rules may occasionally be violated, particularly if X does not belong to the first row of the periodic table. Understanding the logic behind them is obviously better than learning them by rote, because it allows a prediction of when they may fail. Rule 1 expresses a well-known result (S_N2 reactions proceed with inversion of configuration) combined with a generally accepted hypothesis[45] (it is preferable that the nucleophile, the reactive site and the leaving group are in a straight line). These premises are not always valid, as we shall see later. Rules 2 and 3 signify that a *nucleophile*[46] will prefer to approach an unsaturated system in the plane of the π orbitals and from an obtuse angle. These constraints are relatively strong, in the order of at least 5 kcal mol^{-1},[47] so a reaction becomes strongly disfavored if we deviate from the optimal trajectory. In passing, it is worth noting that the scheme

which was originally employed to justify rule 3, is incorrect. Baldwin suggested an obtuse X–C–R angle; the FO approximation requires an obtuse X–C=C. MO calculations[48] show that both angles are, in fact, greater that 90° (p. 76).

[45]Tenud L., Farooq S., Seibl J., Eschenmoser A., *Helv. Chim. Acta*, 1970, **53**, 2059.

[46]Baldwin's rules can *not* be applied to electrophilic cyclizations.

[47]Anh N. T., Elkaïm L., Thanh B. T., Maurel F., Flament J. P., *Bull. Soc. Chim. Fr.*, 1992, **129**, 468.

[48]Eisenstein O., Procter G., Dunitz J. D., *Helv. Chim. Acta*, 1978, **61**, 2538; Strozier R. W., Caramella P., Houk K. N., *J. Am. Chem. Soc.*, 1979, **101**, 1340.

Exercise 15 (D)

Consider the following reaction:

(1) Should it be run under acidic or basic conditions?
(2) Will it be more efficient if X is an electron donor or acceptor?

Answer[49]

(1) Under basic conditions, the reaction will be 5-*endo-trig*. This is forbidden because the conjugated carbonyl group forces the future ring carbon atoms to remain co-planar. The oxygen atom is then too far from the double bond to react. Note that the 5-*endo-trig* below is 'less forbidden' because the chain which will form the ring is more flexible:

Under acidic conditions, the enol form can cyclize through a 5-*exo-trig* transition state:

(2) So X should be a donor, to stabilize the enol.

Exercise 16 (M)

Explain why 5-*exo-trig*, 6-*exo-trig* and 6-*endo-trig* reactions should be allowed whereas 5-*endo-trig* types are forbidden.

---

[49]Johnson C. D., *Acc. Chem. Res.*, 1993, **26**, 476.

Answer

The broken lines in the diagrams show the trace of the plane of the π orbitals. A reaction will occur readily if X lies on this plane and makes an obtuse angle with the C=Y bond. Molecular models show that the carbon backbone is long and flexible enough to satisfy both of these criteria for the *exo* reactions. The 6-*endo* reaction poses problems. If X lies in the π plane, the carbon skeleton has to adopt a boat conformation, leading to a perpendicular attack. However, if X moves slightly out of the π plane, an acceptable compromise can be achieved: the attack trajectory becomes non-perpendicular, with a fair nucleophile–π^* overlap. However, neither condition can be satisfied for a 5-*endo* reaction. Note that a direct application of Baldwin's empirical rules would have masked these subtleties.

Exercise 17 (D)   *Requires at least semi-empirical calculations. MOs are given on p. 276*

The stereochemistry of ketene cycloadditions is remarkable,[50] as exemplified in the following reaction:

(1) Why is the Me (and not the MeO) group adjacent to the ketone function? (*Hint*: the best approach is that leading to the best *total* frontier overlap, cf. p. 144)
(2) The major product has the ketene larger substituents *cis* to the adjacent methoxy group. Why?

Answer[51]

(1) The FO of Me–C_1H=C_2H–OMe are given in the MO Catalog in the Appendix. However, even without calculation, one can predict that its HOMO must resemble that of enol, the conjugative effect of MeO being larger than that of the methyl group. In other words, the larger coefficient is on C_1. It follows that, to ensure the best *molecular* overlap, the ketene central atom must be closer to C_1 than to C_2. Ring closure then occurs by linking C_2 to the ketene terminal atom (**A**):

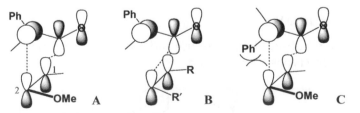

[50]Huisgen R., Mayr H., *Tetrahedron Lett.*, 1975, 2969.
[51]Rey M., Roberts S., Dieffenbacher A., Dreiding A. S., *Helv. Chim. Acta*, 1970, **53**, 417.

(2) Taking **B** as the transition state model (see **39**, p. 77), we can see that approach of the alkene from the phenyl side (**C**) is sterically hindered. To minimize the repulsions, the alkene substituents must be located on the carbonyl side (**A**); therefore, after cyclization, the phenyl group will be *cis* to the MeO group.

Exercise 18 (D) *Requires at least semi-empirical calculations*

Model the reaction between ketene and cyclopentadiene using FO theory, and confirm your model by *optimizing the approach* (do *not* calculate the transition state). Show that a Diels–Alder reaction with the CC bond, although theoretically possible, is improbable.

Answer[52]

The diene HOMO having the largest coefficients at the termini, rule 4 then suggests that one of them should attack the central atom of ketene, making an obtuse angle with C=O. Beware, there is a difficulty here! It is tempting, because of the importance of the ketene LUMO–diene HOMO interaction, to consider that approach **A**, which corresponds to the largest overlap, should be the best. Do not forget that, despite the frequent dominance of frontier control, other factors may intervene. Orientation **A** is in fact disfavored by the electronic repulsion between the oxygen and the diene π system and orientation **B** is a better compromise.

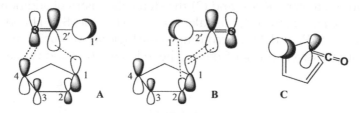

Indeed, when we optimize at the 3–21G level, structure **B** is obtained (even when we start with geometry **A**). **C** is a view from the top of the same structure. The 1–2' distance (3.50 Å) is shorter than the 2–2' distance (3.64 Å), in agreement with the fact that in the diene HOMO, the coefficient of 1 is larger than the coefficient of 2. The 1–2' distance is slightly shorter than the 1'–4 distance (3.74 versus 3.89 Å), but the overlap between 1' and 4 is small, because carbon 4 is located practically in the nodal plane of the 2p orbital of 1' (**C**). This is the reason why the major (and often the only) product is that resulting from the 2 + 2 cycloaddition. See however ref. 52b.

1,2-Asymmetric Inductions

The Cram and Felkin Models

A nucleophilic addition to a prochiral carbonyl compound R'–CO–R gives a racemic mixture of alcohols. If the substituent R' is chiral, the two faces of the carbonyl become

---

[52](a) Huisgen R., Otto P., *Tetrahedron Lett.*, 1968, 4491; (b) Ussing B. R., Hang C., Singleton D. A., *J. Am. Chem. Soc.*, 2006, **128**, 7594.

inequivalent and one of the alcohols is formed in greater quantity than the other. This phenomenon is called *asymmetric induction*. Here, we will only look at 1,2-asymmetric inductions, wherein the chiral carbon is bound directly to the carbonyl center. The structure of the major product can be predicted using a model developed by Cram and Abd Elhafez.[53] It states that if the substituents at the chiral carbon are classed as L (large), M (medium) and S (small), the reactive conformation will have the carbonyl antiparallel to the C*L bond. It is assumed that the oxygen, once complexed with the cation and solvated, becomes very bulky, so it tends to stay as far as possible from the largest group L. The nucleophile then attacks from the less hindered face:

Cornforth *et al.*[54] introduced a variant of the Cram model to be used when one of the substituents is highly electronegative (halogens, etc.). The electronegative group X is oriented *anti* to the oxygen, to minimize dipolar interactions. In other words, X plays the role of the bulky substituent.

Cram's model has been the guiding principle behind almost every study which has appeared subsequently and is still widely employed. Felkin and co-workers[55] pointed out, however, that it cannot satisfactorily explain (1) the enhancement of selectivity with increasing R size and (2) the stereochemistry of addition reactions to cyclohexanones. For example, let us take L = Ph, M = Me, S = H and vary the size of R (from Me to Et, iPr and tBu). According to the logic above, augmenting the size of R will place it progressively further from L. The following transition state:

should become more and more competitive with the Cram structure and selectivity should fall. The opposite is observed experimentally. Reductions with LiAlH$_4$ give diastereomeric ratios which rise steadily from 2.8 (R = Me) to 49 (R = *t*Bu).

Consider now 4-*tert*butylcyclohexanone, a configurationally rigid molecule. The carbonyl plane defines two half-spaces, the lower of which contains only the axial hydrogens at C$_2$ and C$_6$. Even so, the nucleophile generally arrives from above (90% in the reduction by LiAlH$_4$). These results cannot be explained by Cram's model and other factors have been invoked. For instance, Dauben *et al.*[56] suggested that equatorial attack is under 'steric approach control' whereas axial attack is under 'product development control':

---

[53]Cram D. J., Abd Elhafez F. A., *J. Am. Chem. Soc.*, 1952, **74**, 5828.

[54]Cornforth J. W., Cornforth R., Mathew K. K., *J. Chem. Soc.*, 1959, 112.

[55]Chérest M., Felkin H., Prudent N., *Tetrahedron Lett.*, 1968, 2201; Chérest M., Felkin H., *Tetrahedron Lett.*, 1968, 2205.

[56]Dauben W. G., Fonken G. J., Noyce D. S., *J. Am. Chem. Soc.*, 1956, **78**, 2579.

Felkin and co-workers' model correctly predicts the degree of stereoselectivity as a function of the R group and does not require supplementary hypotheses when applied to cyclohexanones. They suggested that torsional repulsions are important, even between bonds only partially formed. To minimize torsional strain, the chiral carbon atom and the carbonyl adopt a staggered conformation in the transition state. Substituent L is *anti* to the nucleophile, which can then attack with minimal hindrance. To explain the preference for **1** over **2**, Felkin and co-workers suggested that M and S interact more strongly with R than O.

This hypothesis would need to hold even for R=H, which is difficult to justify. It can be advantageously replaced by the assumption of non-perpendicular attack.[41] The approaching nucleophile will obviously be much less hindered in **3** than in **4**. Furthermore, as R becomes larger, it tends to push the nucleophile toward the chiral carbon. This causes the difference between M and S to be felt more strongly and the selectivity rises.

An STO-3G study of the reactions (H⁻ + MeCHCl–CHO or EtCHMe–CHO) suggests that the Felkin transition states are more stable than the corresponding Cram structures.[41] This result may be easily explained. During the addition, the strongest interaction occurs between the HOMO of the nucleophile and the LUMO of the substrate. Therefore, the most reactive ketone conformations will be those whose LUMO energy is minimized. To find these conformations, let us formally decompose the molecule into two radicals (RC=O• and the chiral R'• component) and study their subsequent recombination. The recombination of the SOMOs gives two orbitals which are well separated, because the newly formed σ bond is strong (p. 98). As a result, the LUMO of R'–CO–R is the in-phase combination of the fragment LUMOs, whereas the HOMO is the out-of-phase combination of the fragment HOMOs (Figure 6.4).

The LUMO energy minima indicate the points of greatest overlap between the component fragments. They occur when one of the σ bonds lies parallel to the π

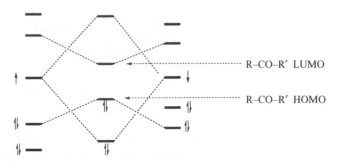

Figure 6.4   The FOs in R'–CO–R built by recombination of the fragment FOs.

system.[57] These conformations are shown in Figure 6.5. The *minimum minimorum* is **5** because σ*(C*L) lies lower than σ* (C*M) or σ* (C*S) (see p. 206).

Once the most reactive substrate conformations are known, it remains to look for the best approach of the nucleophile. An *anti* attack is promoted by a favorable secondary overlap between the nucleophile and σ*(C*L), which is shown by the double arrow. *Syn* attack is disfavored, both by a negative secondary overlap (wavy line) and by the eclipsed relationship between C*L and Nu···C (Figure 6.6). To summarize, the Felkin transition states are favored because they correspond to the best trajectories for attacking the most reactive conformations.

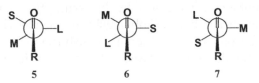

Figure 6.5   The three reactive conformations of R'–CO–R.

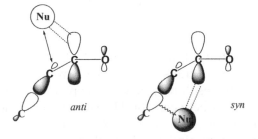

Figure 6.6   Secondary overlaps during syn and anti attacks.

---

[57]The positions of these minima are approximate because it has been assumed that the C*L, C*M and C*S bonds are orthogonal (i.e. they do not interact with one another), which is not strictly true. The O=C–C*–L dihedral angle is not exactly 90° in the most reactive conformation of the isolated ketone, and it is modified still further in the transition state under the influence of the incoming reagent. As Nu–C and C*–L are antiperiplanar and the Nu–C=O angle should be obtuse, the O=C–C*–L dihedral angle is usually smaller than 90°.

Exercise 19 (D)   *Deserves your attention*

An analysis of the conformational profiles of propanal, chloroethanal and 2-chloro-propanal[58] reveals two noteworthy trends:

- The E_{HO} and E_{LU} curves vary symmetrically. When the HOMO energy reaches a maximum, the energy of the LUMO is minimized, and vice versa.
- The E_{HO} curve follows the E_{tot} (total aldehyde energy) curve almost perfectly.

Find an explanation.

Answer

The molecules can be decomposed into (R + CHO). The frontier orbital overlaps of these fragments are at a maximum in the Felkin conformations **5**, **6** and **7**. According to Figure 6.4, these maximum overlaps generate E_{HO} and E_{LU} extrema. Hence the symmetry.

The parallel variations in the total energy and the energy of the HOMO show that (1) the frontier electrons make the largest contribution in the overall binding (which is chemically logical) and (2) the HOMO–HOMO repulsion is larger than the HOMO–LUMO attraction. At first sight, this seems to contradict the FO approximation, which only considers two-electron HOMO–LUMO interactions.

This problem has been partially discussed on p. 114. In the interactions between two fragments of a molecule or between two reagents of a transition state, repulsive terms *must* dominate, otherwise eclipsed configurations would be the rule and activation energies would be negative! We have pointed out that the FO approximation is only valid when the *change* in HOMO–HOMO repulsion is relatively small with respect to the change in the HOMO–LUMO interactions. In conformational problems, the HOMO–HOMO repulsion does vary sensibly, passing through maxima in eclipsed conformations and minima in staggered conformations. The repulsion is enhanced here because the HOMO–HOMO interactions are *intra*molecular, which automatically confers significant overlaps. It follows that the stable conformations are those with minimal HOMO–HOMO interactions. The HOMO–LUMO term is *not* decisive in these problems.

Exercise 20 (D)

(1) Show that the angle of nucleophilic attack in a free carbonyl compound is more obtuse than in its 'ate' complex.
(2) Hence deduce a theoretical rationalization of Seyden-Penne's empirical rule,[59] which states that softer nucleophiles give greater asymmetric induction.

Answer[60]

(1) The non-perpendicular Dunitz–Bürgi attack is due a negative frontier orbital overlap between the nucleophile and the oxygen. In the 'ate' complex, the LUMO

[58]Frenking G., Köhler K. F., Reetz M. T., *Tetrahedron*, 1993, **49**, 3971.
[59]Seyden-Penne J., *Euchem Conference on Chirality*, La Baule, 1972.
[60]Anh N. T., *Top. Curr. Chem.*, 1980, **88**, 145.

coefficient is increased at carbon and decreased at oxygen (p. 61). Hence the unfavorable overlap decreases, the lateral displacement of the nucleophile is diminished and the angle of attack is reduced.

(2) A 'hard' reagent is normally associated with a small cation which is capable of forming a strong complex with the carbonyl; this reduces the angle between the CO and the incoming nucleophile. As the angle diminishes, the difference in energy between the two transition states **3** and **4** is reduced (because the Nu–M and Nu–S distances increase) and the asymmetric induction is lowered.

Exercise 21 (M)   *Deserves your attention*

(1) The angles of nucleophilic attacks on alkenes and alkynes have been calculated[61] (see also ref. 48) to lie in the range 115–130°, definitely larger than the angle of attack on carbonyls (~109°). Why?
(2) The calculated force constants of the transition structures indicate that the deformation of this angle is fairly easy, the cost being about half of the bending of a normal C–C–C or H–C–C angle. Again, this contrasts with the case of the carbonyl group, where the attack angle is relatively rigid.[47] Suggest an explanation.

Answer

(1) The π^*_{CO} orbital is has a larger coefficient at C than at O. The π^*_{CC} orbital has equal coefficients on C_1 and C_2. Therefore, the negative overlap between the nucleophile **Nu** and the distal extremity of the double bond (O or C_2), which is the cause of the Dunitz–Bürgi obtuse angle of attack, is larger with π^*_{CC} than with π^*_{CO}:

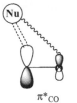 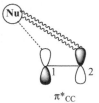

π^*_{CO} $\qquad$ π^*_{CC}

This means that the nucleophile is pushed away more strongly in the alkene case than in the carbonyl case. At the same time, the attractive in phase overlap between **Nu** and C_1 is smaller with π^*_{CC} than with π^*_{CO}, which tends to maintain **Nu** nearer to the vertical position in the carbonyl case. Both factors contribute to give a larger angle for the reaction with alkenes or alkynes.

(2) Consider now the bonding MOs. In the π_{CO} orbital, the larger coefficient is at oxygen, whereas in the π_{CC} orbital the two coefficients are again equal.

[61]Paddon-Row M. N., Rondan N. G., Houk K. N., *J. Am. Chem. Soc.*, 1982, **104**, 7162.

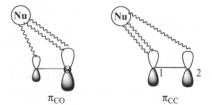

π_{CO} $\qquad$ π_{CC}

To summarize, the **Nu**–O and **Nu**–C_2 interactions are always repulsive, in both the π and π^* orbitals. The interaction of the nucleophile with C_1 is attractive in the π^* orbitals, more strongly for π^*_{CO} than for π^*_{CC}. It is repulsive in the π orbitals, more weakly for π_{CO} than for π_{CC}. As C_1 is nearer to the nucleophile, its influence should dominate. This explains why it is easier to increase the angle of attack for C=C than for C=O.

It is necessary to take into account this difference when examining the stereochemistry of additions to alkenes (see pp. 169–170).

Exercise 22 (E)

The following reaction was used in a synthesis of fumagillin:[62]

(1) conjugated addition
(2) enolate trapping

Two diastereomers are obtained in a 96:4 ratio. What is the stereochemistry of the major product?

Answer

We expect the nucleophile to approach *antiperiplanar* to the lower σ^*_{CO}, giving as major product

The flattening rule

Studies of Felkin's model have shown that the transition state for nucleophilic addition to a carbonyl compound is strongly stabilized when the C_2–X and Nu···C_1 bonds are antiperiplanar.[41] Let us apply this rule to a configurationally rigid cyclohexanone.

---

[62]Taber D. F., Christos T. E., Rheingold A. L., Guzei I. A., *J. Am. Chem. Soc.*, 1999, **121**, 5589.

Our arguments are valid for all angles of attack, but for a clearer drawing we will assume that the nucleophile arrives perpendicular to the carbonyl. Perfect antiperiplanarity is impossible in the normal cyclohexanone geometry, either for equatorial or axial attack. Nonetheless, for the experimentally observed $O=C_1-C_2-H_e$ dihedral angle of $12°7'$,[63] the difference from antiperiplanarity is $21°$ for axial and $45°$ for equatorial trajectories. Flattening the ring (arrows) causes the $Nu\cdots C_1$ and C_2-H_a bonds to move toward antiperiplanarity and axial attack becomes easier:

As the ring flattens, the $OC_1C_2H_e$ dihedral angle increases and the C_1C_2 conformation becomes staggered. Hence, antiperiplanarity, which ensures a good electronic transfer between the nucleophile and the electrophile, also imposes a flattening which automatically minimizes torsional repulsions. Hence the rule[64]: *axial attack is favored by flattened or flexible rings*. The correlation between flexibility of the ketone and the percentage of axial attack was first noted by Suzuki *et al.*[65]

Let us illustrate the flattening rule with a few examples. $LiAlH_4$ reduction of 3-ketosteroids gives 10% β attack, whereas 7-ketosteroids give 55%,[66] results which could not be explained by earlier theories. The B ring of the steroid probably adopts a chair structure, so attack at 7β is equatorial. *Static* torsional repulsions (those occurring in the initial reagent)[55] and steric compression[67] both disfavor equatorial β attack and cannot explain the increase on passing from 3- to 7-substituted compounds. Steric hindrance is also inoperable, the β face being as hindered at C_7 as at C_3. A suggestion[68] has been made that β attacks at C_7 and C_3 are equally favorable, but that 7α attack is hindered by the axial hydrogen 14α. Experimental studies show that this is not the case: $LiAlH_4$ reduction of 1-decalone, which has the same steric environment as a 7-ketosteroid, gives 90–95% α attack[69]:

[63]X-ray structure of *tert*butyl-cyclohexanone (Metras F., personal communication).

[64]Huet J., Maroni-Barnaud Y., Anh N. T., Seyden-Penne J., *Tetrahedron Lett.*, 1976, 159. Also see ref. 41b. Note that antiperiplanarity favors only axial attacks. For the antiperiplanarity of $Nu\cdots C_1$ and C_2C_3 to be achieved for equatorial attacks, the ring must be puckered (arrows in the opposite sense) and may break.

[65]Suzuki T., Kobayashi T., Takegami Y., Kawasaki Y., *Bull. Chem. Soc. Jpn*, 1974, **47**, 1971.

[66]Fieser L. F., Fieser M., *Steroids*, Reinhold, New York, 1959, p. 269.

[67]Schleyer P. v. R., *J. Am. Chem. Soc.*, 1967, **89**, 701; Laemmle J., Ashby E. C., Rolling P. V., *J. Org. Chem.*, 1973, **38**, 2526.

[68]Dauben W. G., Blanz E. B. Jr, Jui J., Micheli R., *J. Am. Chem. Soc.*, 1956, **78**, 3752; Wheeler O. H., Mateos J. L., *Can. J. Chem.*, 1958, **36**, 1049.

[69]Hückel W., Maucher D., Fechtig O., Kurz J., Heinzel M., Hubele A., *Liebigs Ann. Chem.*, 1961, **645**, 115; Grob C. A., Tam S. W., *Helv. Chim. Acta*, 1965, **48**, 1317; Moritani I., Nichida S., Murakami M., *J. Am. Chem. Soc.*, 1959, **81**, 3420.

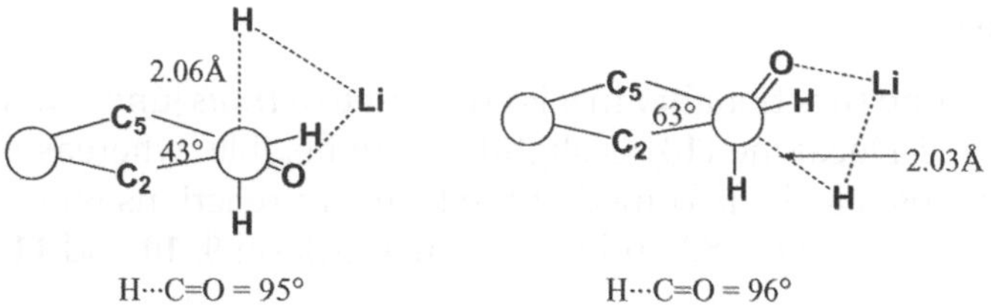

H···C=O = 95°   H···C=O = 96°

Figure 6.7 3–21G transition states for the addition of LiH to cyclohexanone. The values shown correspond to the $C_2C_1C_6C_5$ dihedral angle.

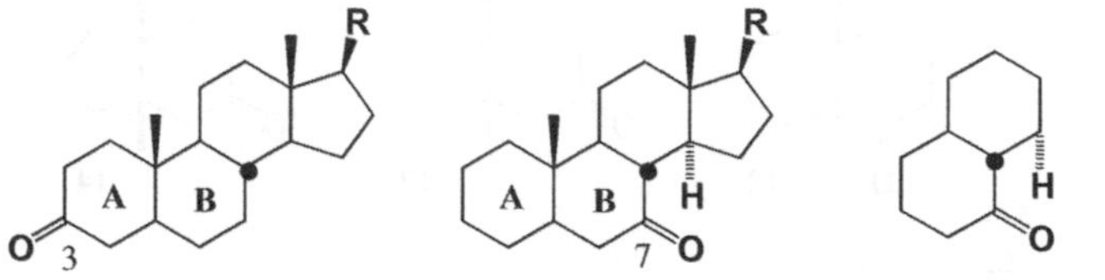

90% α attack   45% α attack   90-95% α attack

The flattening rule rationalizes these data without difficulty. The B ring is linked to two others, so it is less flexible than the A ring. More energy is therefore required to flatten the ring, and axial (α) attack is disfavored. Wheeler and Mateos's studies of reduction rates of 3- and 7-ketosteroids by sodium borohydride[68] support this interpretation. Upon passing from 3- to 7-ketosteroids, β attack rates are essentially unaffected ($k_{7\beta} = 48 \times 10^{-4}$ against $k_{3\beta} = 60 \times 10^{-4}$), but α attack is seriously retarded ($k_{7\alpha} = 64 \times 10^{-4}$ against $k_{3\alpha} = 340 \times 10^{-4}$).

Other applications of the flattening rule may be found in a paper by Wu *et al.*[70] Figure 6.7 (taken from the same paper) shows 3–21G transition states for the axial and equatorial attacks of LiH on cyclohexanone. They have Felkin-type structures, with C_1 and C_6 staggered. As predicted, axial attack is associated with a flattening of the cyclohexanone. The $C_2C_1C_6C_5$ dihedral angle is 43° in the transition state, as against 54° in the cyclohexanone. During equatorial attack, the ring becomes puckered (in order to improve antiperiplanarity between the H···C and C_6C_5 bonds) and the $C_2C_1C_6C_5$ dihedral angle increases to 63°. However, the rigidity of the chair conformation does not allow further puckering and it is therefore impossible to obtain the 45° increase necessary for achieving antiperiplanarity.

Exercise 23 (E)

Arrange the following compounds in order of increasing preference for axial nucleophilic attack:

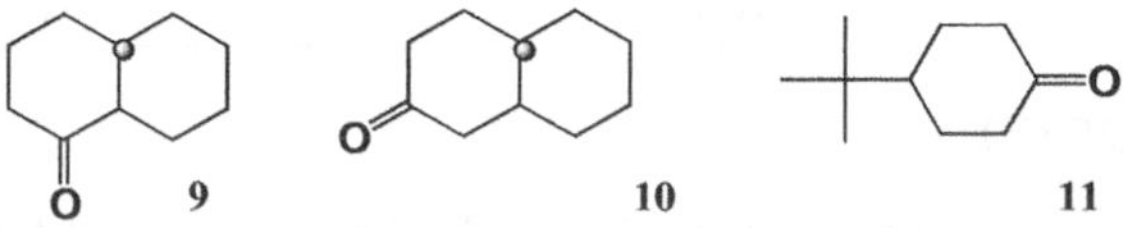

9   10   11

[70]Wu Y-D., Tucker J. A., Houk K. N., *J. Am. Chem. Soc.*, 1991, **113**, 5018. See also: Mukherjee D., Wu Y.-D, Fronczek F. R., Houk K. N., *J. Am. Chem. Soc.*, 1988, **110**, 3328; Coxon J. M., Houk K. N., Luibrand R. T., *J. Org. Chem.*, 1995, **60**, 418.

Answer[65]

The ketone functionality in **9** is adjacent to a *trans* junction, which makes it difficult to flatten. 2-Decalone (**10**) is slightly more flexible, whereas 4-*tert*-butylcyclohexanone is the most easily deformed of the three. In reactions with 2 equiv. of AlMe$_3$, Suzuki *et al.*[65] obtained 60, 82 and 86% axial attack on **9, 10** and **11**, respectively.

Exercise 24 (D)

As Excercise 23, but for compounds:

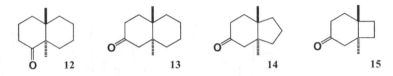

Answer[71]

The molecules in question form a series of *trans* fused bicyclo[4.*n*.,0]ketones, where *n* = 4, 3, 2. As *n* decreases, the dihedral angle forming the junction with the right-hand ring closes and the left-hand ring opens:

Now, if in a ring one dihedral angle (e.g. 6–1–2–3) opens, others elsewhere (e.g. 3–4–5–6) must close. A well-known example of this effect concerns the cyclohexane molecule. In its chair structure, all dihedral angles are 60°, whereas in a boat structure, two of the dihedral angles are zero and the others are greater than 60°.

Consequently, the carbonyl functionality flattens progressively as we move from **13** to **15**.[72] The experimentally observed percentages for axial attack are the following:

	12	13	14	15
LiAlH$_4$	80	85	89	94
NaBH$_4$	78	88	90	94
MeMgI	12	34	42	56

Exercise 25 (E)

Is a 1-ketosteroid more prone to axial attack than a 12-ketosteroid?

---

[71]Casadevall E., Pouet Y., *Tetrahedron Lett.*, 1976, 2841.
[72]Casadevall A., Casadevall E., Moner M., *Bull. Soc. Chim. Fr.*, 1972, 2010; 1973, 657.

Answer[73]

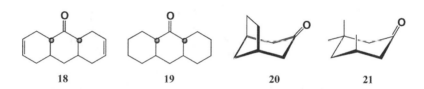

The previous exercise showed that the ketone in **17** must be more puckered than that in **16** because it is located next to a five-membered ring junction. Hence **16** is more prone to axial attack.

Exercise 26 (E)

As Exercise 23, but for the pairs **18–19** and **20–21**.

Answer[74]

In **18**, the dihedral angles about the double bonds in the outermost rings are zero. To compensate, the dihedra forming the junction with the central ring are more open.[75] The reasoning in Exercise 24 then shows that the center ring must be flattened in **18**, so it will suffer more axial attack than **19**. In **20**, the ethylene bridge creates a tie which induces a flattening of the cyclohexanone ring. The experimentally observed percentages of axial attack are:

	18	19	20	21
LiAlH$_4$, THF	51	36	59	24
LiAlH$_4$, ether			70	38
LiAlH$_4$, 3MeOH, THF	44	33		

Exercise 27 (D)

α-Substituted bicyclo[2.2.2]octanones seem to be perfect candidates for testing Cram's model. However, the experimental data are surprising. LiAlH$_4$ reduction occurs from the less hindered face when R = Ph or *t*-Bu, indiscriminately when R = Me and

[73]Ayres D. C., Kirk D. N., Sawdaye R., *J. Chem. Soc. B*, 1970, 505.
[74]Arnaud C., Accary A., Huet J., C. R. *Acad. Sci.*, 1977, **285**, 325; Huet J., personal communication.
[75]Bucourt R., *Top. Stereochem.*, 1974, **8**, 159.

preferentially from the more hindered face when R = Et (68%) and iPr (72%). For the series R = Me, Et, iPr, the larger the R group becomes, the more the hydride attacks from its side![76] Find an explanation.

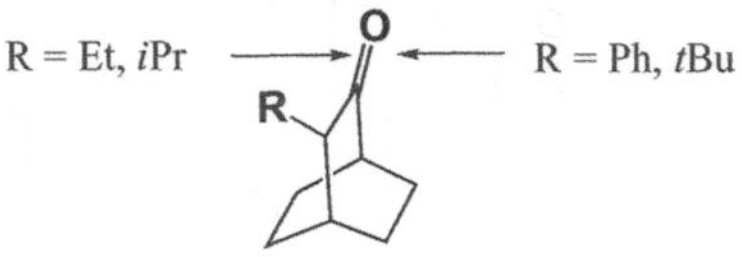

Answer

Our calculations[41] of a Cram-like transition state indicate that the attack at the more hindered face is only slightly disfavored and that small conformational changes could change the preferred trajectory. Lefour has suggested that steric repulsion between R and the *exo*-hydrogen (wavy line) displaces the substituent upwards. The bridge then takes up a Felkin-type conformation. This favors an attack from the same side as R, unless it is very bulky:

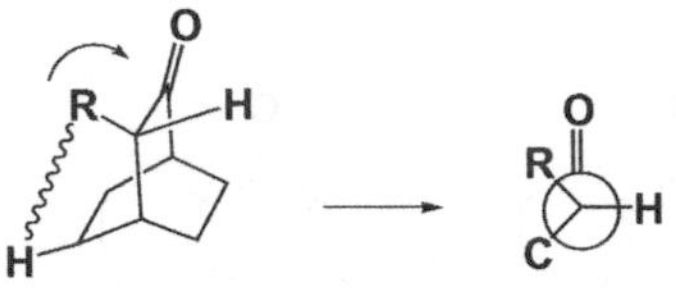

This interpretation does not seem to be applicable to dibenzo derivatives, because Dreiding models show that they are very rigid.

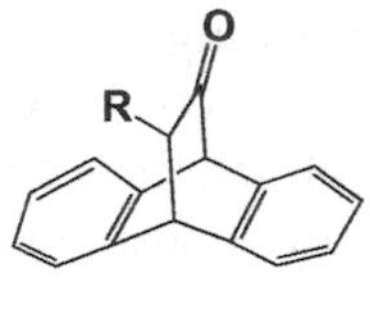

Chérest *et al.*[76] have suggested that the R group may exert an anisotropic inductive effect. It is also possible that in these rigid molecules, a charge control mechanism may favor an attack from the R side.[77]

The importance of being flexible

There is no reason to confine these ideas to six-membered rings. Flexibility, which allows antiperiplanar configuration to be achieved in the transition state, is a crucial factor governing the reactivity of any ketone. To check this idea, we have studied the addition of cyanide ion to cyclopentanone and 3-pentanone.[78] Cyclopentanone can be

---

[76]Varech D., Jacques J., *Tetrahedron Lett.*, 1973, 4443; Chérest M., Felkin H., Tacheau P., Jacques J., Varech D., *Chem. Commun.*, 1977, 372.
[77]Paddon-Row M. N., Wu Y.-D., Houk K. N., *J. Am. Chem. Soc.*, 1992, **114**, 10638; Williams L., Paddon-Row M. N., *Chem. Commun.*, 1994, 353. See also ref. 78.
[78]Anh N. T., Maurel F., Lefour J. M., *New J. Chem.*, 1995, **19**, 353.

flattened easily, but not puckered, so *syn* attack (i.e. from the upper side in the following diagrams, antiperiplanar to the CH bonds) is favored by 2.81 kcal mol^{-1} according to 6–31G* calculations:

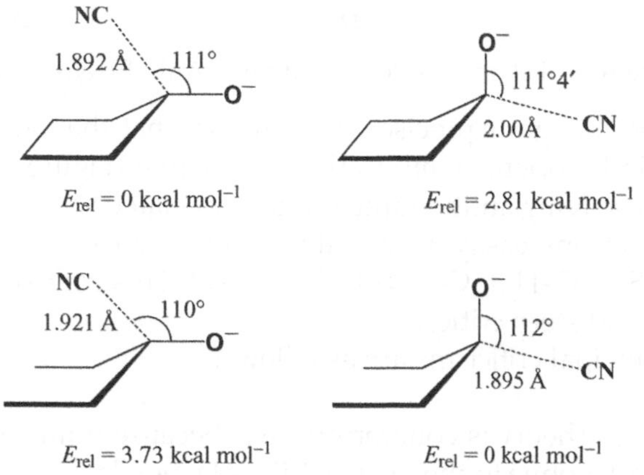

E_{rel} = 0 kcal mol^{-1} E_{rel} = 2.81 kcal mol^{-1}

E_{rel} = 3.73 kcal mol^{-1} E_{rel} = 0 kcal mol^{-1}

3-Pentanone is fairly flexible, so an antiperiplanar attack can occur at either face. The σ^*_{CC} orbital lies lower in energy than σ^*_{CH} so frontier orbital control favors the anti transition state by 3.73 kcal mol^{-1}. Therefore, the flattening rule may be generalized as follows: *antiperiplanar attack and frontier orbital control in general are only important for reasonably flexible ketones.*

Classification of substituents as L, M, S and the Cieplak model

Whether we use the Cram or Felkin model, we have to classify the groups bound to the chiral atom as L, M or S. The classification is usually based on two parameters: the 'effective bulk' of the group and, more importantly, its polarity: chlorine (or even fluorine) is considered bulkier than *tert*-butyl. The 'effective bulk' is difficult to estimate because it is not directly related to steric hindrance, as determined by van der Waals radii or conformational equilibria. For example, although $\Delta G°$ is −3.1 kcal mol^{-1} for a phenyl group and −5.6 kcal mol^{-1} for a *tert*-butyl group,[79] the phenyl 'is almost always considered to be bulkier than an alkyl group'.[80] In fact, Morrison and Mosher[80] have estimated the chances of accurately classifying the relative bulk of two substituents to be 50:50!

We considered that nucleophilic addition is promoted when electron transfer from the reagent to the substrate is facilitated.[41] Consequently, we proposed that X, Y and Z be classified according to their electrophilicity, as measured by the energy of their antibonding orbitals σ^*_{CX}, σ^*_{CY} and σ^*_{CZ}. Thus, the best electron acceptor is considered to be the 'bulkiest group'. This approach uses only one parameter and eliminates the tricky problem of defining 'effective bulk'. See, however, the discussion on pp. 168–169.

---

[79]Eliel E. L., Allinger N. L., Angyal S. J., Morrison G. A., *Conformational Analysis*, Interscience, New York, 1965, p. 44.
[80]Morrison J. D., Mosher H. S., *Asymmetric Organic Reactions*, Prentice-Hall, Englewood Cliffs, N. J., 1971, p. 36, 89.

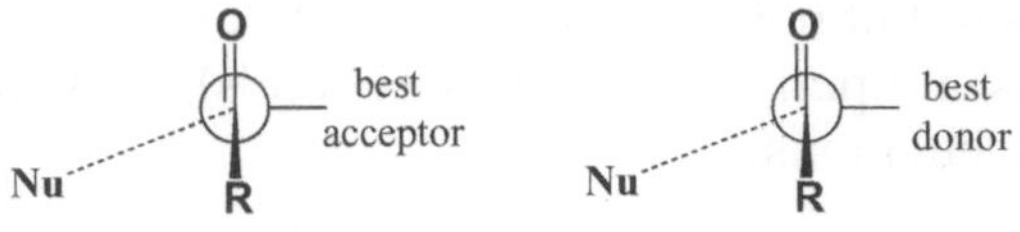

Figure 6.8 Felkin's model according to Anh–Eisenstein[41] (22) and Cieplak[81] (23).

Cieplak[81] suggests precisely the reverse: that the most electron-donating substituent should be oriented *anti* to the nucleophile (Figure 6.8). His rationalization states that the low-lying antibonding orbital associated with the partial Nu···C=O bond can accept electrons easily. Generally, donor character is considered to decrease in the order C–S > C–H > C–C > C–N > C–O. This controversial theory has ardent supporters[82] and stern critics.[83]

The principal criticisms are as follows:

- Cieplak's theory is counterintuitive, because it implies that electron transfer from the electrophile to the nucleophile is favorable.
- The theory rationalizes exceptions well, but is much less reliable for more normal cases. Suppose, for example, that the chiral center has OMe, Et and H substituents. According to Cieplak, C–H is a better donor than C–C or C–O, so we have a situation where L = H. It gives the transition states below:

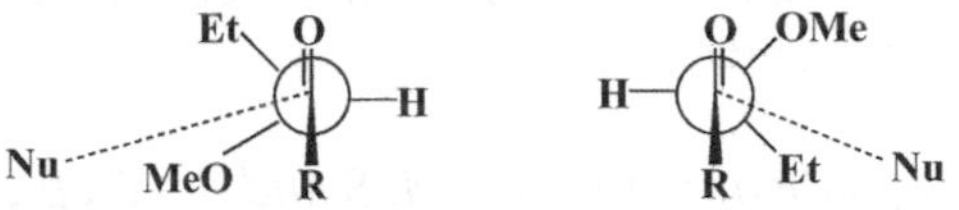

Pauling gives the van der Waals radius of oxygen as 1.40 Å and a methyl group as 2 Å. Consequently, S will be OMe and M will be Et, so **24** should give the major product, particularly since it is favored by dipolar interactions. This prediction is exactly the opposite of the result obtained using Cornforth's model.[54]

- To the partial Nu···C bond correspond *two* orbitals: a low-lying σ^*_{CNu} capable of accepting electrons and a high-lying σ_{CNu}, which is electron releasing. Cieplak takes into account the first but neglects the second; we do the reverse. So which approach

[81]Cieplak A. S., *J. Am. Chem. Soc.*, 1981, **103**, 4540.
[82]E.g.: (a) Cheung C. K., Tseng L. T., Lin M. H., Srivastava S., LeNoble W. J., *J. Am. Chem. Soc.*, 1986, **108**, 1598; 1987, **109**, 7239 (correction); (b) Srivastava S., LeNoble W. J., *J. Am. Chem. Soc.*, 1987, **109**, 5874; (c) Johnson C. R., Tait B. D., Cieplak A. S., *J. Am. Chem. Soc.*, 1987, **109**, 5975; (d) Chung .W. S., Turro N. J., Srivastava S., Li H., LeNoble W. J., *J. Am. Chem. Soc*, 1988, **110**, 7882; (e) Cieplak .A. S., Tait B. D., Johnson C. R., *J. Am. Chem. Soc.*, 1989, **111**, 8847; (f) Lin M. H., LeNoble W. J., *J. Org. Chem.*, 1989, **54**, 998; (g) Xie M., LeNoble W. J., *J. Org. Chem.*, 1989, **54**, 3836; (h) Mehta G., Khan F. A., *J. Am. Chem. Soc.*, 1990, **112**, 6140.
[83]E.g.: (a) Wu Y.-D., Houk K. N., *J. Am. Chem. Soc.*, 1987, **109**, 908; (b) Lodge E. P., Heathcock C. H., *J. Am. Chem. Soc*, 1987, **109**, 3353; (c) Meyers A. I., Sturgess M. A., *Tetrahedron Lett.*, 1988, **29**, 5339; (d) Meyers A. I., Wallace R. H., *J. Org. Chem.*, 1989, **54**, 2509; (e) Wong S. S., Paddon-Row M. N., *Chem. Commun.*, 1990, 456; (f) Coxon J. M., McDonald D. Q., *Tetrahedron*, 1992, **48**, 3353; (g) Coxon J. M., Houk K. N., Luibrand R. T., *J. Org. Chem.*, 1995, **60**, 418; (h) Yamataka H., *J. Phys. Org. Chem*, 1995, **8**, 445.

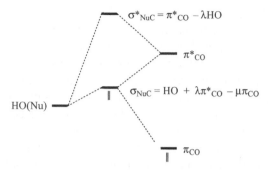

Figure 6.9 A perturbation scheme showing the orbitals of the incipient Nu···C bond.

is better? Two arguments militate against Cieplak. First, it is *chemically* more reasonable to imagine that the nucleophile transfers electrons to its partner than the reverse. Second, a donor substituent *anti* to the incoming nucleophile will tend to weaken the Nu–C bond, therefore prohibiting the bond-making process. Conversely, an acceptor substituent in the *anti* position will strengthen the bond and promote the addition reaction, as we see in Figure 6.9. The σ^*_{CNu} orbital is essentially π^*_{CO}, modified by a small out-of-phase contribution from the HOMO of the nucleophile. Since it is *antibonding* with respect to Nu and C, it will cause the Nu–C bond to weaken when populated. The σ_{CNu} orbital is composed of the nucleophile HOMO mixed in-phase with π^*_{CO} and out-of-phase with π_{CO}. The contribution of the latter is larger, as the nucleophile HOMO and π_{CO} lie close in energy, and this is sufficient to make σ_{CNu} also antibonding between Nu and C. Removal of electron density by a neighboring acceptor then *strengthens* the Nu–C bond. *Ab initio* calculations confirm that the σ_{CNu} and σ^*_{CNu} orbitals are both antibonding with respect to Nu–C. Figure 6.10 shows the three transition states for the $(NC^- + CH_2Cl–CHO)$ system. The Nu-C bond is much shorter in **26** than in **27** or **28**, confirming our prediction that a neighboring acceptor should reinforce (i.e. shorten) the bond which is forming. Structure **26** is also more stable than the other two.

Figure 6.9 suggests[84] that the stabilization due to 'Cieplak factors' is smaller than the 'Anh–Eisenstein' effect. The σ^*_{CNu} level lies higher in energy than π^*_{CO}, so its acceptor properties are smaller than might be imagined. However, the appearance of σ_{CNu} above π_{CO} shows clearly that it has donor character. Calculations confirm this analysis: the lowest energy transition state is **26** (Figure 6.10).

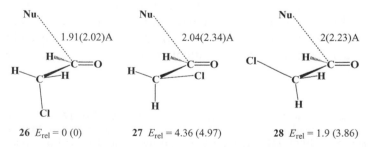

Figure 6.10 Transition states for the reaction of CN^- with $CH_2Cl–CHO$ according to 6–31G* calculations (MP2/6–31G* results in parentheses). All energies in kcal mol^{-1}.

[84]For $(CN^- + H–CHO)$, σ_{CNu} lies at -0.1374 a.u. and σ^*_{CNu} at 0.4432 a.u. (3–21G calculations).

- Calculations by the groups of Paddon–Row,[83] Houk[70,83] and ourselves[78] have shown that the lowest energy transition state has the nucleophile *anti* to the best acceptor. This rule is broken, however, by propanal, [83a,e,85] where the transition state having the methyl group (a better acceptor than H[86]) in the *anti* position is the least stable. This occurs because, methyl being not a good enough acceptor, the energetic gain from having the methyl group in the *anti* position is insufficient to overcome the conformational preference, which favors the methyl group eclipsed by the carbonyl (p. 190)
- The experimental results supporting Cieplak's theory are not entirely convincing.[82] Houk and co-workers[70,83g] have offered an alternative interpretation of these. Furthermore, the rigidity of the molecules used in some of these studies (adamantanone, bicyclo[2.2.1]heptanone) prohibits antiperiplanar attack and introduces a bias against frontier control. Finally, some experimental studies are ambiguous. Thus, *anti* attack at **29** is simultaneously antiperiplanar to the bonds shown with heavy lines, and *syn* to those which are dotted. If this attack is disfavored, is it because it is (a) axial to the cyclohexanone on the left or (b) equatorial to the cyclohexanone on the right?

29

One last comment before finishing. Cieplak's theory is based on the need to provide electrons for the incipient bond, so he invariably orients the best donor *anti* to the incoming reagent, irrespective of whether it is a nucleophile or an electrophile. On the other hand, we are concerned with ensuring the best electron transfer. Thus, *the best acceptor on an electrophile and the best donor on a nucleophile should be oriented anti to the reaction partner.*

Karabatsos's model

Karabatsos[87] introduced a variant of Cram's model based on the following assumptions: (1) the transition states for addition to carbonyls are reactant-like; (2) the reactive conformations are then the most stable ones, which have a neighboring σ bond eclipsing the carbonyl group (cf. p. 188); and (3) the nucleophile approaches from the less hindered side. Assumption (2) is questionable: even in the reactant-like transition state, the stable and reactive conformations may be completely different. Karabatsos's model was the first to draw attention on the importance of conformational factors in asymmetric induction.

---

[85]Frenking G., Köhler K. F., Reetz M. T., *Tetrahedron*, 1991, **47**, 8991, 9005.
[86]Cieplak considers CH to be a better donor than CC. See also: Macaulay J. B., Fallis A. G., *J. Am. Chem. Soc.*, 1990, **112**, 1136; for the opposite point of view: Rozeboom M. D., Houk K. N., *J. Am. Chem. Soc.*, 1982, **104**, 1189 and ref. 83e. In fact, methyl is both a better donor and a better acceptor than H: the σ_{CC} and σ^*_{CC} orbitals are higher and lower in energy than σ_{CH} and σ^*_{CH}, respectively.
[87]Karabatsos G. J., *J. Am. Chem. Soc.*, 1967, **89**, 1367.

Which Factors Control Asymmetric Induction?

Felkin and co-workers[55] were the first to recognize the importance of torsional effects, i.e. the tendency of all bonds, either fully or partially formed, to become staggered in the transition state. The groups of Houk, Paddon-Row,[70,83,88] Frenking *et al.*[89] and Kahn and Hehre,[90] among others, have pointed out the importance of charge control. Karabatsos[87] and later Frenking *et al.*[85,89] emphasized conformational control. Dipolar interactions play a pivotal role in Cornforth's model.[54] Evan and co-workers' models[91] combine the Felkin and Cornforth models. We have stressed the importance of antiperiplanarity, of which the flattening rule[41,64] is a direct consequence. Finally, steric factors should not be overlooked.[83b] No doubt each of these factors may play a role: the problem is to know which influence will dominate under a given set of conditions.

Based on published results, the following inferences seem reasonable:

- Chelation,[92] when it exists, appears to have the strongest influence.
- Torsional effects and non-perpendicular attacks are second in importance. Computational estimates indicate that rotational barriers involving torsional interations with partially formed bonds may be as large as ~ 3 kcal mol^{-1}.[93] To the best of our knowledge, all recent models incorporate non-perpendicular attacks and torsional effects.
- Dipolar and antiperiplanar effects are the next critical factors. Antiperiplanarity (and its corollary, the flattening rule) fully displays its influence only in flexible molecules. The existence of the flattening rule seems to confirm the significance of antiperiplanarity. Houk and co-workers[94] have also pointed out the general occurrence of antiperiplanar effects in nucleophilic, electrophilic and radical additions. So far, almost all the evidence against a significant role of antiperiplanarity is drawn from work with either fluorine-containing compounds[70,85] or rigid molecules such as adamantanones,[82] norbornanones[88b] or bridged biaryl ketones.[95] In compounds lacking flexibility, frontier interactions cannot be maximized.

That dipolar effect may be substantial is shown by a series of results which agree better with the Cornforth than with the Felkin model.[96] Some very interesting results

---

[88](a) Wong S. S., Paddon-Row M. N., *Chem. Commun.*, 1991, 327; (b) Wong S. S., Paddon-Row M. N., *Aust. J. Chem.*, 1991, **44**, 765; (c) Paddon-Row M. N., Wu Y.-D., Houk K. N., *J. Am. Chem. Soc.*, 1992, **114**, 10638; (d) Williams L., Paddon-Row M. N., *Chem. Commun.*, 1994, 353 and reference therein.
[89]Frenking G., Köhler K. F., Reetz M. T., *Tetrahedron*, 1993, **49**, 3971, 3983; 1994, **50**, 11197.
[90]Kahn S. D., Hehre W. D., *J. Am. Chem. Soc.*, 1986, **108**, 7399.
[91](a) Evans D. A., Dart M. J., Duffy J. L., *Tetrahedron Lett.*, 1994, **35**, 8537; (b) Evans D. A., Dart M. J., Duffy J. L., Yang M. G., Livingston A. B., *J. Am. Chem. Soc.*, 1995, **117**, 6619; (c) Evans D. A., Dart M. J., Duffy J. L., Yang M. G., *J. Am. Chem. Soc.*, 1996, **118**, 4322; (d) Evans D. A., Allison B. D., Yang M. G., Masse C. E., *J. Am. Chem. Soc.*, 2001, **123**, 10840; (e) Evans D. A., Siska S. J., Cee V. J., *Angew. Chem. Int. Ed.*, 2003, **42**, 1761.
[92]Reetz M. T., *Angew. Chem. Int. Ed. Engl.*, 1984, **23**, 556; Reetz M. T., *Acc. Chem. Res.*, 1993, **26**, 462; Mengel A., Reiser O., *Chem. Rev.*, 1999, **99**, 1191.
[93]Paddon-Row M. N., Rondan N. G., Houk K. N., *J. Am. Chem. Soc.*, 1982, **104**, 7162.
[94]Caramella P., Rondan N. G., Paddon-Row M. N., Houk K. N., *J. Am. Chem. Soc.*, 1981, **103**, 2438.
[95]Fraser R. R., Kong F., Stanciulescu M., Wu Y.-D., Houk K. N., *J. Org. Chem.*, 1993, **58**, 4440.
[96]Rousch W. R., Adam M. A., Walts A. E., Harris D. J., *J. Am. Chem. Soc.*, 1986, **108**, 3422; Hoffmann R. W., Metternich R., Lanz J. W., *Liebigs Ann.*, 1987, 2395. See also ref. 91e.

are reported in a recent paper[97]: DFT calculations indicate that in the addition of enolboranes to CH_3–CHX–CHO, when X = F, OMe and Cl the Cornforth model is favored whereas for X = PMe_2, SMe and NMe_2 the Felkin model is favored. We have pointed out earlier that one of the surest (and simplest) ways to determine the σ^* level is by looking at the strength of the σ bond (p. 98). Now, bond strengths are CF = 116 kcal mol^{-1}, CO = 85.5 kcal mol^{-1}, CCl = 78 kcal mol^{-1}, CN = 72.8 kcal mol^{-1}, CS = 65 kcal mol^{-1} and CP = 63 kcal mol^{-1}.[98] It is expected that as the CX bond becomes weaker, the Felkin model will fare better and better. This is indeed borne out nicely by the calculations. The energy difference between the Cornforth and the Felkin transition states is −2.4 kcal mol^{-1} for X = F, −1.7 kcal mol^{-1} for X = OMe, −0.2 kcal mol^{-1} for X = Cl,. 0.8 kcal mol^{-1} for X = N, 3.3 kcal mol^{-1} for X = S and 3.5 kcal mol^{-1} for X = P. A crude rule can now be formulated: *If the CX bond strength is smaller than 78 kcal mol^{-1}, the Felkin transition state will be favored over the Cornforth transition state.*

- Charge control becomes dominant mostly in compounds containing fluorine or in rigid systems. This can be expected from the foregoing discussion. More interestingly, electrostatic interactions appear to play a prominent role in silicon compounds.[99]

Note that a frontier interaction such as HOMO(1)–LUMO(2) automatically implies electron transfer from molecule (1) to molecule (2), so that electrostatic factors are already partially taken into account in frontier control.

- Conformational control comes to the fore in the absence of highly polar substituents (as in propanal and α-methylcyclohexanones) and also in reactions with very early (reactant-like) or very late (product-like) transition states. Note that steric control is already incorporated in all models which assume that preferential attacks should come from the less hindered side.[100]

Other factors seem to have little effect. The concept of orbital distortion has appealed to many chemists, including the present author.[101] It is based on the idea that an asymmetric environment can induce a pyramidalization of an sp^2 carbon atom. This causes the π^* orbital to become larger on one face of the molecule than the other,

[97]Cee V. J., Cramer C. J., Evans D. A., *J. Am. Chem. Soc.*, 2006, **128**, 2920.

[98] Huheey J. E., *Inorganic chemistry*, 3rd edn., Harper and Row, New York, 1983, p. A-37.

[99]Wong S. S., Paddon-Row M. N., *Aust. J. Chem.*, 1991, **44**, 765; Fleming I., Hrovat D. A., Borden W. T., *J. Chem. Soc., Perkin Trans. 2*, 2001, 331.

[100]Axial attacks on cyclohexanones come from the more hindered side. This is the reason why Dauben *et al.*[56] introduced the concepts of 'product development control' and 'steric approach control' and Schleyer the 'compression effect' (Schleyer P. v. R., *J. Am. Chem. Soc.*, 1967, **89**, 701; Laemmle J., Ashby E. C., Rolling P. V., *J. Org. Chem.*, 1973, **38**, 2526).

[101]Anh N. T., Eisenstein O., Lefour J. M., Dâu M. E. T. H., *J. Am. Chem. Soc.*, 1973, **95**, 6146; Klein J., *Tetrahedron Lett.*, 1973, 4307; Klein, J., *Tetrahedron*, 1974, **30**, 3349; Eisenstein O., Klein J., Lefour J. M., *Tetrahedron*, 1979, **35**, 225; Liotta C. L., *Tetrahedron Lett.*, 1975, 519, 523; Inagaki S., Fujimoto H., Fukui K., *J. Am. Chem. Soc.*, 1976, **98**, 4054; Ashby E. C., Noding S. A., *J. Org. Chem.*, 1977, **42**, 264; Giddings M. R., Hudec J., *Can. J. Chem.*, 1981, **59**, 459; Rondan N. G., Paddon-Row M. N., Caramella P., Houk K. N., *J. Am. Chem. Soc.*, 1981, **103**, 2436; Houk K. N., Rondan N. G., Brown F. K., Jorgensen W. L., Madura J. D., Spellmeyer D. C., *J. Am. Chem. Soc.*, 1983, **105**, 5980; Jeffrey G. A., Houk K. N., Paddon-Row M. N., Rondan N. G., Mitra J., *J. Am. Chem. Soc.*, 1985, **107**, 321; Frenking G., Köhler K. F., Reetz M. T., *Angew. Chem. Int. Ed.*, 1991, **30**, 1146; Huang X. L., Dannenberg J. J., Duran M., Bertrán J., *J. Am. Chem. Soc.*, 1993, **115**, 4024; Huang X. L., Dannenberg J. J., *J. Am. Chem. Soc.*, 1993, **115**, 6017.

thus favoring attack at the former. However, the carbon s orbital coefficient is very small and, consequently, in most cases orbital distortion can be neglected. Houk and co-workers[102] also concluded that orbital distortions are less important than torsional effects. Moreover, orbital distortion is a static index, i.e. an index based on the isolated substrate, and there is no guarantee that under the influence of the approaching reagent, the substrate will always pyramidalize preferentially in the expected direction.

Care must be taken in generalizing gas-phase computational results, which tend to overemphasize charge control while underemphasizing frontier control. Indeed, in the absence of solvent, coulombic forces, which vary as the inverse square of the distance, are stronger at large distances than frontier interactions, which diminish exponentially with the separation. The reaction then becomes charge controlled with an early transition state in which the reagents may still be in their most stable conformations. The role of the conformational effect is then somewhat overstated. This is especially true in older calculations in which O or Cl were replaced by F and Na^+ and K^+ were modeled by Li^+.

Generally, one should be cautious when interpreting computational results. First, the aforementioned effects are not independent. Clearly, antiperiplanarity can occur only in staggered conformations. Also, electrostatic factors and antiperiplanarity are interrelated through molecule flexibility: the more rigid the molecule, the more important is charge control and the less significant is antiperiplanarity control. We have pointed out that frontier control includes charge transfer. Second, stereoinduction is a small phenomenon, of the order of 1–2 kcal mol^{-1}. It is therefore difficult to attribute a change specifically to a particular factor. In fact, for every set of results, we can have several interpretations, as shown in the following examples.

Wu and Houk[103] found that in the addition of NaH to propionaldehyde, the *anti* transition structure is less stable than the *inside*[104] transition structure. This was interpreted in terms of the donating effect of the methyl group: an *anti* methyl group is expected to destabilize an electron-rich transition state. Now, the relative energies (in kcal mol^{-1}, 3–21G calculations) for the propanal conformers are 0.00 (*inside*), 1.44 (*outside*) and 1.51 (*anti*) and for the transition structure conformers 0.00 (*inside*), 1.77 (*outside*) and 1.94 (*anti*). These figures show that Wu and Houk's interpretation is correct because the *anti* transition structure is destabilized by ~0.43 kcal mol^{-1}. However, this destabilization is small and the main effect is conformational. That this reaction is under conformational control is not really surprising: the calculated transition states are early and therefore reactant-like (the $H^-\cdots C$ distance is 2.015 Å for the *inside* conformer and 2.029 Å for the *anti* conformer, about twice the CH bond distance). Remember that in 'normal' transition structures, the incipient bond has ~1.4 times the equilibrium bond length.

Another kind of difficulty is encountered in the case of a silyl group adjacent to the carbonyl function. Fleming *et al.*[99] calculated the gas-phase reactions of LiH with 2-silylacetaldehyde, 2-silylpropionaldehyde, 2-trimethylsilylacetaldehyde and

---

[102]Wu Y.-D., Houk K. N., Paddon-Row M. N., *Angew. Chem. Int. Ed.*, 1992, **31**, 1019.

[103]Wu Y.-D., Houk K. N., *J. Am. Chem. Soc.*, 1987, **109**, 908.

[104]The *inside* position is synclinal with respect to the insaturation (i.e. the O=C–C–Me dihedral angle is acute) and the *outside* position is anticlinal with respect to the insaturation.

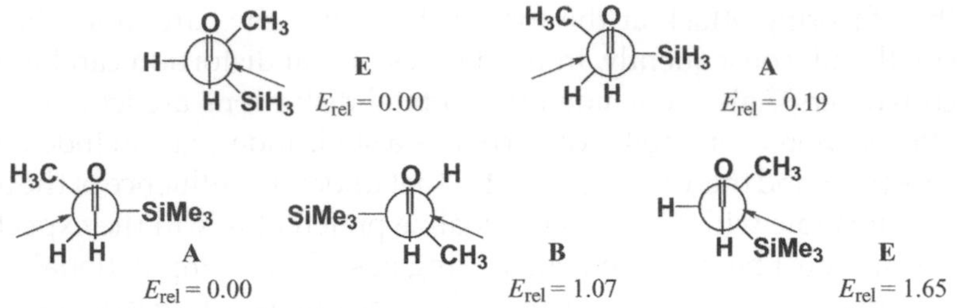

Figure 6.11 MP2/6–31G*//HF/6–31G* transition structures for 2-silylpropionaldehyde and 2-trimethylsilylpropionaldehyde with LiH. Relative energies in kcal mol⁻¹.

2-trimethylsilylpropionaldehyde. Five of the most stable transition structures are shown in Figure 6.11. They are labeled as in the Fleming paper.

Let us recall that in these compounds, the methyl substituent prefers the *inside* position. For the silyl or trimethylsilyl group, the *anti* position is best, the *inside* position second best and the *outside* position always the unfavorable position. So, compared with **E**, transition structure **A** is conformationally favored, sterically favored (the nucleophile approaches from the less hindered side) and also electronically favored (the σ^*_{CSi} orbital is much lower in energy than the σ^*_{CH} orbital). And yet **E** is slightly lower in energy for silylpropionaldehyde (Figure 6.11, top row). An attraction between the incoming hydride and the positively charged Si atom is the simplest explanation. When SiH₃ is replaced by the bulkier SiMe₃ group, **E** becomes less stable by 1.65 kcal mol⁻¹ (Figure 6.11, bottom row). The authors concluded that steric effects must be responsible for this reversal. This is certainly true, but it may be added that electronic factors reinforce steric effects. In fact, Si–C is a relatively weak bond (76 versus 82.5 kcal mol⁻¹ for C–C and 78 kcal mol⁻¹ for C–Cl), so the σ^*_{CSi} orbital is low-lying, despite the fact that its energy is raised by the Si electropositivity. When SiH₃ is replaced by SiMe₃, the acceptor character of the SiC bond is even more marked, which favors **A**.

We would like to draw attention here to a misconception concerning the so-called 'Anh–Eisenstein rule'. When we suggested[41] that a substituent X should be put on the *anti* position if σ^*_{CX} is low-lying, this does *not* mean that an electronegative X will always prefer the *anti* position, or an electropositive Y will avoid it. The energy level of σ^*_{CX} depends on the electronegativity of X (p. 13) *and* on the strength of the CX bond (p. 98). For example, despite the fact that fluorine is very electronegative, the CF bond is also very strong, so the *outside* position would normally be the best for F (Cornforth model). On the other hand, although Si is electropositive, SiH₃ – and a *fortiori* SiMe₃ – can compete with Me for the *anti* position. Examination of the two transition structures below supports this statement.

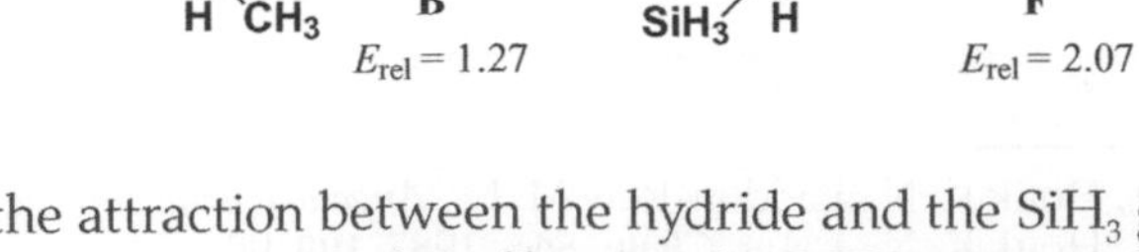

F is favored by the attraction between the hydride and the SiH₃ group (which more than compensates the advantage of **B** of having the SiH₃ group *anti*: compare **E** and **A**, Figure 6.11, top row) and also by having the most electronegative group Me in the *anti*

position. Moreover, **B** is disfavored by the repulsion between the incoming hydride and the methyl group. And yet **B** is more stable by 0.8 kcal mol^{-1}, a fact best explained by the *electron-attracting* character of the Si–C bond.

Let us now return to Cee *et al.*'s paper.[97] We have interpreted their results as showing how increasing electron transfer from the nucleophile to the substrate shifts the reaction control from the Cornforth model to the Felkin model. It must be noted that Cee *et al.* considered this transfer to be unimportant, for in the *anti* transition structure the C–X bond length is about the same as in the isolated substrate. We would like to suggest a possible explanation for this contradiction. In the isolated reagent, two partners are present: the C=O group, which functions as a donor, and the CX bond, which acts as an acceptor. In the transition state, there are three partners, C=O, CX and the nucleophile, which, being electron rich, plays a dominant role. Both C=O and CX then behave as acceptors and electron exchange between them becomes negligible. However, this tentative rationale does not agree entirely with the Morokuma model (see below) and more studies are needed to clarify this point.

Some Recent Models

To illustrate further how various authors appraise the relative importance of the controlling factors, some recent models for stereoinduction are discussed below. They all assume staggered conformations and nonperpendicular attack.

Houk model for electrophilic additions to alkenes[105]

This model assumes *nonperpendicular attacks* (more precisely acute angles of attack, cf. p. 172) and *staggered conformations*. As the electrophile now approaches the double bond near its center, the *inside* position becomes the more hindered position, contrary to the case of nucleophilic addition to carbonyls:

The *inside* position is also disfavored by steric interaction with the circled **H**:

[105]Paddon-Row M. N., Rondan N. G., Houk K. N., *J. Am. Chem. Soc.*, 1982, **104**, 7162; Houk K. N., Rondan N. G., Wu Y.-D, Metz J. T., Paddon-Row M. N., *Tetrahedron*, 1984, **40**, 2257; Houk K. N., Paddon-Row M. N., Rondan N. G., Wu Y.-D., Brown F. K., Spellmeyer D. C., Metz J. T., Li Y., Loncharich R. J., *Science*, 1986, **231**, 1108; Wu Y.-D., Houk K. N., *J. Am. Chem. Soc.*, 1987, **109**, 908; Fleming I., Lewis J.J., *J. Chem. Soc., Perkin Trans. 1*, 1992, 3257.

Therefore, if only steric factors are to be taken into consideration, the largest group should occupy the *anti* position, the medium group the *outside* position and the small group the *inside* position. If only electronic effects are considered, then a donor allylic group D should be *anti* to maximize electron transfer from the donating high-lying σ_{CD} orbital to the transition state LUMO. The *outside* position is second best, and D avoids the *inside* position where the σ_{CD} overlap with π^* will be negligible. An attractor A will prefer the *inside* or *outside* position. These rules merely reflect the best way to facilitate electron transfer from the nucleophile to electrophile: an attractor (donor) substituent of the chiral carbon of the electron-poor (electron-rich) compound should be *anti* to the electron-rich (electron-poor) reagent.

Electrostatic interactions become prominent in reactions with rigid alkenes, for example 7-isopropylidenebenzonorbornene[106] or benzobicyclo[2.2.2]octadiene.[107]

Morokuma model for conjugated additions

Dorigo and Morokuma[108] postulated that for conjugated additions, a substituent on the γ-carbon atom has the following effect: *An electron-withdrawing group in the anti position is disfavored because it destabilizes the transition structure by removing electron density from the enal fragment; an electron-donating group is favored in this position both for steric reasons and because it stabilizes the enal fragment.*

There remains the problem of choosing between the *inside* and *outside* positions. This is determined by the relative magnitude of the repulsion between the *inside* substituent and the enal fragment on the one hand, and the repulsion between the *outside* substituent and the incoming nucleophile on the other:

According to Bernardi *et al.*,[109] for (*E*)-enoates with two γ-hydrocarbon substituents, the dominant interaction is with the nucleophile. Therefore, the smallest substituent S should be put in the *outside* position. For (*Z*)-enoates, S should be put in the *inside* position.[110] For γ-alkoxy-(*E*)-enoates, a Felkin transition state may rationalize organolithium[111] and alkoxyde[112] additions, but does not explain the results for

[106]Wu Y.-D., Li Y., Na J., Houk K. N., *J. Org. Chem.*, 1993, **58**, 4625.
[107]Paquette L. A., Bellamy F., Wells G. J., Böhm M. C., Gleiter R., *J. Am. Chem. Soc.*, 1981, **103**, 7122.
[108]Dorigo A. E., Morokuma K., *J. Am. Chem. Soc.*, 1989, **111**, 6524.
[109]Bernardi A., Capelli A. M., Gennari C., Scolastico C., *Tetrahedron: Asymmetry*, 1990, **1**, 21.
[110]For experimental results pertaining to this problem, see: Kruger D., Sopchik A. E., Kingsbury C. A., *J. Org. Chem.*, 1984, **49**, 778; Yamamoto Y., Nishii S., Ibuka T., *Chem. Comm.*, 1987, 1572; Idem, *J. Am. Chem. Soc.*, 1988, **110**, 617; Honda Y., Hirai S. M., Tsuchiashi G., *Chem. Lett.*, 1989, 255.
[111]Tatsuta K., Amemiya Y., Kanemura Y., Kinoshita M., *Tetrahedron Lett.*, 1981, 3997.
[112]Mulzer J., Kappert M., Huttner G., Jibril I., *Angew. Chem. Int. Ed.*, 1984, **23**, 704; Lopez Herrera F. J., Pino Gonzales M. S., *Tetrahedron*, 1986, **42**, 6033.

cuprate[113] and alkylcopper[114] reactions. For γ-alkoxy-(Z)-enoates, a modest selectivity is predicted.[115]

Although some significant advances have been made, the problem of conjugated addition stereoselectivity is still far from being completely clarified. For an illustrative example, see ref. [116].

Radical additions to alkenes

Theoretical studies[117] have shown that the transition states for additions of nucleophilic and electrophilic radicals to alkenes are remarkably similar. They all involve *nonperpendicular attacks* (attack angles between 97° and 110°, depending on the calculation level) and *staggered arrangement* with respect to the incipient bond. There is a tendency (more marked in nucleophilic additions[118]) toward *trans* bending of the alkene, with greater pyramidalization of the carbon atom under attack. Giese *et al.*[119] found that 1,2-stereoinduction in radical addition to alkenes can be rationalized by the Felkin model.

Evans electrostatic models for 1,2-and 1,3-asymmetric inductions[91]

The Evans electrostatic model for 1,2-asymmetric induction incorporates Dunitz–Bürgi attack and torsional effects into the Cornforth model:

For reactions not under chelating conditions, Evans and co-workers proposed a merger of the Felkin (1,2) model and a polar (1,3) model. Thus, C_α is arranged according to the Felkin model with the L group *anti* to the nucleophile. The $C_\alpha C_\beta$ bond is staggered and the polar substituent X on C_β is oriented in the direction opposite to the C=O group:

[113]Rousch R. W., Lesur M. B., *Tetrahedron Lett.*, 1983, 2231; Salomon R. G., Miller D. B., Raychaudhury S. R., Avasthi K., Lal K., Levison B. S., *J. Am. Chem. Soc.*, 1984, **106**, 8296.

[114]Yamamoto Y., Nishii S., Ibuka T., *Chem. Comm.*, 1987, 464.

[115]Experimental results: Larcheveque M., Tamagnan G., Petit Y., *Chem. Comm.*, 1989, 31.

[116]Yamamoto Y., Chounan Y., Nishii S., Ibuka T., Kitahara H., *J. Am. Chem. Soc.*, 1992, **114**, 7652.

[117]Delbecq F., Ilavsky D., Anh N. T., Lefour J. M., *J. Am. Chem. Soc.*, 1985, **107**, 1623; Houk K. N., Paddon-Row M. N., Spellmeyer D. C., Rondan N. G., Nagase S., *J. Org. Chem.*, 1986, **51**, 2874; Sosa C., Schlegel H. B., *J. Am. Chem. Soc.*, 1987, **109**, 4193; Gonzalez C., Sosa C., Schlegel H. B., *J. Phys. Chem.*, 1989, **93**, 2435; Zipse H., He J., Houk K. N., Giese B., *J. Am. Chem. Soc.*, 1991, **113**, 4324.

[118]Strozier R. W., Caramella P., Houk K. N., *J. Am. Chem. Soc.*, 1979, **101**, 1340.

[119]Giese B., Damm W., Roth M., Zehnder M., *Synlett*, 1992, 441.

The inside alkoxy effect

This effect has been proposed by Stork and Kahn[120] for the dihydroxylation of γ-hydroxyenones and by Houk *et al.*[121] for 1,3-dipolar cycloadditions of allylic ethers. In the preferred transition state, the largest group should be *anti* to the oxidant, while the C–O bond takes up the *inside* position. The rationale suggested by Houk *et al.* is the following. In electrophilic attack on an allylic ether, the π bond becomes electron deficient. Therefore, the donor substituent, usually an alkyl group R, should be put in the *anti* position. The alkoxy substituent, which is electron withdrawing through its σ^*_{CO} orbital, will avoid the *anti* position. When CO is *inside*, it is near the plane and overlap of σ_{CO} with π is minimized. This model, but not the Kishi model[122], accounts for the increase selectivity with increasing size of R[123]:

However, a computational study[124] shows that the Kishi model controls the stereoselectivity for (Z)-alkenes. Note also that in the Diels–Alder reactions of hexachloropentadiene with chiral alkenes, the *inside* alkoxy' effect is attributed to electrostatic repulsion of the oxy group in the[125] *outside* position with the chlorine atom of hexachloropentadiene in the 1-position.

6.2.2 Electrophilic Additions

An electrophilic addition generally proceeds in two steps.[126] In the first, the most important interaction occurs between the LUMO of the electrophile and the HOMO of the alkene. Orbital overlap is largest when the electrophile attacks at the center of the π bond:

Sometimes,[127] such cyclic intermediates can be isolated:

[120]Stork G., Kahn M., *Tetrahedron Lett.*, 1983, **24**, 3951.
[121]Houk K. N., Moses S. R., Wu Y.-D., Rondan N. G., Jäger V., Schohe R., Fronczek F. R., *J. Am. Chem. Soc.*, 1984, **106**, 3880.
[122]Cha J. K., Christ W. J., Kishi Y., *Tetrahedron*, 1984, **40**, 2247.
[123]Evans D. A., Kaldor S. W., *J. Org. Chem.*, 1990, **55**, 1698.
[124]Haller J., Strassner T., Houk K. N., *J. Am. Chem. Soc.*, 1997, **119**, 8031.
[125]Haller J., Niwayama S., Duh H.-Y., Houk K. N., *J. Org. Chem.*, 1997, **62**, 5728.
[126]March J., *Advanced Organic Chemistry*, 4th edn, John Wiley & Sons Inc., New York, 1992, p. 734.
[127]Slebocka-Tilk H., Ball R. G., Brown R. S., *J. Am. Chem. Soc.*, 1985, **107**, 4504.

Arigoni and co-workers[128] used isotopic labeling experiments to show that the cyclization of linalool to α-terpineol occurs at the center of the double bond rather than at its extremities. In principle, the intermediate cation **30** could adopt an elongated conformation **31** leading to **33** (which is not found) or the folded conformation **32**, where attack of the carbocation at the center of the double bond gives the observed product **34**.

As the electrophile becomes smaller, it is less able to bridge the lobes lying at each end of the double bond. Thus, cyclic transition states become less favored along the series iodonium > bromonium > chloronium.[129] For example, a theoretical (3–21G) study of the addition of Br_2, Cl_2 and F_2 to ethylene[130] gave the transition states below (figures in italics refer to overall charges).

Very similar geometries are observed for Br_2 and Cl_2. The three-membered ring structure is clearly visible and the departure of the X^- anion is already in progress. Point charges are developing and the Br–Br and Cl–Cl bonds are lengthened from their equilibrium distances by 37 and 43%, respectively.[131] A subsequent S_N2 attack of the displaced X^- on the cyclic carbocation will give the experimentally observed *trans-addition* product:[132]

[128]Godtfred S., Obrecht J. P., Arigoni D., *Chimia*, 1977, **31**, 62.
[129]Fahey R. C., *Top. Stereochem.*, 1968, **3**, 273; Hassner A., Boerwinkle F., Levy A. B., *J. Am. Chem. Soc.*, 1970, **92**, 4879.
[130]Yamabe S., Minato T., Inagaki S., *Chem. Commun.*, 1988, 532.
[131]Huheey J. E., *Inorganic Chemistry*, 3rd edn., Harper and Row, New York, 1983, pp. 258 and A40.
[132]Freeman F., *Chem. Rev.*, 1975, **75**, 439.

The case of fluorine is different. The F–F bond is barely polarized and only slightly elongated (by 21%). The more rectangular form of this transition state implies *cis* addition. Experiments also suggest that this is the case.[133] There are probably two reasons behind the change in mechanism. First, heterolytic rupture would generate F^+, which is extremely unfavorable because F is the most electronegative element of all. Second, fluorine is very small (having a covalent radius of 0.71 Å, compared with 0.37 Å for H 0.77 Å for C 0.99 Å for Cl and 1.14 Å for Br), so it does not easily bind to both carbons of the ethylene simultaneously to form a cyclic three-membered intermediate.

Exercise 28 (M)

Is this four-centerd transition state forbidden by the Woodward- Hoffmann rules?

Answer

No more than a hydroboration is! The transition state for a Woodward–Hoffmann-allowed reaction is aromatic in character, whereas a forbidden one is antiaromatic. However, the rules of aromaticity only apply when each atom of the annulene uses one, and only one, AO to bind to its neighbors. This is not the case here: each fluorine atom uses two AOs, one for bonding to the other F atom and one for the incipient bond to C.

Protons are even smaller than F^+, so they should give open transition states and nonstereoselective additions, which have been confirmed experimentally.[126] A loss of stereoselectivity can even be observed during the bromination of certain substituted alkenes. The HOMO, and hence the bromonium ion, are dissymmetric and an equilibrium can be established with the open form. In some cases, such as the benzylic cation below, this form dominates.

Experiments have shown that the bromination of *cis*- and *trans*-phenylpropenes in CCl_4 is non-selective.[134] For stilbene, *trans* addition occurs in solvents having low dielectric constants, but the reaction loses stereoselectivity if ε rises above 35.[135]

---

[133]Rozen S., Brand M., *J. Org. Chem.*, 1986, **51**, 3607.

[134]Fahey R. C., Schneider H., *J. Am. Chem. Soc.*, 1968, **90**, 4429; Rolston J. H., Yates K., *J. Am. Chem. Soc.*, 1969, **91**, 1469, 1477, 1483.

[135]Buckles R. E., Miller J. L., Thurmaier R. J., *J. Org. Chem.*, 1967, **32**, 888; Heublein G., Lauterbach H., *J. Prakt. Chem.*, 1969, **311**, 91; Ruasse M. F., Dubois J. É., *J. Am. Chem. Soc.*, 1975, **97**, 1977; Bellucci G., Bianchini R., Chiappe C., Marioni F., *J. Org. Chem.*, 1990, **55**, 4094.

6.2.3 Application to the Aldol Addition

Transition state structures are not always easy to calculate. One approach allowing qualitative models to be derived for complex reactions links *geometric constraints* to the chemical characteristics of a reaction. If the constraints are sufficiently numerous, few transition states can satisfy them all simultaneously.

Thus, the aldol reaction[47,136] can be viewed as a nucleophilic addition to an aldehyde (obtuse angle of attack) or an electrophilic addition to an enolate (acute angle). When these preferences are taken together, it can be deduced that the *syn* approach is more favorable than the *anti* approach (which requires two obtuse angles). Note that *syn* terminology signifies an acute $O=C\cdots C=C$ dihedral angle, whereas *anti* indicates that it is obtuse:

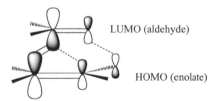

A choice must now be made between *syn–cis* and syn–trans orientations (where *cis* and *trans* indicate the relative configurations of the two oxygen atoms with respect to the C=C double bond):

syn–cis   syn–trans

Chelation by M^+ favors *syn–cis*. Search for the best overlap between the enolate HOMO and the aldehyde LUMO leads to the same conclusion,[137] as the dotted secondary interactions are attractive:

LUMO (aldehyde)

HOMO (enolate)

It may come as a surprise to see favorable secondary interactions between two negatively charged oxygen atoms. This is less unreasonable than it may seem:

- Under experimental conditions, the charges are partially neutralized by the cation.
- Bond formation between atoms having the same charge is well known. Mulzer

---

[136]For a recent review on aldol reactions, see Mahrwald R. (Ed.), *Modern Aldol Reactions*, Wiley-VCH, Weinheim, 2004.
[137]Maximum frontier overlap thus encompasses the first two geometric constraints. We have detailed these, however, to show the preferential angles of attack.

et al.[138] note that diazoalkanes react with carbonyl compounds to give 1,3-dipolar adducts, despite repulsive interactions between the electronegative N and O atoms. We are only considering secondary interactions, which is clearly a much less stringent requirement.

- The cyclic HOMO–LUMO interaction bears a striking resemblance to the transition state for 1,3-dipolar cycloadditions.

Next, we can introduce some substituents. (Z)-Enolates generate diastereomeric transition states **35** and **36**, whereas (*E*)-enolates give **37** and **38**.

To choose between these transition states, a new factor needs to be introduced. Calculations modelling addition reactions to aldehydes[139] have shown that the bulk of the R group forces the incoming nucleophile to approach the aldehyde from the side bearing the hydrogen:

In other words, the C···Nu bond lies closer to CH than to CR when projected on to the carbonyl plane. This can be translated as a rotation of the aldehyde shown by the arrows in structures **35–38**. This rotation diminishes the R–R_2 repulsion in **35** and increases the R–R_1 repulsion in **36**. If these repulsions were of similar magnitude *before the rotation*, the *erythro* transition state **35** will be favored over the *threo* precursor **36**. The R–R_2 distance in **35** being smaller than R–R_1 in **36**, this condition is only met when $R_2 \ll R_1$. Hence the prediction that (Z)-enolates will give *erythro* aldols when R_2 is significantly less bulky than R_1. The percentage of *threo* product should rise as the size of R_2 increases. This hypothesis agrees nicely with experimental studies conducted by Fellman and Dubois[140]:

---

[138]Mulzer J., Brünstrub G., Finke J., Zippel M., *J. Am. Chem. Soc.*, 1979, **101**, 7723.
[139]Thanh B. T., Thesis, Université Paris XI (Orsay), 1988.
[140]Fellman P., Dubois J. É., *Tetrahedron*, 1978, **34**, 1349.

R$_2$	*erythro : threo*
Me	*erythro*
iPr	48 : 52
tBu	20 : 80

The proportion of *threo* product also rises as a function of R when R$_1$ and R$_2$ are of comparable bulk. Increasing the size of R disfavors **35**:

R	*rythro : threo*
Me	48 : 52
tBu	29 : 71

Finally, if R$_2 \ll$ R$_1$, the bulk of R will have little effect upon the *erythro:threo* ratio and the *erythro* product will dominate. The *erythro* aldol is always obtained in the reaction

irrespective of substituent at the aldehyde (R= Me, Et, iPr, tBu). These conclusions follow straightforwardly from our model[47], but they are much less easy to obtain from other theories. For example, it has been suggested that[141] a 1,3-diaxial interaction between R and R$_1$ destabilizes the Zimmerman–Traxler transition state **40** with respect to **39** (these are the analogs of **36** and **35**, respectively).

Obviously, this conclusion should hold when R$_1$ = R$_2$. The first series of examples show that this is clearly not the case. Furthermore, increasing the bulk of R should disfavor **40**. This fails to explain the second series, where the reverse trend is observed when R$_1$ and R$_2$ are similar in size.

Our analysis can also be applied to (*E*)-enolates, which give rise to transition states **37** and **38**. The major product will have *threo* stereochemistry when R$_2 <$ R$_1$ and an *erythro* configuration when R$_2 >$ R. (*E*)-enolates are less markedly affected by the bulk of R$_2$ and R$_1$ than their Z counterparts. Increasing hindrance from R$_1$ (and, to a lesser extent, R) increases the proportion of *threo* aldol. For a detailed discussion, see ref. 47.

---

[141]Zimmerman H. E., Traxler M. D., *J. Am. Chem. Soc.*, 1957, **79**, 1920.

6.3 Substitution Reactions

6.3.1 Bimolecular Electrophilic Substitutions

S_E2 reactions generally occur with retention of configuration. In some cases, this stereochemistry has been directly proved.[142] It can also be deduced from observations such as the very easy substitutions which occur in neopentyl compounds (whose rear face is sterically blocked),[143] and at bridging atoms.[144] According to March,[145] inversion of configuration *only* occurs when the leaving group side is severely hindered.

Why should there be such a preference for retention of configuration? During a 'front attack' the electrophile E^+ overlaps simultaneously with the reaction center C and the leaving group X, leading to a cyclic transition state (**41**). Being a two-electron species, this structure is aromatic. The transition state for attack from the rear face resembles an allyl cation, so it is less stable. An FO analysis provides the same conclusions. The strongest interaction occurs between the LUMO of the electrophile and the HOMO of the substrate (i.e. the σ_{CX} bonding orbital), so the best overlap occurs during a 'front attack' (**42**).

6.3.2 Bimolecular Nucleophilic Substitutions

Vinylic S_N2 reactions are discussed on p. 198. Here, we restrict ourselves to nucleophilic susbstitutions at saturated centers. Aromaticity criteria then strongly disfavor retention of configuration. Indeed, the transition state leading to retention, **43**, resembles a four-electron (two from the nucleophile and two from the CY bond) antiaromatic annulene.

At first sight, FO theory may seem to provide no useful insight here. The strongest FO interaction involves the HOMO of the nucleophile and the LUMO of the substrate which, in this case, is the CY antibonding orbital. σ^*_{CY} is a carbon sp³ AO, mixed out-of-phase with one of the leaving group AOs. The two large lobes point toward each other (**44**).[146] In a back-side attack leading to inversion, the nucleophile overlaps with the small lobe of the carbon AO, which does not seem very promising. In a front-side attack (leading to retention), the favorable overlap with the major lobe of the carbon

[142]Winstein S., Traylor T. G., *J. Am. Chem. Soc.*, 1955, **77**, 3741; Gielen M., Nasielski J., *Ind. Chim. Belge*, 1964, **29**, 767; Jensen F. R., *J. Am. Chem. Soc.*, 1960, **82**, 2469; Jensen F. R., Nakamaye K. L., *J. Am. Chem. Soc.*, 1966, **88**, 3437.

[143]Hughes E. D., Volger H. C., *J. Chem. Soc.*, 1961, 2359.

[144]Winstein S., Traylor T. G., *J. Am. Chem. Soc.*, 1956, **78**, 2597; Schöllkopf U., *Angew. Chem.*, 1960, **72**, 147.

[145]March J., *Advanced Organic Chemistry*, 4th edn, John Wiley & Sons, Inc., New York, 1992, p. 573.

[146]Certain semi-empirical calculations (extended Hückel, CNDO, etc.) orient the large lobes away from each other in the σ^* orbital. Such an orbital is nonbonding rather than antibonding, because the two AOs hardly interact.

AO is offset by an out-of-phase overlap with the AO of the leaving group. Salem had to resort to numerical calculations before concluding that inversion is favored.[147]

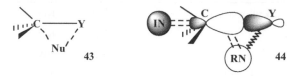

43   44

This apparent indetermination is not a failure of FO theory, but rather an illustration of its ability to provide more balanced conclusions, as opposed to the rules of aromaticity which can sometimes be a little too 'categorical'. The absence of any overwhelming drive for either inversion or retention implies that by a judicious choice of the reaction conditions it is possible to obtain S_N2 reactions *with retention of configuration*.[148] To promote them, the overlap of the nucleophile with the carbon atom must be increased, whereas its overlap with the leaving group must diminish. This may be achieved by reducing the mixing coefficient in the σ^*_{CY} orbital:

$$\sigma^*_{CY} = N\left(\varphi_C + \frac{\langle \varphi_Y | P | \varphi_Y \rangle}{E_C - E_Y} \varphi_Y\right)$$

where, E_C and E_Y represent the energies of the AOs φ_C and φ_Y. The mixing coefficient is inversely proportional to their energy gap, so increasing this gap will tend to increase the contribution of φ_C to σ^*_{CY} and diminishes that of φ_Y (assuming that the numerator remains unchanged). To increase the denominator, we can raise E_C, lower E_Y, or do both at once. Now, raising E_C amounts to reducing the electronegativity of the reaction center. This is most easily achieved by moving down the relevant column of the periodic table. Hence, all other parameters being equal, the replacement of a carbon atom by silicon (or germanium, tin or lead) will increase the propensity of the reaction to proceed with retention. The question that we need to ask at this stage concerns precisely how 'equal' all other things are, because the replacement of carbon by silicon will introduce other modifications. Fortunately, they also tend to favor retention. Thus, for any given leaving group, the Si–Y bond will be longer than the C–Y, which will distance the leaving group from the nucleophile and reduce their repulsion. Furthermore, the silicon valence orbitals are much more diffuse than those of carbon. So, at long distance, it should be possible to have a reasonable Si–Nu overlap while keeping the Nu–Y repulsions acceptably low. Hence, a frontal approach is facilitated.

The consequences of lowering E_Y are less clear. The replacement of Y = Cl by Y = F brings an increase in electronegativity and a valence orbital contraction, which favor retention, but also a shortening of the Si–Y distance, which promotes inversion. Numerical calculations suggest that the replacement of a given leaving group by a more electronegative homolog *from the same column of the periodic table* will promote

[147]Salem L., *Chem. Bri.*, 1969, **5**, 449.
[148]Anh N. T., Minot C., *J. Am. Chem. Soc.*, 1980, **102**, 103; Anh N. T., *Top. Curr. Chem.*, 1980, **88**, 145; Minot C., *Nouv. J. Chim.*, 1981, **5**, 319.

Table 6.1 Effects of changing the nucleophile and leaving group

Nucleophile	X = H	OMe	SMe or SPh	F	Br or Cl
R'Li (R' = alkyl or aryl)	Ret	Ret	Ret	Ret	Inv
Allyllithium	Ret	Ret	Inv	Inv	Inv
PhCH$_2$Li	Ret	Inv	Inv	Inv	Inv
R'MgX (R' = alkyl or aryl)		Ret	Inv	Inv	Inv
R'MgX (R' = allyl, benzyl)		Inv	Inv	Inv	Inv
LiAlH$_4$		Ret	Inv	Inv	Inv
*t*BuOK or KOH	Ret	Ret		Ret	Inv

retention. Indeed, the replacement of leaving groups such as Cl and SR by F and OR increases the degree of retention in experimental studies.[149] Table 6.1 shows how changing the nature of the nucleophile and the leaving group affects the stereochemical outcome of the reaction below:

Note that H, an inert substituent in carbon chemistry, is a reasonably good leaving group when bound to silicon. This is due to a number of factors. First, the Si–H bond strength is much lower than the C–H bond strength (76 versus 98 kcal mol^{-1}). Second, the Pauling electronegativities of H, C and Si are 2.20, 2.55 and 1.90, respectively, so a heterolysis of the Si–H bond to give Si$^+$H$^-$ is reasonable, but a rupture of C–H giving C$^+$H$^-$ is more difficult. Even so, these arguments do not explain why *retention* of configuration generally accompanies substitution of H, especially given its modest electronegativity and the relatively short Si–H bond length. The most likely reasons are as follows. An overlap with the major lobe of the Si AO will favor a frontal attack. This is obviously not hindered by the H substituent because its compact valence orbitals overlap very weakly with the incoming nucleophile. Additionally, the hydride ion is the only leaving group which has no inner-shell electrons, so interelectronic repulsions are minimized.

Corriu and Lanneau[150] suggested that *harder nucleophiles induce greater retention of configuration*. This empirical rule can be interpreted in the following way. Hard reagents are normally small. Therefore, they have highly contracted valence orbitals which inter-

[149]Corriu R. J. P., Guérin C., Moreau J. J. E., *Top. Stereochem.*, 1984, **15**, p. 87.
[150]Corriu R., Lanneau G., *J. Organomet. Chem.*, 1974, **67**, 243.

Table 6.2 Stereochemistry in reactions of 46 and 47

	46	47
EtLi	63% Ret	100% Inv
nBuLi	82% Ret	59% Inv
Allyllithium	86% Ret	100% Inv
PhCH$_2$Li	99% Ret	100% Inv

Inv, inversion; Ret, retention.

act poorly with the leaving group. Conversely, a soft reagent tends to be a voluminous species having rather diffuse valence orbitals.[151] It will interact significantly with the leaving group, so it favors reactions proceeding with inversion.

Another method exists for promoting the interaction between the nucleophile and the reactive center during a frontal attack. Consider a tetracoordinate silicon atom **45**. As the R$_2$SiR$_3$ angle closes, the R$_1$SiX angle must open. During this process, the four silicon hybrid orbitals become nonequivalent. Those involved in bonds to R$_2$ and R$_3$ have diminished s character, whereas the remainder are richer in s character than a normal sp^3 hybrid.[152] Now hybrid orbitals become more dissymmetric as their s character increases: the large lobe becomes larger and the small lobe shrinks. This favors retention over inversion. The SiC, CC and aromatic CC bond lengths are 1.85; 1.54 and 1.40 Å, respectively, so the intracyclic C–Si–C angle in **46** should be smaller than the tetrahedral value. This suggests that **46** will react with more retention of configuration than **47**, which is precisely the result obtained by experiment[153] (Table 6.2).

6.4 The Limitations of Rule 4

An absurd situation would exist if there were no repulsions: activation energies would be negative and eclipsed conformations would be preferred (Exercise 19, p. 153). Rule 4 neglects repulsion and deals exclusively with *attractive* HOMO–LUMO terms, so it can only be approximate. Nonetheless, the approximation is excellent: rule 4 is easy to use and generally gives predictions which agree with experiment. The reason is obvious for electrophilic additions: the system is electron deficient, so interelectronic repulsions are usually small enough to be neglected. In other reaction classes, interactions between occupied orbitals tend to provoke also the outcome predicted by rule 4. Thus, the adoption of antiperiplanarity in asymmetric induction reactions (which maximizes the HOMO–LUMO terms) automatically leads to a staggered transition state (pp. 152 and 156) and *minimization* of torsional repulsions. In a nucleophilic addition, the minimization of overlap between the HOMO of the nucleophile and the π_{CO} orbital also favors attack from an obtuse angle:

---

[151]Pearson R. G., *J. Chem. Educ.*, 1968, **45**, 581, 643.
[152]The angle between two valence orbitals increases with increasing s character. Two pure p orbitals make an angle of 90° with each other. The angle becomes 109°28′ for two sp^3, 120° for two sp^2 and 180° for two sp orbitals.
[153]Ref. 149, p. 115.

As both HOMO–LUMO attraction *and* HOMO–HOMO repulsion promote approach from an obtuse angle, this explains why this trajectory is so highly favored. According to calculations,[47] an additional 5 kcal mol^{-1} are required to provoke a nucleophilic attack from an acute angle, whereas only 1 kcal mol^{-1} suffices to switch an electrophilic attack from acute to obtuse (the electrophilic case is controlled by HOMO–LUMO interactions alone). It is the repulsions between the occupied orbitals which displace the nucleophile from the π^*_{CO} plane during an addition to an aldehyde (p. 176).

Hence Rule 4 usually gives the best trajectory. However, the best trajectory provides only a *static* and highly idealized description of a reaction. In reality, reagents collide in a random fashion and react when (a) they have sufficient energy *and* (b) their orientation vaguely resembles the ideal geometry. It is conceivable that imperfect trajectories will make up a significant proportion of reaction 'events': they may not have the lowest energy, but they occur much more often. Concepts such as 'reaction funnels', 'reaction cones', 'reaction windows', 'accessible volumes' and 'permissible deviations' have been introduced[154] in an attempt to rationalize these nonoptimal approaches.

Finally, we should remember that the ease with which a reaction proceeds is controlled by the free energy of activation, $\Delta G^{\ddagger}$. Frontier orbital interactions provide information concerning the transition state potential energy, which usually gives a reasonable approximation of $\Delta G^{\ddagger}$. Nonetheless, as Menger[140c] has pointed out, we need to add a kinetic energy parameter to obtain the internal energy, a PV term to obtain the enthalpy and finally an entropic contribution to obtain the free enthalpy. So, bear in mind that we have neglected a significant number of terms.

Exercise 29 (D)   *Deserves your attention*

(1) Find the frontier orbitals of methylene CH$_2$ *without recourse to calculations*. Assume that the carbon has sp^2 hybridization, so that the carbene is bent.

(2) CH$_2$ reacts with ethylene to form cyclopropane. The symmetrical approach shown below appears to be the simplest. It also respects the 'principle of least motion', often invoked in the literature. Use FO analysis to show that this mechanism is improbable and to find the best approach. Assume that the carbene has a singlet configuration.

[154](a) Schneider S., Lipscomb W. N., Kleiner D. A., *J. Am. Chem. Soc.*, 1976, **98**, 4770; (b) Wipke W. T., Gund P., *J. Am. Chem. Soc.*, 1976, **98**, 8107; (c) Menger F. M., *Tetrahedron*, 1983, **39**, 1013; (d) Anh N. T., Thanh B. T., Thao H. H., N'Guessan Y. T., *New J.Chem.*, 1994, **18**, 489.

Answer

(1) Since two of the hybrid orbitals are already involved in the CH bonds, only two valence orbitals remain. These are an sp^2, which mixes slightly with the hydrogens to give the HOMO, and a pure p orbital: the carbene LUMO. We can be sure that these are the frontier orbitals because they are essentially nonbonding whereas for the CH bonds, being strong, their orbitals are either firmly bonding or firmly antibonding (p. 98).

HOMO   LUMO

(2) In a 'least motion' approach, the HOMO–LUMO overlap is zero, so the reaction is forbidden (rule 1, p. 47):

A lateral approach allows positive overlaps, so the reaction is allowed. As the reaction progresses, the carbene rotates (see arrows).

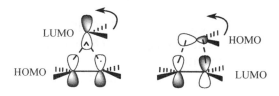

Remark 1: This cyclopropanation process is *not* a cycloaddition, but a nonlinear cheletropic reaction (p. 72).

Remark 2: The popularity of the principle of least motion is possibly due to the similarity of its name with the principle of least *action*, and a misunderstanding. For most chemists, applying this principle involves finding the pathway which causes the least change in the positions of the reacting nuclei. This approach is simple, seductive, easy to remember… and often erroneous! Such a trajectory minimizes the energy required for the displacement of atoms but offers no guarantee that the attractive terms will be maximized. Cutting your expenditure is not the only way to get rich; increasing your salary can be useful too! Every antarafacial reaction violates this intuitive interpretation of the principle of least motion. A careful examination of articles defining the principle of least motion in mathematical terms shows that these (rather complex) definitions usually have *nothing* to do with the idea of minimizing atomic displacement.

Exercise 30 (D)

We might expect conjugate additions to isophorone to be difficult, given that C_4 is disubstituted (p. 109). Indeed, treatment of this compound with $(EtO)_2P(O)CHCN^-$ M^+ (20 °C, 1 h, THF) gave no reaction when M = K. The corresponding lithium reagent, M = Li, gave 30% of a 1,2-addition product and no 1,4 addition.[155] However, the conjugate addition did occur when the reagent was changed to $PhCHCN^-$ Li^+. At −70 °C, the yield was 45% in THF but only 10% in a THF–HMPA (4:1) mixture.[156] Explain these trends.

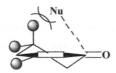

isophorone

Hint: $PhCHCN^-$ Li^+ forms a weakly ion pair in THF. Why is this?

Answer[157]

The influence of the cation reveals electrophilic assistance, which is more efficient with Li than K. The observed 1,2-addition is therefore 'normal'. However, the nonperpendicular Dunitz–Bürgi attack of the nucleophile on the carbonyl function will be sterically hindered by the *gem*-dimethyl group:

This hindrance is accentuated for $PhCHCN^-$, which adopts a planar, rather bulky structure to delocaliz its charge. Therefore, conjugate addition is promoted. Finally, HMPA is a basic solvent which traps the cation. As this effect is found to inhibit the reaction, we can deduce that this conjugate addition is under complexation control and that $PhCHCN^-$ Li^+ exists as a weak ion pair in THF.

Exercise 31 (M)

A comparison of the epoxidation of **48**, **49**, **50**, **51** and **52** by *m*-chloroperbenzoic acid was performed. It showed very high α stereoselectivity for **48–50**, good β selectivity for **52** and no selectivity at all for **51**. Interpret these data.

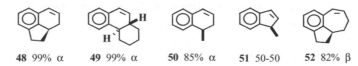

48 99% α   **49** 99% α   **50** 85% α   **51** 50-50   **52** 82% β

[155]Deschamps B., *Tetrahedron*, 1978, **34**, 2009.
[156]Sauvetre R., Seyden-Penne J., *Tetrahedron Lett.*, 1976, 3949.
[157]Lefour J. M., Loupy A., *Tetrahedron*, 1978, **34**, 2597.

Answer[158]

Felkin's model applies also to electrophilic additions (p. 169). We need only place the best donor substituent antiperiplanar to the incoming electrophile. In **48**, the pentagonal ring is almost planar, so the circled hydrogen atom is axial. Therefore, the β hydrogen on the adjacent carbon is also axial. The rule of antiperiplanar attack then dictates preferential α epoxidation. Structure **49** has a *trans* junction, with axial angular hydrogens: α epoxidation is again favored. The most stable conformation of **50** has the methyl group in an equatorial position. This is also a reactive conformation because of an axial allylic hydrogen. Again, α attack dominates. However, **50**, being flexible, can exist in a second conformation having an axial methyl group. Since this minor configuration is also fairly reactive (the α allylic hydrogen is axial), a loss of selectivity follows.

Structure **51** is a planar, rigid molecule. The allylic bonds are symmetrically disposed about the molecular plane and antiperiplanar attack is impossible. The diagram above shows why **52** favors β-epoxidation. The seven-membered ring is fairly flexible, which allows competing reactions with the minor conformer. A diminution of selectivity is to be expected.

Verify the conformational analyses above with molecular models (or, preferably, by computer modeling). It is important to be able to make qualitative predictions of such results; dihedral angle analysis makes this possible.[159] The technique is not particularly difficult; the potential of this tool will amply repay the investment.

Exercise 32 (D)   *Deserves your attention*

The stereochemistry of nucleophilic additions to cycloheptanone has been studied by quantum mechanical calculations.[160] Employing a Monte Carlo search, 1000 conformations were generated and minimized. The five lowest conformers, labelled **A**, **B**, **C**, **D** and **E**, are shown below with their relative energies. The global minimum **A** is a chair with the carbonyl on the two-carbon stern fragment. The twist conformer **B** is almost as stable. The three remaining conformers can be roughly characterized as chairs with the carbonyl located at the bow.

For each conformer, two directions of attack are considered: 'top' and 'bottom'. Can you, *just by inspecting* **A**, **B**, **C**, **D** and **E**, predict some of the lowest and highest transition structures?

Hint. Reread the, in Section 6.2.1 above, the sub-sections *The flattening rule* (p. 155) and *The importance of being flexible* (p. 160).

[158]Martinelli M. J., Peterson B. C., Khau V. V., Hutchinson D. R., Leanna M. R., Audia J. E., Droste J. J., Wu Y.-D., Houk K. N., *J. Org. Chem.*, 1994, **59**, 2204.
[159]Bucourt R., *Top. Stereochem.*, 1974, **8**, 159; Toromanoff E., *Tetrahedron*, 1980, **36**, 2809.
[160]Ando K., Condroski K. R., Houk K. N., Wu Y.-D., Ly S. K., Overman L. E., *J. Org. Chem.*, 1998, **63**, 3196.

A
$E_{rel} = 0.00$

B
$E_{rel} = 0.34$

C
$E_{rel} = 1.32$

D
$E_{rel} = 1.93$

E
$E_{rel} = 2.24$

top

bottom

Answer

For steric reasons, top attack on **A**, **C**, **D** and **E** will give rise to high-energy transition states. Assuming that cycloheptanone is flexible enough for antiperiplanarity to exert its influence fully, we expect that the more adaptable conformer will lead to the most stable transition state. The best choice would then be **B**. The second best is more difficult to pick. **A** is more flexible (carbonyl is located at the stern) than **C** (carbonyl at the bow), but has only one CC bond antiperiplanar to the incoming nucleophile. **D** is certainly less favorable than **C** because its 'vertical' CC bonds are tilted more to the left, so that more distortion is required in order to reach antiperiplanarity. Finally, **E** is believed to be the least favorable conformer as it is less flexible and has only one antiperiplanar CC bond.

The energies of the nine transition structures are given below (top addition to **A** results in a conformational change to give conformer **B**).

	A	B	C	D	E
Top		0.00	3.58	1.79	4.15
Bottom	0.29	0.00	0.01	1.27	2.37

7 Some Structural Problems

Stable conformations

Rule 5 *The most stable conformations in a neutral molecule are those which minimize HOMO–HOMO interactions between the constituent fragments. For ions, the most stable configurations are those maximizing HOMO–LUMO interactions.*

Reactive conformations

Rule 6 *The most reactive conformations are those having the strongest HOMO–HOMO and LUMO–LUMO interactions between their constituent fragments.*

Beware: these are the reactive conformations of an *isolated molecule*. They may be modified in the presence of a reagent.

Structural anomalies

Rule 7 *When two fragments combine, an XY bond will shorten (lengthen) if the HOMO of the fragment containing it is antibonding (bonding) between X and Y. It will also shorten (lengthen) if the LUMO is bonding (antibonding) between X and Y.*
Rule 8–*Abnormal valence angles may be found if they increase HOMO–LUMO interactions between molecular fragments*

7.1 Principle of the Method

Strictly, frontier orbital theory applies only to bimolecular processes. Therefore, in unimolecular reactions and/or in structural problems, the molecule is formally split into two fragments, the recombination of which is treated as a bimolecular reaction. However, this ingenious artifice is a rather crude approximation, to be used with caution.

There is a fundamental difference between the treatment of structural problems and that of bimolecular reactions. Usually, study of bimolecular reactions only requires an

Frontier Orbitals   Nguyên Trong Anh
© 2007 John Wiley & Sons, Ltd

evaluation of the HOMO–LUMO interactions between the partners. In a structural study, the repulsive interactions between the fragments must also be considered, especially the HOMO–HOMO interaction (cf. Exercise 19, p. 153), which is crucial in conformational analysis (rule 5). Structural analyses are less reliable than reactivity studies, even though they take more interactions into account (p. 234).

7.2 Stable Conformations

7.2.1 Aldehydes, Alkenes and Enol Ethers

Ethanal and Propene

staggered conformation **1**   eclipsed conformation **2**

Microwave spectroscopy[1] shows that the staggered conformation (**1**) of ethanal is less stable than the eclipsed conformation (**2**) by approximately 1 kcal mol^{-1}. Hehre and Salem[2] analysed this molecule as the union of a CHO with a methyl group. The FOs of the CHO group are π_{CO} and π^*_{CO}. The methyl group orbitals[3] are presented in Figure 7.1. Only π_{Me} and π^*_{Me}, the two orbitals on the left, have the correct symmetry to interact with π_{CO} and π^*_{CO}.

Three interactions occur when the CHO and Me fragments combine (Figure 7.2). The first (1) takes place between the occupied π_{Me} and π_{CO} orbitals. It is a destabilizing

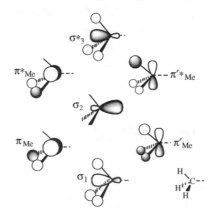

Figure 7.1   The orbitals of the methyl group.

[1]Kilb R. W., Lin C. C., Wilson E. B., *J. Chem. Phys.*, 1957, **26**, 1695.
[2]Hehre W. J., Salem L., *Chem. Commun.*, 1973, 754. Also see: Dorigo A. E., Pratt D. W., Houk K. N., *J. Am. Chem. Soc.*, 1987, **109**, 6591.
[3]To obtain these orbitals without calculations, see: Anh N. T., *Introduction à la Chimie Moléculaire*, Ellipses, Paris, 1994, pp. 319–321.

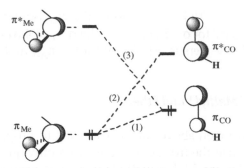

Figure 7.2 The main interactions between CH_3 and CHO.

four-electron interaction which increases with the overlap. The positive secondary overlap (wavy lines in the diagram), absent in **2**, thus destabilizes the staggered conformation **1**:

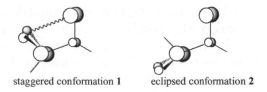

staggered conformation **1**　　　eclipsed conformation **2**

(2) and (3) are two-electron interactions, so an in-phase overlap will confer stability. However, secondary overlaps are both out-of-phase in the staggered conformation **1**, again disfavoring it with respect to **2**:

interaction (2)　　　　　　interaction (3)

A similar reasoning shows that propene is also more stable in its eclipsed form:

more stable than

Exercise 1 (E)

Show that $SiMe_3$ is a σ-donor and π-acceptor with respect to alcanes. *Indications*: Bond energies: H–C = 98 kcal mol^{-1}, H–Si = 76 kcal mol^{-1}, C–C = 82.6 kcal mol^{-1}, C–Si = 76 kcal mol^{-1}. Electronegativities (Pauling scale): H = 2.20, C = 2.55, Si = 1.90.

Answer

As Si is electropositive with respect to carbon, an Si–C bond is polarized Si^+C^-; Si is therefore a σ-donor. The Si–C bond being weaker than the C–H bond by 22 kcal mol^{-1}, this means that π_{SiMe_3} lies higher than π_{Me} and $\pi^*_{SiMe_3}$ lies below π^*_{Me} (cf. p. 98). $SiMe_3$ is thus both a better π-donor and a better π-acceptor than Me. However, as Si bears a partial positive charge, its orbitals are all lowered (cf. Rule, p. 103). The acceptor effect is therefore dominant.

According to 6–31G* calculations, the σ_{SiC} orbital in silaethane lies at −12.33 eV (versus −13.86 eV for σ_{CC} in ethane), the σ^*_{SiC} at 4.31 eV (versus 11.24 eV for σ^*_{CC}), the π_{SiH_3} at −12.75 eV (versus −13.28 eV for the π_{Me}) and the $\pi^*_{SiH_3}$ at 5.37 eV (versus 6.61 eV for the π^*_{Me}).

Propanal and Methyl Vinyl Ether

The approach given above can be extended to propanal and methyl vinyl ether. For propanal, we use the frontier orbitals of an ethyl group in place of π_{Me} and π^*_{Me} and refer back to Figure 7.2. For methyl vinyl ether, we replace π_{Me} and π^*_{Me} by the frontier orbitals of OMe, and π_{CO} and π^*_{CO} by π_{CC} and π_{CC}.

The CH group orbitals of methylene

We deal here only with the CH group orbitals and do not consider the orbitals corresponding to the two other bonds of the carbon atom. Four AOs intervene in the CH bonds, the two hybrid carbon orbitals and the two *s* hydrogen orbitals. To find their linear combinations requires solving an equation of the fourth degree. We can get around the problem by taking the symmetry combinations of the carbon hybrids:

symmetrical combination   antisymmetrical combination

Each combination can only mix with a hydrogen combination having the same symmetry. We have thus substituted a four-AO problem by two two-AO problem. The latter are solved using the theory of heteronuclear diatomic molecules and the results are shown in Figure 7.3.

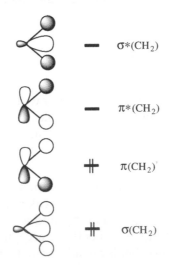

$\sigma^*(CH_2)$

$\pi^*(CH_2)$

$\pi(CH_2)$

$\sigma(CH_2)$

Figure 7.3   CH group orbitals of CH_2.

The ethyl group can be built by combining the orbitals of the methyl (Figure 7.1) and methylene (Figure 7.3) groups. Its HOMO is an out-of-phase combination of the CH_3 and CH_2 HOMOs, whereas its LUMO is an in-phase combination of the fragments LUMOs (cf. p. 152):

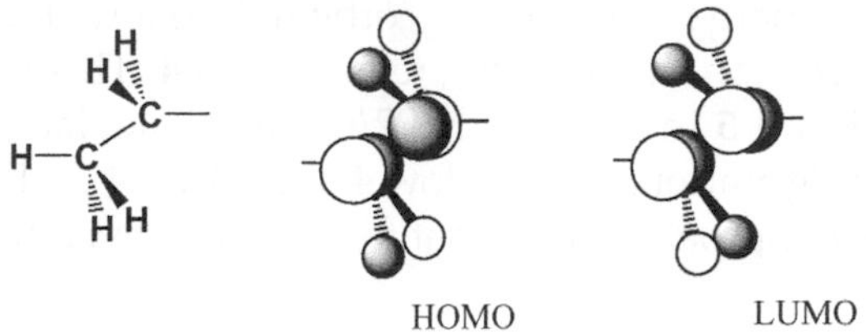

HOMO   LUMO

Consider now three conformations of propanal, **3** and **4**, where the carbonyl eclipses the CMe and CH bonds, respectively, and **5**, which has the CMe and CO bonds *trans* to each other. Experiments show that **3** is the preferred configuration.[4]

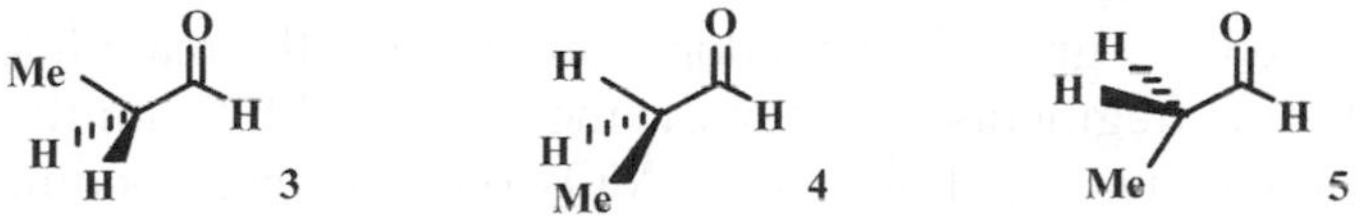

Let us first compare **3** with **5**. The type (1) destabilizing interaction between π_{CO} and the ethyl group HOMO increases with the overlap. It is attenuated by a secondary overlap (dotted) in conformation **3** so this form is favored:

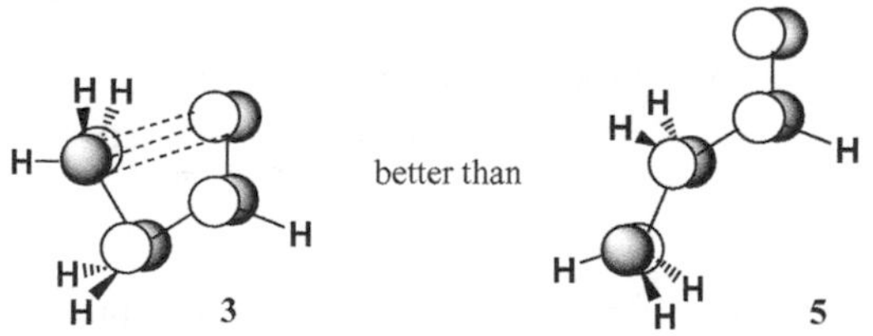

better than

The two-electron interactions HOMO(Et)–π^*_{CO} and LUMO(Et)–π_{CO} [type (2) and type (3) respectively] are stabilizing. The in-phase secondary overlaps again favor conformation **3**.

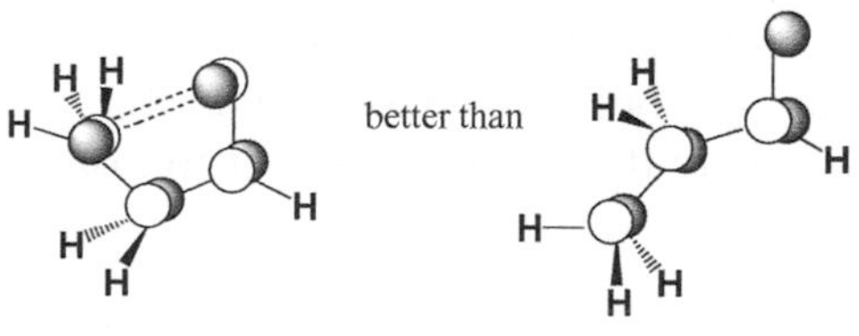

better than

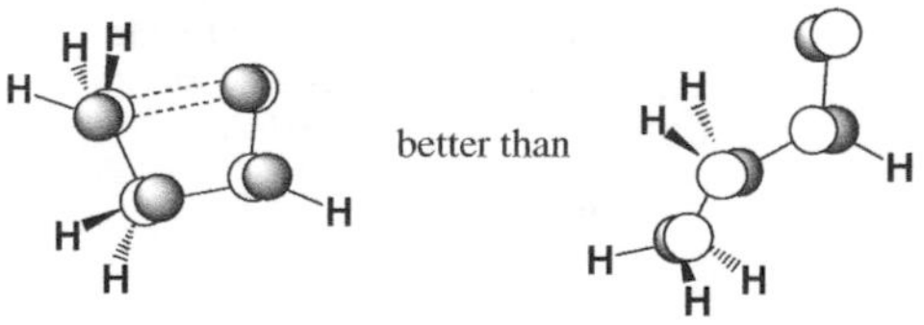

better than

[4]Karabatsos G. J., *J. Am. Chem. Soc.*, 1967, **89**, 1367 and references therein.

In **4**, the CMe bond lies almost parallel to the π system, so the HOMO of the ethyl fragment can be approximated here by the σ_{CMe} orbital. According to rule 5, the relative stability of **4** depends on the value of the σ_{CMe}–π_{CO} interaction compared with that of the HOMO(Et)–π_{CO} interaction in **3** and **5**. The HOMO(Et), being a π-type orbital, is higher in energy than the σ_{CMe} orbital. It follows that the HOMO(Et)–π_{CO} repulsion in **5** is larger than the σ_{CMe}–π_{CO} repulsion in **4**. This conformation should therefore lie between **3** and **5** in stability. MP2/6–31G* calculations[5] confirm that **3** is the lowest energy configuration; it lies below **4** and **5** by 1.4 and 2.2 kcal mol^{-1}, respectively.

A similar approach shows that for methyl vinyl ether:

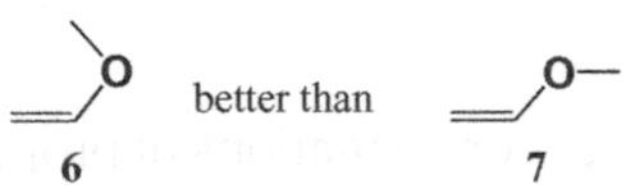

Optimizing at the 6–31G* level, it is found that **6** is more stable than **7** by 2.01 kcal mol. The methoxy group being a donor substituent, the main interaction between the C=C and MeO fragments is that of the methoxy HOMO with the π^*_{CC} orbital. Mixing these two fragment orbitals forms two MOs, one occupied bonding combination HOMO + λ π^*_{CC} and one unoccupied antibonding combination π^*_{CC} – μHO. This is shown as

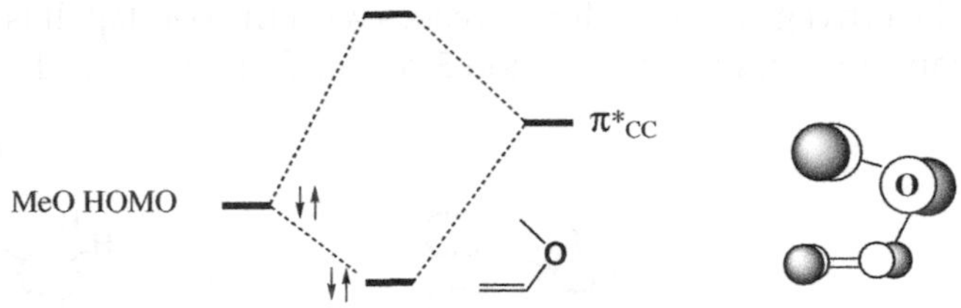

The physical significance of this scheme is that two electrons, belonging initially to the methoxy HOMO, are after interaction partially delocalized into the π^*_{CC} orbital. This HOMO is – just like the ethyl HOMO – antibonding between O and Me, so withdrawing electrons from it will strengthen the O–Me bond, which should then shortens. The π^*_{CC} orbital is antibonding, so populating it will weaken – and therefore lengthen – the C=C bond (rule 7). Finally, in the occupied HOMO + $\lambda\pi^*_{CC}$ MO, the overlap between C_2 and O is positive (i.e. bonding), so the C_2O bond should shorten. This shortening is more marked as the HOMO–LUMO interaction becomes stronger. Let us now compare the calculated structures of **6** and **7**:

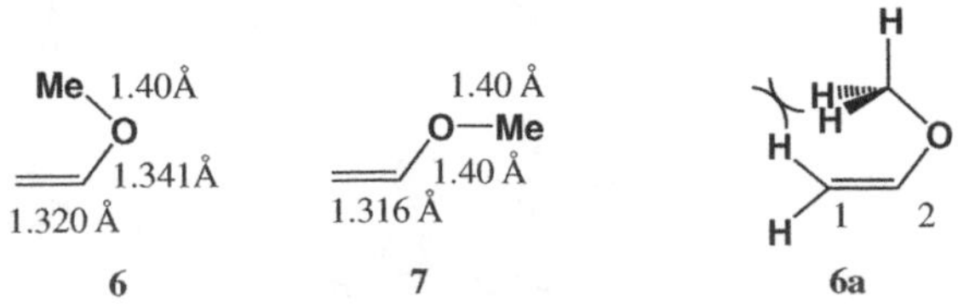

---

[5]Frenking G., Köhler K. F., Reetz M. T., *Tetrahedron*, 1993, **49**, 3971.

All our predictions are borne out, except for the O–Me bond shortening. This does not occur because of the steric hindrance between the inside hydrogen of C_1 and the methyl group (cf. **6a**). To relieve this interaction, the angles in **6** widen. The C=C–O angle has a value of 122°83 in **7** and of 128°59 in **6**. Similarly, the values for the C_2OMe angle are 116°10 and 118°21, respectively. The C_2O bond length, which decreases from 1.40 Å in **7** to 1.34 Å in **6**, is a clear indication that the frontier interaction is more favorable in **6**.

In each of the above analyses, the frontier orbital interactions shown in Figure 7.2 all favor the eclipsed configurations (**2**, **3** and **6**). This is no longer true for higher homologs such as butanal, higher aldehydes and alkyl vinyl ethers. The interactions between filled orbitals become more numerous and tend to favor conformations analogous to **5** or **7** with zigzag carbon skeletons. These four-electron interactions are the molecular orbital expression of steric hindrance.

Exercise 2 (E)

Conformations **A** and **B** of methyl formate have been compared by 6–31G* calculations. Would you expect **A** to be more stable by 2 or 6 kcal mol^{-1}?

A   **B**

Answer

Methyl formate is considered as resulting from the formal union of C=O and MeO. The difference with the methyl vinyl ether case is the replacement of a C=C bond by a C=O bond. Now, the π^*_{CO} is lower in energy and has a larger coefficient at carbon. The interaction with the MeO HOMO will be stronger. Indeed, calculations give **A** more stable than **B** by 6.27 kcal mol^{-1}. Comparison of the bond lengths and angles confirms that the favorable MeO HOMO–π^* interaction is enhanced, compared with the methyl vinyl ether case.

Me 1.419 Å	1.411 Å
O O=C–O = 125°75	O—Me O=C–O = 123°19
O= 1.316 Å C–O–Me = 116°83	O= 1.323 Å C–O–Me = 117°70
1.183 Å	1.177 Å
A	**B**

Note that although the C–O–Me angle in **A** is narrowed by 0°9, the O=C–O angle widens by 2°56′. Note also that, due to steric repulsion, the Me–O bond is lengthened instead.

Exercise 3 (E)

Of the two conformations **A** and **B** of acrolein, which one is more stable?

^{1}O 4
    2 3
   A   **B**

Answer

As in the methyl formate case, the major interaction is that of the π^*_{CO} with the HOMO of the substituent. However, the HOMO of MeO is antibonding between O and Me, whereas the HOMO of C=C is bonding between the two carbons. So:

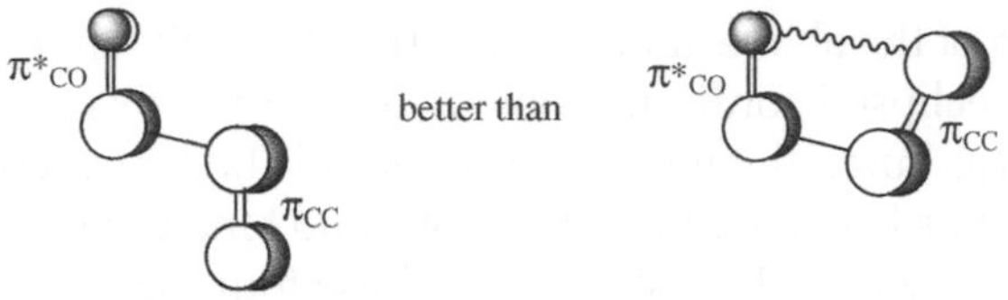

because in conformation **A**, there is an out-of-phase overlap (wavy line) diminishing the stabilization. According to 6–31G* calculations, **B** is more stable than **A** by 1.66 kcal mol^{-1}. That the π^*_{CO}–π_{CC} interaction favors **B** is shown by the shorter 2–3 bond. The 1–2–3 and 2–3–4 angles open in **A** to diminish both the steric hindrance and the unfavorable secondary 1–4 overlap.

	A	B
2–3 bond length	1.486 Å	1.478 Å
1–2–3 angle	121°46	121°26
2–3–4 angle	124°37	123°84

Exercise 4 (M)

Why is mesityl oxide more stable in its cisoid than its transoid form?

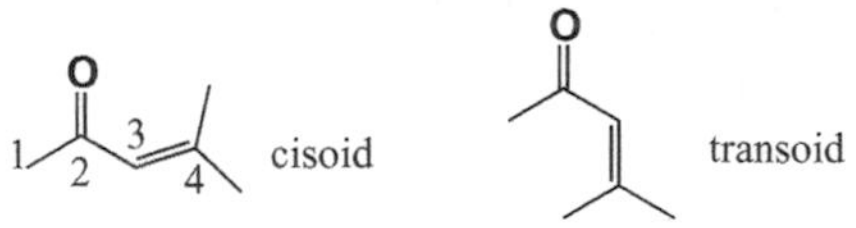

Hint. According to 6–31G* calculations, the cisoid and transoid forms of methyl vinyl ketone have practically the same energy and the same bond lengths and angles. Justify this result by comparing the MOs of formaldehyde and acetaldehyde given on p. 249.

Answer

The fact that the cisoid and transoid forms have the same energy, angles and bond lengths strongly suggests that the frontier interactions are negligible in methyl vinyl ketone. This results from two congruent effects. The methyl substituent raises the π^*_{CO}, increasing the π^*_{CO}–π_{CC} energy gap and diminishes its importance. The coefficient at the central carbon atom in the π_{CO} orbital becomes smaller, so the π_{CO}–π_{CC}^* also decreases.

In mesityl oxide, the two new methyls increase the frontier interactions which favor the transoid form, but this cannot compensate for the strong steric repulsion between C_1 and the methyl located at C_4:

	cisoid	transoid
2–3 bond	1.488 Å	1.490 Å
1–2–3 angle	114°68'	133°29'
2–3–4 angle	127°98'	128°55'

Chloroethanal and 2-Chloropropanal

Chloroethanal contains two polar bonds, C–Cl and C=O, so dipole–dipole interactions must be included in its conformational analysis. Let us look at conformations **8**, **9** and **10**, which are the analogs of **3**, **4** and **5**.

The frontier orbital interactions which favored conformation **3** in propanal also stabilize conformation **8**. However, **8** is destabilized by dipole–dipole interaction, whereas conformations **9** and **10** are stabilized by other factors. Thus, the CCl bond is highly polar and moderately strong (78 kcal mol^{-1}), so it has a low-lying σ^*_{CCl} orbital. This permits a stabilizing π_{CO}-σ^*_{CCl} interaction in conformation **9**.

Conformation **10** minimizes repulsions between the C–Cl and C=O dipoles. Therefore, **9** and **10** should be reasonably close in energy to **8**. Frenking *et al.* calculations[5] show that stability increases in the order **8** < **9** < **10**. However, the most stable configuration is a structure related to **10** which has an O=C–C–Cl dihedral angle of 156°: it lies 0.8 kcal mol^{-1} below **8**.

Combining the results for propanal and chloroethanal suggests that the conformational stability of 2-chloropropanal decreases in the order **13** < **12** < **11**. This order has been confirmed by calculations.[5]

7.2.2 Conformations of Some Ions

Conformational studies are easier for ions than for molecules. In a cation, the LUMO is very low-lying, so it interacts strongly with the HOMOs of the substituents. In the same way, the HOMO in an anion lie at very high energy, so it interacts strongly with the substituents' LUMOs. In both cases, it is possible, to a first approximation, to neglect HOMO–HOMO interactions, which considerably simplifies the analysis. This approach gives better results for cations, which are electron-deficient species, than for anions.

Walsh Orbitals

Walsh orbitals[6] are the MOs corresponding to the cyclopropane C–C bonds. We begin by supposing that carbon orbitals are sp^2 hybrid. Indeed, many studies show that cyclopropane is more like ethylene than propane or cyclobutane. For example, it adds bromine slowly, whereas cyclobutane is inert. Hydrogenation of cyclopropane occurs at 120°and cyclobutane at 200°. Asymmetric CH_2 stretching modes appear in the IR spectrum of cyclopropane at 3050 and in that of ethylene at 3080 cm^{-1}. Methyl cyclopropyl ketone shows a carbonyl stretch at 1695 cm^{-1}. Finally, the HCH angle is 116°6′ in ethylene, and 118° in cyclopropane.

Two of these carbon sp^2 hybrids are used to form CH bonds, so one sp^2 and one pure p orbital remain to bind to the neighboring carbons. The sp^2 and p orbitals are orthogonal so, to a first approximation, the six MOs of the cyclopropane skeleton can de divided into two groups of linear combinations of the sp^2 and p orbitals, respectively (Figure 7.4).

The Cyclopropylcarbinyl Cation[7]

Two limiting conformations can be drawn for the cyclopropylcarbinyl cation, C_3H_5–CH_2^+. Structure **14** has the vacant p orbital lying in the cyclopropane plane whereas **15** has them orthogonal.

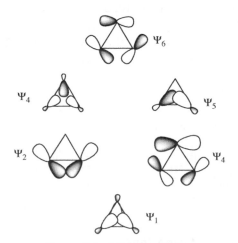

Figure 7.4   Walsh orbitals.

[6]Walsh A. D., *Nature*, 1947, **159**, 712; *Trans. Faraday Soc.*, 1949, **45**, 179.
[7]Hoffmann R., in *XXIIIrd International Congress of Pure and Applied Chemistry*, Vol. 2, Butterworths, London, 1971, p. 233.

The strongest interaction between the C_3H_5 and CH_2^+ fragments will occur between the p orbital and the cyclopropane HOMO (Figure 7.4). Only one of the two degenerate cyclopropane HOMOs has a finite coefficient at the substituted carbon. This two-electron stabilizing interaction increases with the overlap, so **14** is the most stable configuration. This prediction is fully vindicated by experiment.[8]

positive overlap   zero overlap

Substituted Ethyl Ions[9]

We will state the following rule without proof:[10]

Rule: *Replacing any atom by a more electronegative (or electropositive) atom X has no effect on the energy of MOs having a zero coefficient at X. The fall (or rise) in the energy of the remaining orbitals will be proportional to the value of their coefficient at X.*

Applying this rule to the methyl group (Figure 7.1), we can derive the orbitals of CH_2X, where X is an electronegative atom. The π_{Me} and π^*_{Me} orbitals are unchanged. All of the others are lowered in energy, with π'_{Me} and π'^*_{Me} more strongly affected (see Figure 7.5, central column). The loss of degeneracy will induce strong conformational preferences for CH_2X-CH_2. Figure 7.5 shows the perturbation schemes for the union of CH_2X and CH_2^+ in two different conformations. In (A), the vacant p orbital can only interact with π'_{Me}, whereas in (B), its only overlap is with π_{Me}. As the latter lies at higher energy, the interaction is stronger, so (B) is the more stable structure.

The problem of the anion is slightly more complicated (Figure 7.6). The p orbital is doubly occupied, so four interactions must be considered: the two-electron stabilizing (2) and (4) and the four-electron destabilizing interactions (1) and (3). Fortunately, their effects are complementary. π'_{Me} and π'^*_{Me} lie at lower energy than π_{Me} and π^*_{Me}, so (2) provides more stabilization than (4). Additionally, repulsion is lower in (1) than in (3). Consequently, (A) should be more stable than (B) and conformational preferences are inverted upon going from cations to anions.

The previous study might seem to be a theoretician's amusement, of no practical value. This is not true, as we will see when analyzing why vinylic S_N2 reactions generally proceed with *retention of configuration*.

---

[8]Pittman C. U. Jr, Olah G. A., *J. Am. Chem. Soc.*, 1965, **87**, 2998; Kabakoff D. S., Namanworth E., *J. Am. Chem. Soc.*, 1970, **92**, 3234.

[9]Hoffmann R., Radom L., Pople J. A., Schleyer P. v. R., Hehre W. J., Salem L., *J. Am. Chem. Soc.*, 1972, **94**, 6221.

[10]This result is obvious in diatomics: compare the perturbation schemes giving the π MOs of ethylene and formaldehyde. However, the rule is valid only if X and the replaced atom give rise to bonds of similar energy. If we replace, for example, O by S, the occupied MOs are indeed raised, but the antibonding MOs may be lowered because β_{CS} is much smaller than β_{CO}.

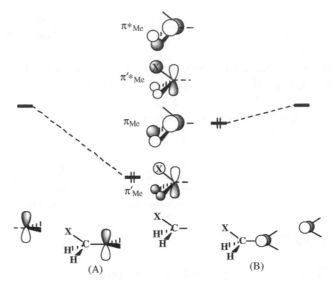

Figure 7.5 Conformational analysis of the $CH_2X-CH_2^+$ cation.

The stereochemistry of vinylic S_N2 reactions[11]

For ease of discussion, we will divide the reaction into two steps. The first is the nucleophilic addition of Y^- to give an anionic intermediate; the second is the departure of the leaving group X^-, which occurs after the intermediate has changed conformation. The conclusions that we draw remain valid for a concerted process. Rule 4 (p. 129) indicates that Y^- will approach along the plane of the π^* orbital, giving initially intermediate **16**:

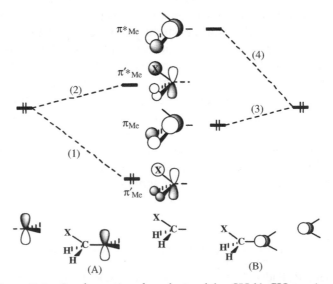

Figure 7.6 Conformational analysis of the $CH_2X-CH_2^-$ anion.

[11]Stohrer W. D., *Tetrahedron Lett.*, 1975, 207. See also: Miller S. I., *Tetrahedron*, 1977, **33**, 1211 and references therein.

$$Y^- \; + \; \underset{X}{\overset{H}{\longmapsto}} \!\!\!\!=\!\!\!\!\underset{H}{\overset{H}{\longmapsto}} \; \longrightarrow \; \underset{Y}{\overset{X \overset{H}{\cdots}}{\diagup}} \!\!\!\!\!\!\!\! \overset{\ominus \cdots H}{\underset{H}{\diagup}} \quad 16$$

However, this structure is not favorable for the departure of the leaving group X. Two factors dictate that the most reactive conformation will have the CX bond eclipsed by the p orbital of the carbanion. First, the exothermic formation of the π bond can then occur simultaneously with the CX bond rupture, thus diminishing the activation energy. Second, the CX bond will be weakened in conformation **17**, so it will cleave more easily. Imagine **17** as resulting from the union of the CHXY and CH_2^- fragments. As they combine, σ^*_{CX} interacts with the p orbital of the CH_2^- portion. This gives rise to two new orbitals, one bonding and the other antibonding. The bonding combination $p + \lambda \, \sigma^*_{CX}$ will be occupied by the electrons previously associated with CH_2^-. Physically, this means that electrons have been transferred from the p lone pair into the antibonding σ^*_{CX} orbital, thus weakening the C–X bond.

In principle, transformation of **16** into **17** could involve rotating X upwards (**18**) or downwards (**20**). Configuration (**20**) can only be reached with difficulty because it passes through the high-energy conformation **19**, whose p orbital lies orthogonal to CX. Hence the preferred profile passes through **18**, which means that the reaction proceeds with retention of configuration (Figure 7.7).

Figures 7.6 and 7.7 are molecular orbital rationalizations of a familiar concept in organic chemistry: an anionic species will be stabilized by an adjacent acceptor. Replacing the anionic doublet by the lone pair of heteroatom leads to the anomeric and gauche effects (see below).

At first sight, our treatment may appear to be needlessly complicated. However, perturbation methods (which can be fairly rapid after a little practice) offer two important advantages over empirical rules. They are general (for example, the anomeric effect or the stabilization of cyclobutadienes by transition metals can be treated in the same way) and provide more information, particularly on structural distortions which, in turn, can help us predict the system's reactivity.

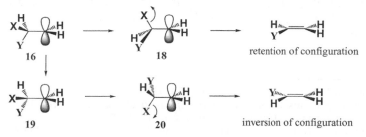

Figure 7.7 Stereochemistry of vinylic $S_N 2$ substitutions.

7.2.3   The Anomeric Effect

Lone Pairs in Ethers

The existence of two different ionization potentials in ethers, R–O–R, shows that their lone pairs are not equivalent. This experimental phenomenon is most easily rationalized by supposing that the oxygen is sp^2 hybridized, as shown in **21**.[12] One of the lone pairs has then sp^2 character and the other is a p orbital. When their electron densities are combined, a 'rabbit ear' structure (**22**) is obtained. This often-used representation is incorrect (because it implies an equivalence of the lone pairs) but convenient, especially when the molecule is deformed slightly to make the oxygen perfectly tetrahedral. Below, we will see how the stability of an ether is increased when one of the lone pairs lies antiparallel to a low-lying neighboring σ^*_{CX} orbital. This is easily seen using representation **22**, but less obvious with **21**. We will use whichever of these representations is better adapted to the problem at hand.

The Anomeric Effect

Substituents in cyclohexanes prefer an equatorial orientation. Nonetheless, an electronegative substituent X situated α to the oxygen in a tetrahydropyran will prefer an axial position. This is the *anomeric effect*.

Several interpretations of this phenomenon have appeared,[13] but we will use Salem and co-workers' version of a theory proposed by Altona and co-workers.[14] An anomeric effect only arises when X is localized on the carbon adjacent to the oxygen, which suggests that the effect results from an interaction between O and $C_\alpha X$. More precisely, X being always electronegative, this implies an interaction between one of the oxygen lone pairs and the low-lying antibonding σ^*_{CX} orbital. We will assume that the ring has a chair configuration, so that its bonds are staggered. This automatically minimizes torsional repulsions, which are neglected in our approximate analysis.

---

[12]Sweigart D. A., *J. Chem. Edue.*, 1973, **50**, 322.

[13]Deslongchamps P., *Stereoelectronic Effects in Organic Chemistry*, Pergamon Press, Oxford, 1983, Chap. 2; Juaristi E., Cuevas G., *Tetrahedron*, 1992, **48**, 5019.

[14]Romers C., Altona C., Buys H. R., Havinga E., *Top. Stereochem.*, 1969, **4**, 39; David S., Eisenstein O., Hehre W. J., Salem L., Hoffmann R., *J. Am. Chem. Soc.*, 1973, **95**, 3806.

When X is equatorial (**23**), the p lone pair interacts mainly with σ^*_{CH} and the sp^2 lone pair with σ^*_{CX}. For axial orientation (**24**), the principal interactions are sp^2–σ^*_{CH} and p–σ^*_{CX}. The smallest HOMO–LUMO gap appears in the last of these combinations, so **24** is favored.

Structure **22** provides a simpler but less rigorous explanation of the anomeric effect. The largest stabilization occurs when the one of the oxygen lone pairs lies parallel to σ^*_{CX}. This only happens when X is axial (**26**).

Applications

The p–σ^*_{CX} interaction gives rise to two new orbitals. Only the lower, p + λ σ^*_{CX} (**27**), is occupied. It is O–C_α bonding and C_α–X antibonding, so the OC_α bond is shortened and the $C_\alpha X$ lengthened. An X-ray structure confirms this analysis.[15]

In physical terms, this implies donation from the p lone pair into the antibonding σ^*_{CX} orbital. The oxygen atom becomes electron depleted and less basic: any protonation will be at the sp^2 lone pair because attack at the p orbital will diminish the anomeric stabilization.[16] Conversely, the electron transfer increases the basicity of the X atom. It becomes more easily protonated, which further weakens the CX bond and encourages the X group to leave. Deslongchamps has shown that this phenomenon has important consequences in acetal chemistry (X = OR), a detailed discussion of which appears in ref. 13.

A Warning

We have considered the anomeric effect simply as an electron transfer from the oxygen lone pair into the CX bond. However, X is usually a heteroatom, so a full analysis

---

[15]Jeffrey G. A., Pople J. A., Radom L., *Carbohydr. Res.*, 1972, **25**, 117; Bürgi H. B., Dunitz J. D., Shefter E., *Acta Crystallogr., Sect.* B, 1974, **30**, 1517.
[16]We saw a similar situation when dealing with 'ate' complexes (Exercise 3, p. 59).

must include the possibility of donation from an X lone pair into the C–O antibonding orbital. In theory, nine *gauche* conformations are possible for an acyclic acetal and 27 for an orthoester. Symmetry and 'classical' conformational arguments allow a number of these to be rejected immediately, but analyzing such species is not always simple.

The energy associated with a 'pure' anomeric effect (i.e. one not influenced by other factors) has been estimated at ~1.4 kcal mol^{-1}.[17] Putting an OR group axial confers a destabilization of ~0.8 kcal mol^{-1}, so the energy difference between **28** and **29** is only about 0.6 kcal mol^{-1}. Therefore, the anomeric effect is small. Bonds will only be significantly weakened when several anomeric effects are superimposed or when they are reinforced by protonation.

28   **29**

Exercise 5 (E)

Find the more stable isomer in each of the pairs (**A, A'**) and (**B, B'**) below.

A   **A'**   **B**   **B'**

Answer[18]

Two anomeric effects operate in **A**: the donation of the shaded oxygen lone pair into CS and of the shaded S lone pair into CO. In **A'**, only the donation from the shaded sulfur occurs. Thus **A** is the more stable isomer. A similar argument shows that **B** is more stable than **B'**.

A   **A'**

Differentiating between the (**A, A'**) and (**B, B'**) pairs involves comparing the relative strength of two anomeric effects: the donation of oxygen electron density into CS and S electron density into CO. The σ^*_{CO} and σ^*_{CS} energies should be similar because the electronegativity of oxygen (which lowers the energy) is offset by the greater CO bond strength (which tends to raise it). This being the case, the element having the higher-lying lone pair (i.e. the sulfur) will produce the stronger anomeric effect. This

[17]Descotes G., Lissac M., Delmau J., Duplan J., *C. R. Acad. Sci. C*, 1968, **267**, 1240; Beaulieu N., Dickinson R. A., Deslongchamps P., *Can. J. Chem.*, 1980, **58**, 2531.
[18]Eliel E. L., Giza C. A., *J. Org. Chem.*, 1968, **33**, 3754; Zefirov N. S., Shekhtman N. M., *Dokl. Akad. Nauk SSSR*, 1967, **177**, 842; 1968, **180**, 1363.

argument is supported by experimental data which give equilibrium constants of **A/A'** = 65/35 and **B/B'** = 90/10.

Exercise 6 (E)

Will the following compound be more stable with R' placed axially or equatorially?

Answer[19]

When R' is axial, the nitrogen lone pair lies antiparallel to the CO and the conformation is stabilized by an anomeric effect.

7.2.4 The Geminal Effect

When two functional groups are borne by the same atom, they interact to cause either stabilization or destabilization. This is the *geminal* effect.[20] If the substituents are both strong σ-acceptors and π-donors [for example a *gem*-dimethoxy–$C(OMe)_2$–], a stabilization occurs.[21] This is easy to understand. Oxygen, being electronegative, is a σ-acceptor and σ^*_{CO} is low in energy and is π-donor thanks to its lone pair. Therefore, we have here a double anomeric effect:

If the substituents are both σ- and π-acceptors, there is destabilization.[22] FO interpretation of this conclusion is not so obvious. Clearly, there is no longer electron donation from one substituent to the σ* orbital of the other and therefore no more stabilization, but the reason of the destabilization is not apparent (steric hindrance?).

[19]Allingham Y., Cookson R. C., Crabb T. A., Vary S., *Tetrahedron*, 1968, **24**, 4625; Riddell F. G., Lehn J. M., *J. Chem. Soc. B*, 1968, 1224; Booth H., Lemieux R. U., *Can. J. Chem.*, 1971, **49**, 777.
[20]Benson S. W., *Angew. Chem. Int. Ed. Engl.* 1978, **17**, 812; Schleyer P. v. R., Clark T., Kos A. J., Spitznagel G. W., Rohde C., Arad D., Houk K. N., Rondan N. G., *J. Am. Chem. Soc.*, 1984, **106**, 6467; Leroy G., *J. Mol. Struct.*, 1985, **120**, 91
[21]Schleyer P. v. R., Kos A. J., *Tetrahedron*, 1983, **39**, 1141; Schleyer P. v. R., Jemmis E. D., Spitznagel G. W., *J. Am. Chem. Soc.*, 1985, **107**, 6393; Reed A. R., Schleyer P. v. R., *J. Am. Chem. Soc.*, 1987, **109**, 7362.
[22]Wu Y.-D., Kirmse W., Houk K. N., *J. Am. Chem. Soc.*, 1990, **112**, 4557.

7.2.5 The Gauche Effect

Wolfe[23] defines the *gauche* effect as the tendency of molecules to adopt a structure maximizing the number of *gauche* interactions between adjacent lone pairs and/or polar bonds. Heteroatoms are assumed to be sp^3 hybridized, thus having equivalent lone pairs (**22**). This phenomenological definition is unsatisfactory because it groups three different situations under the same title.

Two Adjacent Lone Pairs

Let us compare the *trans* (**30**) and *gauche* (**31**) configurations of hydrogen peroxide. For the sake of clarity, the p lone pairs (perpendicular to the plane of the page) are omitted from **30**. In **31**, they are shown only as dotted lines.

Conformation **30** is destabilized by two severe repulsions: the lone pairs are degenerate *and* they overlap very strongly. That eclipsed lone pairs overlap well seems obvious, but the interactions are no smaller in the *trans* case, as you can convince yourself by factorizing the *cis* and *trans* hybrids into their components.[24] Conformation **31** is destabilized less, as each repulsion involves a non-degenerate interaction between a p lone pair and the sp^2 hybrid of the adjacent oxygen. Furthermore, these orbitals are staggered, so their overlaps are relatively small.

The experimental value of the dihedral angle[25] in HOOH is 111°. The interactions between adjacent lone pairs are repulsive, but the other *gauche* effects result from attractive influences (see below). Had we use the 'rabbit ear' structure **22**, with equivalent lone pairs, the experimental structure would be hard to explain.

A Lone Pair Adjacent to a Polar Bond

This is simply the anomeric effect viewed in a different way! Consider fluoromethanol (**32**). It has an HOCF dihedral angle of 60°,[26] so the CF bond will be perfectly antiparallel to one of the lone pairs if the oxygen atom is represented by **22**. However, care is needed if we depict the oxygen by **21**. Conformation **33** is *not* the best: the p–σ^*_{CF}

---

[23]Wolfe S., *Acc. Chem. Res.*, 1972, **5**, 102; Zefirov N. S., *Tetrahedron*, 1977, **33**, 3193; Juaristi E., *J. Chem. Educ.*, 1979, **56**, 438.
[24]Hoffmann R., Imamura A., Hehre W. J., *J. Am. Chem. Soc.*, 1968, **90**, 1499.
[25]Olovsson I., Templeton D. H., *Acta Chem. Scand.*, 1960, **14**, 1325 ; Hunt R. H., Leacock R. A., *J. Chem. Phys.*, 1966, **45**, 3141.
[26]Wolfe S., Rauk A., Tel L. M., Czismadia I. G., *J. Chem. Soc. B*, 1971, 136.

interaction is maximized but the sp^2–σ^*_{CF} interaction is zero. Stabilization increases when the CH_2F group is rotated slightly clockwise (**33** → **34**) to allow σ^*_{CF} to interact with both of the oxygen lone pairs.[27]

A counter-clockwise rotation would bring the CF bond *synclinal* to the sp^2 lone pair. The *synclinal* overlap is weaker than the *anticlinal*,[28] because of the negative overlap between sp^2 and the fluorine AO in σ^*_{CF}. We have used this argument in our treatment of the Walden inversion (p. 179). Pushing the analogy a little further, we can equate the sp^2–σ^*_{CF} interaction to an S_N2 attack of the sp^2 orbital on the CF bond. This attack must proceed with inversion (because the reaction occurs at a carbon center), so the sp^2 orbital will be best oriented antiparallel to CF.

Two Adjacent Polar Bonds

Gauche effects between two adjacent polar bonds are relatively weak, except when the conformation is controlled by hydrogen bonds (FCH_2CH_2OH, CH_2OH–CH_2OH, etc.). For example, the *gauche* conformation of dichloroethane is only favored over the *trans* by 1.2 kcal mol^{-1} in the gas phase, and both conformations have the same energy in solution.[29] The simplest rationalization[30] seems to be the following. In *trans* conformation **35**, the main interactions are σ_{CX}–σ^*_{CY} and σ^*_{CX}–σ_{CY}. In the *gauche* conformation **36**, they are σ^*_{CX}–σ_{CH} and σ_{CH}'–σ^*_{CY}. The HOMO–LUMO gaps are smaller in **36** so this conformer is favored.

This argument is only molecular orbital parlance for the idea that the best acceptor CX prefers to be oriented opposite to the best donor (CH rather than CY). It cannot, however, explain why the *gauche* conformation is slightly favored in $FCH_2C_2H_5$.[30a]

---

[27]This is the same problem as the complexation of C=O by M$^+$ (p. 58). The C=O—M angle must be obtuse if the cation is to interact with both lone pairs.
[28]Gavezzotti A., Bartell L. S., *J. Am. Chem. Soc.*, 1979, **101**, 5142.
[29]Morino Y., *J. Mol. Struct.*, 1985, **126**, 1.
[30](a) Radom L., Lathan W. A., Hehre W. J., Pople J. A., *J. Am. Chem. Soc.*, 1973, **95**, 693; (b) Baddeley G., *Tetrahedron Lett.*, 1973, 1645.

7.3 Reactive Conformations

The reactive conformations of any given molecule will depend on the nature of the incoming reagent and the direction of its approach. This means that transition state geometries are needed to obtain the *true* reactive conformations. Transition states are difficult to calculate, so we will study only the reactive forms of an *isolated* molecule.

Rule 2 (p. 47) states that the most reactive conformations have the highest-lying HOMO and/or the lowest-lying LUMO. The perturbation scheme in Figure 6.4 (p. 152) shows that the antibonding combination of the fragment HOMOs generates the molecular HOMO. The LUMO is the bonding combination of fragment LUMOs. Hence, to identify the most reactive conformations, we need the configurations having the smallest HOMO–LUMO gap, and the greatest HOMO–HOMO and LUMO–LUMO overlap for the fragment FOs.

Let us picture the ketone CXYZ–CO–R as a union of CXYZ with CO–R. Its most reactive conformations are those corresponding to minima on the LUMO energy curve, i.e. those where the LUMO–LUMO interactions are strongest (rule 6). The LUMO of the CO–R fragment is π^*_{CO} whereas the LUMO of CXYZ is a combination of σ^*_{CX}, σ^*_{CY} and σ^*_{CZ}. The largest overlaps will occur when the carbon p orbital is eclipsed by CX, CY or CZ. Thus, the minima on the LUMO curve are given by the Felkin conformations **37–39**.

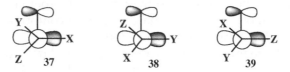

To prove this, let us decompose the R–CXYZ LUMO into a component parallel to π^*_{CO} and a second one perpendicular to π^*_{CO}. In **37–39**, the parallel component is mainly σ^*_{CX}, σ^*_{CY} and σ^*_{CZ}. If, for example, σ^*_{CX} is the lowest-lying orbital, it will have the strongest interaction with π^*_{CO}, and **37** is the most reactive conformation. The relative energies of σ^*_{CX}, σ^*_{CY} and σ^*_{CZ} can be predicted using the guidelines below:

- The σ^*_{CX} energy falls as the electronegativity of X increases (cf. p. 13).
- The σ^*_{CX} energy also falls as the CX bond weakens (cf. p. 98).
- The HOMO–LUMO gap will fall as the number of atoms in the substituent X, Y or Z rises.[31] This rule follows directly from Figure 6.4 (p. 152). Thus, the HOMO of the ethyl group, being an antibonding combination of the HOMOs of CH_2 and CH_3, lies higher in energy than the HOMO of the methyl group. Similarly, the LUMO of the ethyl group is lower-lying than that of the methyl group.

Occasionally, these guidelines may contradict each other: F is the most electronegative element, but σ^*_{CF} orbitals lie relatively high in energy because the C–F bond is

---

[31]When the substituent is a conjugated group, this result can be immediately verified, using Coulson's equations (p. 18).

particularly strong. In any case, these guidelines only localize approximately the reactive conformations. For example, even when σ^*_{CX} is the lowest-lying orbital, **37** is not necessarily the optimum conformation, Rotating R–CXYZ slightly allows π^*_{CO} to overlap simultaneously with σ^*_{CX} and σ^*_{CY}. This may be a better structure, in much the same way as **34** is better than **33**. Finally, note that rule 6 is based purely upon frontier orbital considerations; sometimes other factors are important. For example, when X = Cl and Y and Z are relatively apolar groups, the strong dipolar stabilization associated with a conformation such as **40** may make it competitive with **37–39**.

40

7.4 How to Stabilize Inherently Unstable Species

In principle, this is a simple process: we determine the causes of instability in the target molecule and eliminate them by appropriate transformations (substitution, complexation, etc.). Indeed many 'abnormal' molecules with varied structures have now been made.[32] For each problem, several solutions are possible. There is not enough space to illustrate every conceivable approach here, so we will merely consider four representative molecules: norcaradiene (p. 226), cyclobutadiene, trimethylenemethane and carbene.

7.4.1 Cyclobutadiene

The first attempts to prepare cyclobutadiene (CB) were made around 1870,[33] but its antiaromaticity was only understood by Hückel in 1932. The prediction that CB could be stabilized by complexation was published in 1956[34] and the first cyclobutadienes were made – as transition metal complexes – 3 years later.[35]

Orgel and Longuet-Higgins' analysis can be explained in the following way. Figure 7.8 shows that two π electrons of CB are non-bonding. They make no contribution to molecular stability but still introduce interelectronic repulsions. Thus, they confer thermodynamic instability. Furthermore, Hund's rule requires that Ψ_2 and Ψ_3 both contain a single electron. This adds kinetic instability because diradical structures are extremely reactive.[36]

---

[32]Vögtle F., *Fascinating Molecules in Organic Chemistry*, John Wiley & Son Inc., New York, 1992.
[33]Cava M. P., Mitchell M. J., *Cyclobutadiene and Related* Compounds, Academic Press, New York, 1967; Bally T., Masamune S., *Tetrahedron*, 1980, **36**, 343; Maier G., *Angew. Chem. Int. Ed.*, 1988, **27**, 309.
[34]Longuet-Higgins H. C., Orgel L. E., *J. Chem. Soc.*, 1956, 1969.
[35]Hubel W., Braye E. H., *J. Inorg. Nucl. Chem.*, 1959, **10**, 250; Criegee R., Schröder G., *Liebigs Ann. Chem.*, 1959, **623**, 1.
[36]We have implicitly assumed that cyclobutadiene is square. If it distorts and adopts a rectangular structure, its frontier orbitals become nondegenerate (Jahn–Teller effect) but the orbital separation will remain small and the molecule will still be highly reactive.

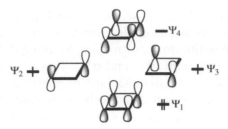

Figure 7.8 The π MOs of cyclobutadiene.

Let us turn to the frontier orbitals of the $Fe(CO)_3$ group.[37] The three lowest-lying metal orbitals, d_{xy}, d_{xz} and d_{yz}, form a triply degenerate set which accommodates six of the eight iron valence electrons. Only two electrons remain; the doubly degenerate d_{z^2} and $d_{x^2-y^2}$ orbitals shown below accept one each:

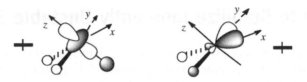

These orbitals have the correct symmetry for mixing with Ψ_2 and Ψ_3. Figure 7.9 shows the interaction of Ψ_3 with $d_{x^2-y^2}$ in $CBFe(CO)_3$. This two-electron (stabilizing) interaction removes the thermodynamic instability. Furthermore, the single electrons in the parent fragments are paired in the complex, so the kinetic instability is also eliminated.

Cyclobutadienes can also be stabilized by other methods. In 1958, Roberts suggested the stabilization of *free* cyclobutadiene by the incorporation of suitable substituents.[38] His hypothesis was confirmed 10 years later by Gompper and Seybold's synthesis of **41**.[39] It is relatively easy to show that the presence of alternating donor and acceptor substituents will confer stability upon cyclobutadiene.[40]

EtO$_2$C \qquad NH$_2$

NH$_2$ \qquad CO$_2$Et

41

7.4.2 Trimethylenemethane

Free trimethylenemethane (TMM) is also a highly reactive species. It has no Kekulé resonance form and can only be observed by trapping at low temperatures in a neutral

[37](a) Albright T. A., Burdett J. K., Whangbo M. H., *Orbital Interactions in Chemistry*, John Wiley & Sons, Inc. New York, 1985, p. 339; (b) Anh N. T., *Introduction à la Chimie Moléculaire*, Ellipses, Paris, 1994, p. 319.

[38]Roberts J. D., *Chem. Soc. Spec. Publ. No. 12*, 1958, 111.

[39]Gompper R., Seybold G., *Angew. Chem. Int. Ed. Engl.*, 1968, **7**, 824.

[40]Anh N.T. Introduction à la Chimie Moléculaire, Ellipses, Paris, 1994, p. 258.

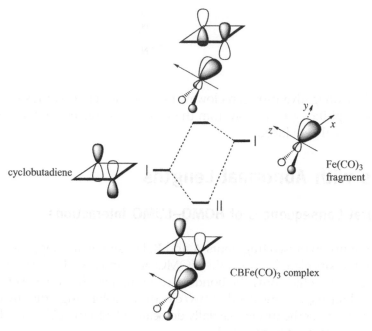

Figure 7.9 The interaction of Ψ_3 with $d_{x^2-y^2}$ in $CBFe(CO)_3$.

matrix. Its MOs are given in Figure 7.10. Two of them are singly occupied nonbonding orbitals which have the same symmetry as those of cyclobutadiene. In this light, the stability $TMMFe(CO)_3$ is unsurprising.

7.4.3 Stable Carbenes

Carbenes, having a low-lying vacant p orbital and a lone pair, can react with nucleophiles and electrophiles alike. Introducing very bulky substituents is one obvious way to diminish its reactivity. A more interesting solution is electronic stabilization. For example, the following carbene has indefinite stability in the solid state:[41]

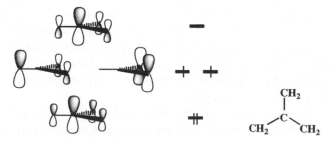

Figure 7.10 Molecular orbitals of trimethylenemethane.

[41]Arduengo A. J. III, Rasika Dias H. V., Harlow R. L., Kline M., *J. Am. Chem. Soc.*, 1992, **114**, 5530.

The electronegative nitrogens lower the lone pair energy, thus decreasing the carbene nucleophilicity. Donation from their lone pairs raises the p orbital and the electrophilicity is also reduced.

7.5 Bonds with Abnormal Lengths

7.5.1 Structural Consequences of HOMO–LUMO Interactions

Consider a union of two fragments **A** and **B**. For simplicity, suppose that the principal interaction takes place between the HOMO of **A** and the LUMO of **B**. It generates two new orbitals, of which only the bonding combination HOMO + λLUMO is occupied. In physical terms, this means that two electrons 'belonging' initially to the HOMO of **A** have been, after the union, partially delocalized into the LUMO of **B**. Thus, electron transfer has occurred from **A** to **B**.

Now, consider the influence of this process upon an IJ bond localized within the **A** fragment. If the overlap between I and J is positive in the **A** HOMO, then the HOMO(**A**)–LUMO(**B**) interaction will withdraw bonding electron out of IJ. The IJ distance increases as a consequence. If the overlap between I and J is negative in the **A** HOMO, the antibonding IJ electron density will diminish and the IJ bond shortens. Consider now a KL bond in fragment **B**. If the overlap between K and L is positive in the **B** LUMO, the electron transfer will increase the bonding electron density between K and L reduce the KL distance. If the overlap is negative, the bond will lengthen. Rule 7 on p. 187 encompasses all these results.

It follows from this analysis that *any substituent, whether donor or acceptor, weakens the bond to which it is attached.* Consider an acceptor substituent. Let **M** be the bond and **N** the substituent. The HOMO(**M**) is the σ bonding orbital, which is depopulated by the **M**–**N** interaction. A similar reasoning shows that a donor will also weaken the bond because it will partially populate the σ* antibonding orbital.

We may now wonder whether two acceptors, two donors or a donor and an acceptor will weaken most a bond. Two like substituents seem intuitively less able to weaken the bond than the mixed combination. This can be justified more rigorously by introducing the substituents sequentially. Let the first substituent be a donor. It will cause the energies of the σ and σ* orbitals to rise (p. 66), so they will interact more readily with the LUMO of an acceptor.

Is it better to place a substituent at each end of the bond, or to place both substituents upon the same atom? A stepwise approach again provides the answer. If a donor **D** is present, the σ bonding orbital becomes dissymmetric, with the larger coefficient at the nonsubstituted atom:

Rule 3 (p. 87) states an acceptor **A** (which can be viewed as an 'electrophile') then prefers the nonsubstituted carbon atom. 'Arrow pushing' could have predicted a heterolytic cleavage:

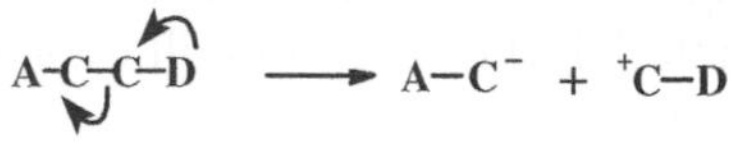

$$\text{A--C--C--D} \longrightarrow \text{A--C}^- + {}^+\text{C--D}$$

Our approach shows that this substitution pattern also favors homolytic ruptures.

Exercise 7 (E)

Explain why a donor-substituted σ bond adopts the structure shown in **42**, with the bigger coefficient on the nonsubstituted atom.

Answer

In MO theory, a σ and a π bond are both just two overlapping orbitals and a monosubstituted bond is similar to an allyl system. Did you realize that **42** *must* have the same structure as the HOMO of an enol (p. 29)?

7.5.2 Applications to Nucleophilic Additions

Additions and additions–eliminations

When a nucleophile Z^- attacks a carbonyl compound R–CO–X, a tetrahedral intermediate **43** is formed. If the σ^*_{CX} orbital lies at high energy (σ^*_{CH}, σ^*_{CC}, etc.), it does not interact strongly with O^-, the CX bond is not weakened and the intermediate alcoholate remains stable until it is hydrolysed. However, when X = OR', NH_2, Cl or O–CO–R', the low-lying σ^*_{CX} will interact strongly with O^-. The oxygen being charged, the electron transfer into σ^*_{CX} will far exceed a simple anomeric effect. Consequently, the CX bond may cleave spontaneously, giving **44**. This is why aldehydes and ketones undergo simple additions, whereas esters, amides, acid chlorides and acid anhydrides give addition–elimination reactions.[42]

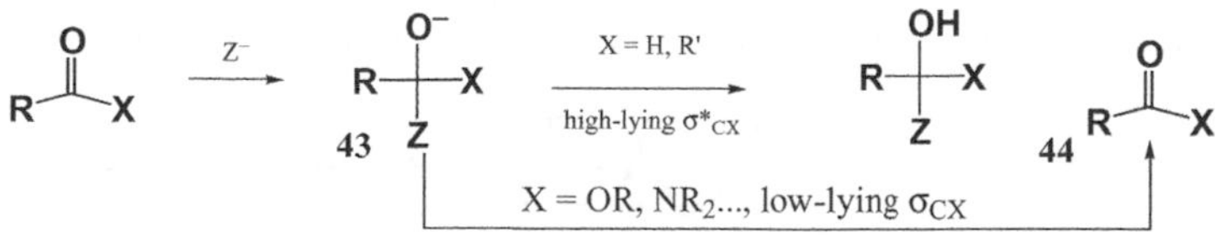

Reversible and Irreversible Reactions. Multistep Mechanisms

Intermediate **43** can also be formed by the addition of X^- to R–CO–Z. It is obvious from the discussion above that additions will be reversible if the reagent is an alkoxide, an amide or a carboxylate, etc. Hydrogenations and alkylations are usually irreversible.

---

[42]Obviously, we can also rationalize these facts by saying that RO, Cl, RCO_2, etc. are better leaving groups than H or R. However, it is always useful to look at a problem from different angles. For example, the present treatment can explain much more easily why the Meerwein–Ponndorf reaction is reversible (see below).

This approach is fairly general. Let us assume that a reaction product contains a bond which is sufficiently weak to break spontaneously. If this is the bond formed by the incoming reagent, the reaction will be reversible. If it is one of the bonds present in the starting material, the reaction will evolve to give a new product and we have a multi-step mechanism.

7.5.3 Substituent Effects

Fragmentations, Enolizations and Related Reactions

Cleaving CC bonds

The stability of **43** is a function of the σ^*_{CX} orbital energy. In the previous section, we accepted that a σ^*_{CO} orbital is sufficiently low-lying to allow spontaneous CO bond cleavage, but that σ^*_{CC} is high enough that a CC bond is normally stable. This over-simplified explanation requires some elaboration. The σ^*_{CO} orbital does lie lower in energy than σ^*_{CC}, but not by very much. Oxygen is certainly more electronegative but the CO bond is stronger (85.5 kcal mol^{-1} versus 82.5 kcal mol^{-1} for the C–C bond, cf. p. 98). This suggests that it should be possible, using more vigorous conditions, to break CC bonds. Experiments confirm this. Alkylation of a ketone by methyllithium is irreversible because the newly formed CC bond (bold in **45**) is weakened only by a single alkoxide donor. Aldol reactions are reversible because the CC bond is sapped by the simultaneous presence of a donor and an acceptor (**46**).

In general, C–C bonds which undergo fragmentation reactions[43] are substituted by a donor at one end (e.g. O or N) and an acceptor at the other (σ^*_{CX}):

Obviously, the fragmentation occurs most easily when the CC bond is at its weakest, i.e. when the donor lone pair and CX are both antiperiplanar to CC:

[43]Grob C. A., *Angew. Chem. Int. Ed. Engl.*, 1969, **8**, 535; Fisher W., Grob C. A., Sprecher G. v., Waldner A., *Tetrahedron Lett.*, 1979, 1905; Gleiter R., Stohrer W. D., Hoffmann R., *Helv. Chim. Acta*, 1972, **55**, 893.

Exercise 8 (E)

If we require compound **48**, should we use **47a** or **47b** as the starting material?

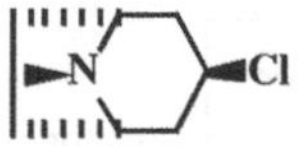

| 47a | 47b | 48 |

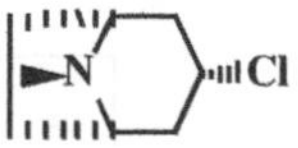

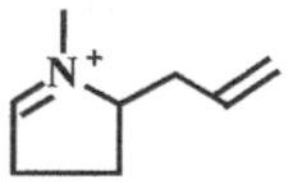

Answer[43]

Only **47a** satisfies the stereoelectronic requirements:

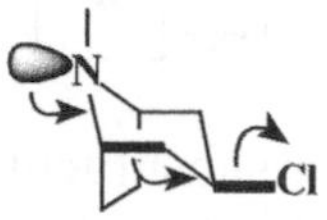

Exercise 9 (M)

Explain the following transformation:

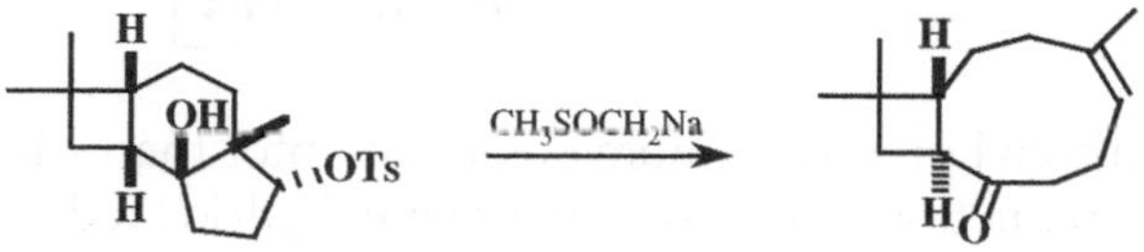

$$CH_3SOCH_2Na$$

Answer[44]

The reaction begins with a fragmentation induced by the proton abstraction, followed by an epimerization α to the ketone:

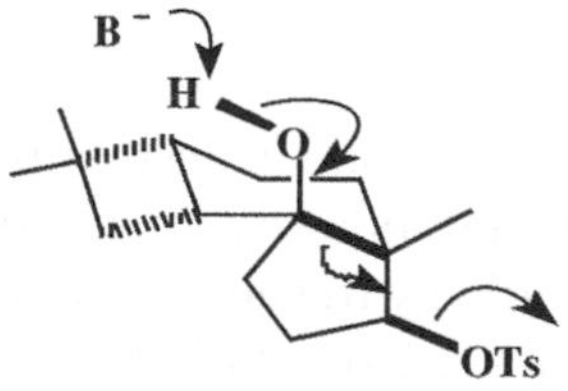

Use molecular models to check that only the Z double bond is formed. Stereoelectronically controlled fragmentations have often been used to provide stereodefined routes to large rings or unsaturated compounds.

Cleaving CH bonds

Substituents on both extremities are necessary to break a CC bond, *a fortiori* to cleave the inherently stronger CH bond. However, hydrogen is monovalent and cannot bear a substituent, so an external reagent is required to play this role. Consequently, *homolytic and heterolytic cleavages of CH bonds are usually bimolecular processes.* When

---

[44]Corey E. J., Mitra R. B., Uda H., *J. Am. Chem. Soc.*, 1964, **86**, 485.

the carbon atom bears an acceptor, the CH bond can be cleaved by an electron donor (i.e. a base) to give C^-H^+. Enolizations fall into this category:

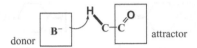

as do expulsions of an α-proton to stabilize carbocations:

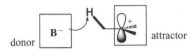

If the carbon has a donor substituent, we need an electrophile to attack the hydrogen. This provokes a hydride transfer (cf. the Cannizzaro and Meerwein–Ponndorf reactions):

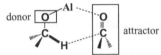

The carbonyl group is a more potent acceptor than AlH_3 or BH_3, so the Meerwein–Ponndorf reduction is reversible but metal–hydride reductions are not.

Exercise 10 (D)   *Deserves your attention*

Seyden-Penne and co-workers[45] studied the Horner–Emmons reaction at conjugated carbonyls. They found that 1,4-additions are irreversible whereas 1,2-additions are reversible, more so with Li than K enolates. Explain these results.

Answer

In 1,2-addition, the C–Nu bond is destabilized by O^- (a strong donor) and the double bond (simultaneously a donor and an acceptor). In the conjugate addition, the C–Nu bond is destabilized only by the enolate, a moderate donor.

The 1,2-product exists as two equilibrating forms, one where the cation is associated with O^- (which is stable because the donor power of the alkoxide is attenuated) and one where it is associated with the nucleophile, a leaving group. The latter form undergoes the cleavage, which is promoted better by Li^+ than K^+.

Exercise 11 (E)

The thermal ring opening of bicyclo[3.2.0]hepta-3,6-dien-2-one **A** to give tropolone **B** is forbidden (disrotatory opening of a cyclobutene). So, in spite of the strain in the starting material, it only occurs at 350 °C.

[45]Deschamps B., Lefebvre G., Redjal A., Seyden-Penne J., *Tetrahedron*, 1973, **29**, 2437; Redjal A., Seyden-Penne J., *Tetrahedron Lett.*, 1974, 1733.

Much milder conditions can be employed when substituents are introduced at the ring junctions.[46] Comment the results given in the table below. Each reaction has a half-life of 2 min at the given temperature T.

	R_1	R_2	$T(°C)$
(a)	H	H	350
(b)	NHMe	CN	−75
(c)	NH_2	CN	−50
(d)	NHMe	CO_2H	−44
(e)	OH	CHO	−42
(f)	NHMe	CO_2Me	−33
(g)	OH	CN	0
(h)	NH_2	Ph	4
(i)	OH	CO_2H	10
(j)	OH	CO_2Me	10
(k)	OH	Ph	55

Answer

The donor character of the R_1 substituents increases along the series OH < NH_2 < NHMe. The oxygen lone pair lies at $\alpha + 2\beta$ and the nitrogen at $\alpha + 1.5\beta$. The NHMe lone pair energy is raised by the methyl group. The acceptor character of the R_2 substituents increases approximatively as follows (with the LUMO energy in parentheses): Ph $(\alpha - \beta)$ < CO_2Me $(\alpha - 0.796\beta)$ < CO_2H $(\alpha - 0.789\beta)$ < CN $(\alpha - 0.781\beta)$ < CHO $(\alpha - 0.618\beta)$.

Any bond is weakened by the presence of any substituent, donor or acceptor. The effect is reinforced when one end is substituted by a donor and the other an acceptor

The Cope Reaction

Substitutions at positions 3 and 4. The anionic oxy-Cope reaction

Assuming a concerted mechanism, the Cope transition state can then be modeled by two allyl fragments in interaction:

---

[46]Kobayashi T., Hirai T., Tsunetsugu J., Hayashi H., Nozoe T., *Tetrahedron*, 1975, **31**, 1483.

The rearrangement comprises the breaking of the 3–4 bond (cost ≈ 80 kcal mol^{-1}), the shifting of the double bonds (cost ≈ 8 kcal mol^{-1}) and the formation of the 1–6 bond (liberating ~80 kcal mol^{-1}). Therefore, the activation energy comes mostly from the 3–4 bond breaking.[47] Any substituent put at position 3 or 4 will then weaken the 3–4 bond and facilitate the reaction.

Table 7.1   Activation energies of some Cope reactions (B3LYP/6–31G* calculations)

Activation energy (kcal mol^{-1})	Calculated[48]	Experimental
1,5-Hexadiene	33.2	33.5 ± 0.5[49]
3-Phenyl-1,5-hexadiene	28.4	28.1 ± 0.4[50]
1,3-Diphenyl-1,5-hexadiene	30.2	30.5 ± 0.2[51]
1,4-Diphenyl-1,5-hexadiene	29.2	29.9 ± 0.2[52]
2,4-Diphenyl-1,5-hexadiene	26.7	24.6 ± 0.8[50]
1,3,5-Triphenyl-1,5-hexadiene	29.2	27.8 ± 0.2[51]
1,3,4,6-Tetraphenylhexadiene	19.1	21.3 ± 0.1[53]

The effect is naturally more pronounced if the substituent is charged. The best known example is the so-called 'anionic oxy-Cope' reaction,[54] where an O$^-$ at position 3 induces a rate acceleration of 10^{10}–10^{17}-fold, a result generally attributed to the 3–4 bond weakening.[47,55] The lone pairs of HN$^-$ lie at even higher energies than those of O$^-$. The 3–4 bond breaking then becomes so easy that the 'anionic amino-Cope' reaction occurs by a stepwise mechanism.[56]

As shown in Table 7.1, B3LYP/6–31G* calculations reproduce adequately the activation energies of Cope reactions. These calculations confirm (Table 7.2, $\Delta H^\ddagger$ in kcal

[47]Steigerwald M. L., Goddard W. A. III, Evans D. A., *J. Am. Chem. Soc.*, 1979, **101**, 1994; Delbecq F., Anh N. T., *Nouv. J. Chim.*, 1983, **7**, 505. In some late transition states, the formation of the 1–6 bond can reduce appreciably the activation energy. We shall return to this point later.

[48]Hrovat D. A., Chen J., Houk K. N., Borden W. T., *J. Am. Chem. Soc.*, 2000, **122**, 7456.

[49]Doering W. v. E., Toscano V. G., Beasley G. H., *Tetrahedron*, 1971, **27**, 5299.

[50]Dewar M. J. S., Wade L. E., *J. Am. Chem. Soc.*, 1977, **99**, 4417.

[51]Doering W. v. E., Yang Y., *J. Am. Chem. Soc.*, 1999, **121**, 10112.

[52]Doering W. v. E., Birladeanu L., Sarma K., Teles J. H., Klärner F.-G., Gehrke J. -S., *J. Am. Chem. Soc.*, 1994, **116**, 4289.

[53]Doering W. v. E., Birladeanu L., Sarma K., Blaschke G., Scheidemantel U., Boese R., Benet-Bucholz J., Klärner F.-G., Gehrke J.-S., Zinny B. U., Sustmann R., Korth H.-G., *J. Am. Chem. Soc.*, 2000, **122**, 193.

[54]Evans D. A., Golob A. M., *J. Am. Chem. Soc.*, 1975, **97**, 4765.

[55]Evans D. A., Baillargeon D. J., Nelson J. V., *J. Am. Chem. Soc.*, 1978, **100**, 2242; Ahlgren G., *Tetrahedron Lett.*, 1979, 915; Gajewski J. J., *J. Am. Chem. Soc.*, 1991, **113**, 967.

[56](a) Gajewski J. J., Gee K. R., *J. Am. Chem. Soc.*, 1991, **113**, 967; (b) Sprules T. J., Galpin J. D., Macdonald D., *Tetrahedron Lett.*, 1993, **34**, 247; (c) Scialdone M. A., Meyers A. I., *Tetrahedron Lett.*, 1994, **35**, 7533; (d) Lee J. K., Gajewski J. J., *J. Org. Chem.*, 1996, **61**, 9422; (e) Yoo H. Y., Houk K. N., Lee J. K., Scialdone M. A., Meyers A. I., *J. Am. Chem. Soc.*, 1998, **120**, 205.

Table 7.2 Influence of substituents on the activation energy and transition structure of the Cope reaction

Substituent	$\Delta H^{\ddagger}$ (calc.)	3–4 bond	1–6 bond
None	33.2	1.965	1.965
3-O⁻	9.9	2.330	3.280[55e]
3-CN	29.3	2.082	2.131[57]
3-CH=CH$_2$	28.1	2.079	2.150[57]
3-C$_6$H$_5$	28.4	2.062	2.122[48]
1-CN	35.5	2.131	2.082
1-CH=CH$_2$	37.1	2.150	2.079
1-C$_6$H$_5$	36.2	2.122	2.062
2-CN	28.0	1.825	1.825
2-CH=CH$_2$	30.4	1.779	1.830
2-C$_6$H$_5$	30.4	1.821	1.837

mol⁻¹, bond lengths in Å) that a substituent at position 3 lowers the activation energy. The transition states become looser and dissymmetric: compared with the parent transition state, the 3–4 bond has lengthened and the 1–6 bond even more. Note that the reaction is facilitated whether the substituent is a donor (O⁻), an acceptor (CN) or a conjugating group (vinyl, phenyl).

Substitutions at other positions

A donor (acceptor) substituent at position 1 will decrease the 1-coefficient and increase the 2-coefficient in the π_{12} orbital (in the $\pi_{12}{}^*$ orbital). It will therefore (slightly) assist the 3–4 bond breaking and impede the 1–6 bond making. An unsaturated substituent stabilizes the starting material, but the conjugation is lost when the double bond shifts into the 2–3 position, so the net result will be an increase of the activation energy. Table 7.2 shows that a cyano group at position 1 raises the activation energy by 2.3 kcal mol⁻¹, a vinyl group by 3.9 kcal mol⁻¹ [57] and a phenyl by 3 kcal mol⁻¹ [48]. Note that a Cope reaction interconverts a 3-substituted-1,5-hexadienediene and a 1-substituted-1,5-hexadiene. Therefore, these two compounds have the same transition state. Only the atom numbering differs: the 1–6 bond of the 1-substituted derivative becomes the 3–4 bond of the 3-substituted derivative.

The 3-substituted transition state is earlier than the 1-substituted transition state. For example, when the substituent is a cyano group, the 3–4 bond is shorter (2.082

---

[57]Hrovat D. A., Beno B. R., Lange H., Yoo H.-Y., Houk K. N., Borden W. T., *J. Am. Chem. Soc.*, 1999, **121**, 10529.

versus 2.131 Å), the 1–6 bond is longer (2.131 versus 2.082 Å) and the 5–6 bond is also longer (1.402 versus 1.393 Å). Let us recall that 3-substitution accelerates but 1-substitution slows the Cope reaction. Indeed, a 1-substituted hexadiene is stabilized by conjugation and reacts less readily. On the other hand, a 3-substituted hexadiene is activated by the 3–4 bond weakening. We can also consider this difference as an indication that Cope reactions are facilitated more by the 3–4 bond breaking than the 1–6 bond making. This is easily understood by looking at the curve giving the bond energy as a function of the bond length. The energy cost for breaking a bond is higher at the beginning, when the bond stretching starts. Afterwards, it becomes easier and easier. Conversely, the energy gain obtained in bond making is initially rather small and builds up later on. In other words, *any effect favoring the 1–6 bond making is felt only if the transition state is sufficiently late.*

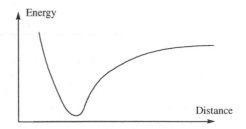

Analysis of 2-substitution is much simpler. A donor (attractor) substituent at position 2 will decrease the C_2 coefficient at and increase the C_1 coefficient in the π_{12} ($\pi_{12}{}^*$) orbital. It will therefore weaken slightly the 3–4 bond and favor significantly the 1–6 bond making. Moreover, conjugation is not destroyed by the double bond shifting from the 1–2 to the 2–3 position. Hence, overall, 2-substitution is expected to facilitate the Cope reaction. According to B3LYP/6–31G* calculations, a phenyl group at position 2 lowers the activation energy by 2.8 kcal mol^{-1},[48] a CN group by 5.2 kcal mol^{-1} [57] and a vinyl group by 2.8 kcal mol^{-1}.[57] Note that the activation energy for 2-cyano-1,5-hexadiene is lower than that for 3-cyano-1,5-hexadiene. In line with the remark above, its transition state is a late one.

Because the 3–4 bond is less weakened and the 1–6 bond formation is facilitated, 2-substituted transition states are more symmetrical than 1- or 3-substituted transition states. This conclusion can also be reached using another argument. If, following Hrovat *et al.*,[57] we assume that the concerted transition state is a resonance between the limiting forms **A** in which the 3–4 and 1–6 bonds are completely made, and **B** in which these bonds are completely broken:

it is clear that substitutions at positions 1, 3, 4 and 6 will stabilize the **B** form and lead to loose and dissymmetric transition states. On the other hand, 2- or 5-

substitutions will stabilize the **A** form and give tighter and more symmetrical transition states.[58]

Polysubstitutions

From the analyses above, we can qualitatively predict whether two substituents will reinforce or oppose each other (see Exercise 13, p. 220).

In some cases, however, the result is far from obvious. We have seen that a vinyl group at position 1 or 6 raises the activation energy by 3.9 kcal mol^{-1}. Hanna and coworkers[59] have observed that compound **C** does not undergo the anionic oxy-Cope reaction. If, however, the isopropyl group is replaced by an isopropenyl group (**C** → **D**), the reaction, instead of being hindered by the terminal double bond, occurs readily and with good yields.

A likely explanation is that the anionic oxy-Cope transition state is polarized and can be modeled by an acrolein (free or as an 'ate' complex) interacting with an allylic anion. The isopropenyl group stabilizes the transition state by delocalizing the negative charge. If this argument is valid, then an attractor substituent put at position 4 or 6 will accelerate the reaction and a donor substituent at the same positions will impede it (see Exercise 12).

---

[58]We should not push this model too far however. Consider, for example, a 1-substituted hexadiene. In the limiting form **B**, the radical site at 1 is stabilized. Yet the transition state is late with the incipient 1–6 bond shorter than the partial 3–4 bond: apparently, the allyl radical 1–2–3 prefers to react by its *less* reactive site!

[59]Gentric L., Hanna I., Ricard L., *Org. Lett.*, 2003, **5**, 1139.

In order to check this interpretation, B3LYP/6–31G*//3–21G calculations of compounds **E–I** were performed and the results are shown below. The 3–4 and 1–6 bond lengths (in Å) are those of the 3–21G transition states. The activation enthalpies (kcal/mol^{-1}) are indicated under the formulae.

$$\text{E } 32.25 \text{ kcal mol} \qquad \text{F } 32.58 \qquad \text{G } 33.20$$

$$\text{H } 34.31 \qquad \text{I } 31.10$$

Comparison of **E** and **F** shows that a 6-vinyl group raises the activation enthalpy by only 0.33 kcal mol^{-1} instead of the 3.9 kcal mol^{-1} found for the Cope reaction. This is consistent with a delocalization stabilization of the negative charge on the fragment 4–5–6. On the other hand, an ethyl substituent at position 6 (**G**) destabilizes the negative charge and should render the rearrangement more difficult. This is indeed the case. These results rationalize Hanna and co-workers' observations, if we take **F** and **G** as models for **D** and **C**, respectively. The effects should be enhanced if Et is replaced by a better donor (e.g. methyl vinyl ether) and the vinyl group by an aldehyde, which is a much better acceptor. These expectations are fulfilled (**H** and **I**). However, more precise calculations are still needed to confirm our interpretation fully.

Exercise 12 (E)

The anionic oxy-Cope reaction is accelerated when a thiomethoxy group is put at positions 4 or 6. A methoxy group at the same positions slows down the reaction. Explain.

Answer[60]

The transition state can be modeled by an acrolein interacting with an allylic anion. A thioalkoxy substituent at position 4 or 6 stabilizes the allylic anion and favors the reaction. In fact, the 3–4 bond rupture becomes so easy that the reaction then occurs by a stepwise mechanism. MeO, being a donor, destabilizes the anion and impedes the reaction.

Exercise 13 (M)

Predict the outcomes of 1,3-, 1,4-, 2-4-, 2,5-, 3,3- and 3,4-disubstitutions for the Cope reaction.

---

[60]Haeffner F., Houk K. N., Ravindra Reddy Y., Paquette L. A., *J. Am. Chem. Soc.*, 1999, **121**, 11880.

Answer

Following Hrovat *et al.*,[57] we consider the Cope transition state to be a resonance structure of the limiting forms **A** and **B** (see p. 218). Synergy is then expected for substitutions at positions 1, 3, 4 and 6, which all favor **B**, and for substitutions at positions 2 and 5, which favor **A**. For 1,3-, 1,4-, 2,5-, 3,3- and 3,4-disubstitutions, the net effect should be larger than the sum of monosubstitution effects. Most probably 1,4-disubstitution weakens the 3–4 bond more than 1,3-disubstitution (the two substituents, one direct, the other a vinylogue, are both borne by atom 3).

2- and 4-substituents have opposite effects, the former favoring the resonance form **A** and the latter form **B**. The net result should be less than the sum of two monosubstitutions.

These deductions are in line with the results published for the phenyl[57] and vinyl[48] substituents. The agreement is not so good for the cyano substituent.[48]

Exercise 14 (E)

Compound **A** is stable in the solid state and in hexane at 20 °C. However, it is rapidly transformed into **B** by traces of BF_3.[61] Why does BF_3 promote the conversion of **A** into **B** but not the reverse transformation of **B** into **A**?

Answer

BF_3 reacts with the carbonyl group to form an 'ate' complex of **A**, whose powerful acceptor properties promote the Cope rearrangement. The reverse reaction does not occur because the 'ate' complex of **B** cannot influence the 3 or 4 position.

Exercise 15 (M)

(1) Explain why the four compounds below rearrange at very different rates and predict the effects of the solvent and the cation. Bear in mind that the 'oxy-Cope' transformation of **D** is performed in the presence of crown ethers.[62]

[61]Cookson R. C., Hudec J., Williams R. O., *Tetrahedron Lett.*, 1960, 29.
[62]Berson J. A., Walsh E. J. Jr, *J. Am. Chem. Soc.*, 1968, **90**, 4730; Evans D. A., Golob A. M, *J. Am. Chem. Soc.*, 1975, **97**, 4765.

(2) Temperatures above 140 °C are necessary for transposing **E**. So why does **F** rearrange at −15 °C?[63]

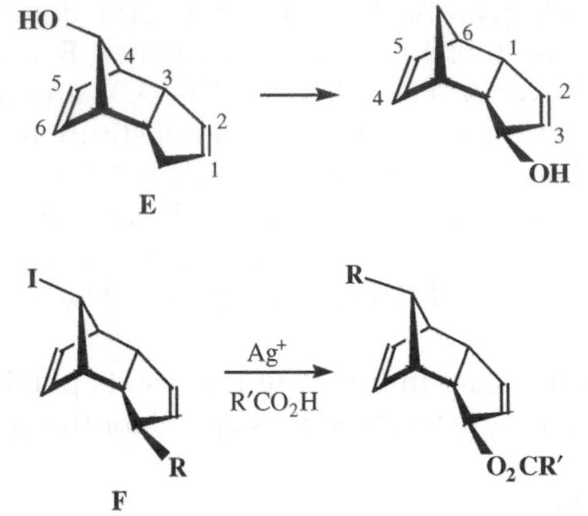

E

F

Ag⁺ / R′CO₂H →

Answer

(1) The reaction is facilitated by weakening of the 3–4 bond, which increases as the donor character of the 4-substituent rises: **B** < **C** < **D**. The rearrangement rate for **C** should also increase along the series Li⁺ < Na⁺ < K⁺ because the interaction with the alkoxide diminishes as the cation increases in size. The same argument suggests that increasing solvent basicity will facilitate rearrangement of **C**. Solvent effects should be nonexistent for **A** and small for **B**. Acidic solvents should disfavor **D**.

(2) The Ag⁺-induced heterolytic cleavage of the C–I bond generates a cationic intermediate which weakens the C_3–C_4 bond. Therefore, the cation rearranges easily, captures a solvent molecule and gives the product.

The Claisen Reaction

By analogy with the Cope reaction, can we predict substituent effects on the Claisen reaction?

A donor substituent at C_4 will weaken noticeably the CO bond and lower the activation energy of the reaction:

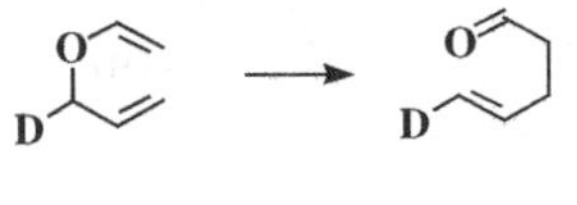

The Claisen transition state can be modeled by an enolate interacting with an allylic cation.[64] Atom 1 will then bear a negative charge and atoms 4 and 6 positive charges.

---

[63]Breslow R., Hoffmann J. M. Jr, *J. Am. Chem. Soc.*, 1972, **94**, 2111.
[64]Coates *et al.* (ref. 66b) used a similar model.

A donor substituent at position 4 or 6 will favor the reaction and an attractor similarly placed will hinder it. Conversely, an attractor (donor) at position 1 stabilizes (destabilizes) the enolate fragment. As atoms 2 and 5 bear no charge, the electronic character of the substituent has little consequence. By analogy with the Cope reaction, we expect that a substituent at these positions also accelerates the Claisen reaction.

These predictions are roughly verified. A CN group at position 6 reduces the rate 10-fold.[65] A methoxy group at position 4 or 6 accelerates the reaction 96- and 9.5-fold, respectively.[66] Generally, a donor substituent at positions 2, 4 and 6 lowers the activation energy. Thus,[67] compounds **A** rearrange at 100 °C whereas compounds **B** react at 25–60 °C:

Some experimental results show that our analysis is incomplete. Thus, a 5-methoxy substituent *slows* the reaction 40-fold.[66,70] If however, the thermodynamic factor[68] (a carbonyl is more stable than an enol ether[69]) is taken into account, most of the 'anomalous' results can be rationalized.[70] Let us clarify this point by discussing the data summarized in Table 7.3 (6–31G* calculations, energies in kcal mol^{-1}, bond lengths in Å). Relative values are given in parentheses. The fourth column gives the variations of activation energy due to the thermodynamic factor (i.e. to the reaction exothermicity), evaluated according to Marcus theory.[70a]

A priori, a 1-OH substituent will have a rather small electronic effect. A donor group placed on the negatively charged part of the transition state is unfavorable, but it populates the σ_{CO}* orbital, which is favorable. According to Yoo and Houk,[70] the electronic effect lowers the activation energy only by 0.4 kcal mol^{-1}. The remaining 2.3 kcal mol^{-1} are due to the thermodynamic factor, an α-hydroxyketone being more stable than an enol ether:

---

[65]Burrows C. J., Carpenter B. K., *J. Am. Chem. Soc.*, 1981, **103**, 6983.

[66](a) Curran D. P., Suh Y., *J. Am. Chem. Soc.*, 1984, **106**, 5002; (b) Coates R. M., Rogers B. D., Hobbs S. J., Peck D. R., Curran D. P., *J. Am. Chem. Soc.*, 1987, **109**, 1160.

[67]Ireland R. E., Mueller R. H., Willard A. K., *J. Am. Chem. Soc.*, 1976, **98**, 2868.

[68]Yamabe S., Okumoto S., Hayashi T., *J. Org. Chem.*, 1996, **61**, 6218.

[69]More precisely, the Claisen reaction comprises a replacement of a C=C double bond (energy = 146–151 kcal mol^{-1}, according to Smith M. B., March J., *March's Advanced Organic Chemistry*, 5th ed., John Wiley & Sons, Inc., New York, 2001, p. 24) by a C=O double bond (173–181 kcal mol^{-1}) and of two C–O bonds (85–91 kcal mol^{-1}) by two C–C bonds (83–85 kcal mol^{-1}). Therefore, it should be exothermic by ~20 kcal mol^{-1}.

[70](a) Yoo H. Y., Houk K. N., *J. Am. Chem. Soc.*, 1997, **119**, 2877; (b) Aviyente V., Yoo H. Y., Houk K. N., *J. Org. Chem.*, 1997, **62**, 6121.

Table 7.3 Calculated (6–31G*) activation energies, reaction energies and transition geometries of some Claisen reactions

Compound	$\Delta H^{\ddagger}$	ΔE_{rxn}	$\Delta\Delta E_{thermo}^{\ddagger}$	3–4 bond	1–6 bond
Parent	46.9 (0)	−21.1(0)	(0)	1.918	2.266
1-OH	44.3 (−2.7)	−26.2(−5.1)	(−2.3)	1.937	2.261
2-OH	38.2 (−9.1)	−39.2 (−18.1)	(−7.8)	1.878	2.334
4-OH	45.9 (−1.0)	−15.2 (+5.9)	(+2.8)	1.998	2.340
5-OH	51.7 (+5.0)	−22.3 (−1.2)	(−0.8)	1.945	2.293
6-OH	46.0 (−0.6)	− 17.9 (3.2)	(+1.5)	2.016	2.330

A carboxylic acid is even more stable than an α-hydroxyketone, which is why the thermodynamic factor accounts for 86% of the activation energy lowering of 2-hydroxyallyl vinyl ether (2-hydroxy-AVE):

highly exothermic

On the other hand, the ease of reaction of 4-hydroxy-AVE is certainly due to the weakening of the 3–4 bond, as the thermodynamic factor is now unfavorable. Similarly, a 6-OH substituent is beneficial for electronic, not thermodynamic, reasons. The most interesting – and apparently still unexplained – problem is the 5-substituted case. By analogy with the Cope reaction, we expect 5-substitution to facilitate also the Claisen reaction.[71] Now, according to RHF/6–31G* calculations, an OH put at position 5 *raises* the activation energy by 4.8 kcal mol^{-1}.[70a] Experimental measures indicate that the rate is reduced 40-fold by a 5-methoxy group[66] and increased 5-fold by a 5-CN group.[65]

We would like to suggest the following explanation. A 5-substituent favors 1–6 bond formation (cf. p. 218). Now, when AVE takes the reactive chair conformation, the 1,2-double bond orbitals are raised by conjugation with the oxygen lone pairs, as confirmed by 6–31G* calculations:

In 5-cyano-AVE, the 5–6 double bond is electron poor and reacts readily with the electron-rich 1–2 double bond, thus facilitating 1–6 bond formation. In 5-methoxy-AVE, both double bonds are electron rich. The 1–6 bond formation is then disfavored and

---

[71]In fact, MNDO and AM1 calculations predict an acceleration: Dewar M. J. S., Healy E. F, *J. Am. Chem. Soc.*, 1984, **106**, 7127 ; Dewar M. J. S., Jie C., *J. Am. Chem. Soc.*, 1989, **111**, 511.

the Claisen reaction rendered more difficult.[72] According to 6–31G* calculations, the HOMO–LUMO energy gap is 13.84 eV in the parent AVE, 12.73 eV in 5-cyano-Ave and 14.05 eV in 5-methoxy-AVE.

Exercise 16 (M)

Why do Lewis acids promote Claisen reactions? Suggest another method for accelerating aza- and thia-Claisen rearrangements.

Answer

Their effect resembles electrophilic assistance (p. 61). Coordination of the oxygen atom to the Lewis acid increases its effective electronegativity, thus lowering the energy of its valence orbitals. Lowering σ^*_{CO} allows a stronger interaction with the neighboring π orbitals. In turn, this facilitates the reaction:

Thus, the thermal rearrangement of **A** only occurs after prolonged heating at 180–200 °C but yields of 88% can be achieved even at −78 °C upon addition of $TiCl_4$.[73] Other Lewis acids (Et_2AlCl, BCl_3) also increase the rate constant by factors of up to 10^{10}.[74]

If this interpretation is correct, onium ions should rearrange more rapidly than their neutral parents. Indeed, sulfonium ion **B** rearranges at room temperature, whereas the parent **C** is unchanged after refluxing in toluene.[75] **D** and **E** rearrange at 250 °C[76] and 80 °C,[77] respectively.

---

[72]Admittedly, the transition state is polarized and the 4–5–6 fragment bears a partial positive charge. However, as the substituent is put at position 5, which is a node for the allyl cation, the donor or attractor character of the substituent has little influence on the stability of the cation.

[73]Narasaka K., Bald E., Mukaiyama T., *Chem. Lett.*, 1975, 1041.

[74]Borgylya J., Madeja R., Fahrni P., Hansen H. J., Schmid H., Barner R., *Helv. Chim. Acta*, 1973, **56**, 14; Sonnenberg F. M., *J. Org. Chem.*, 1970, **35**, 3166.

[75]Bycroft B. W., Landon W., *Chem. Commun.*, 1970, 967.

[76]Hill R. K., Gilman N. W., *Tetrahedron Lett.*, 1967, 1421.

[77]Brannock K. C., Burpitt R. D., *J. Org. Chem.*, 1961, **26**, 3577.

Substituted Cyclopropanes[78]

A monosubstituted cyclopropane will take the form of an isosceles triangle, but will it be elongated or flattened, and will the nature of the substituent influence the geometry?

Let us assume an acceptor substituent, such as an aldehyde. The strongest interaction will occur between the aldehyde LUMO and the cyclopropane HOMO[79] and cause a transfer of electron density from the ring to the substituent. The HOMO + LUMO occupied combination is shown below. The HOMO is bonding between C_1 and C_2 and between C_1 and C_3, but antibonding between C_2 and C_3. Rule 7 then states that the C_1–C_2 and C_1–C_3 bonds will be lengthened and the C_2–C_3 bond shortened. Hence the triangle will be elongated.

The geometries of **49**[80] and **50**[81] confirm this prediction.

In the case where the substituent is a donor, its HOMO interacts mainly with the cyclopropane Ψ_4 orbital (Figure 7.4, p. 196). Ψ_4 is antibonding between C_1 and C_2 and between C_1 and C_3, but bonding between C_2 and C_3. The cyclopropane flattens, but the effect is smaller, Ψ_4 being more antibonding than Ψ_3 is bonding.

The study above has some applications. Normally, the equilibrium between cycloheptatriene (**51**, R = H) and norcaradiene (**52**, R = H) is strongly displaced in favor of the cycloheptatriene. To favor the norcaradiene, the electrocyclic ring opening must be inhibited by strengthening the 1–6 bond. Our analysis shows that this may be achieved by incorporating acceptor groups R. Donor R groups shift the equilibrium to the left.[82]

---

[78]Hoffmann R., *Tetrahedron Lett.*, 1970, 2907; Günther H., *Tetrahedron Lett.*, 1970, 5173; Hoffmann R., Stohrer W. D., *J. Am. Chem. Soc.*, 1972, **94**, 779.
[79]The latter has two degenerate MOs (Figure 7.4, p. 196) but the Ψ_2 orbital does not interfere because it has a zero coefficient at the substituted carbon.
[80]Fritchie C. J. Jr, *Acta Crystallogr.*, 1966, **20**, 27.
[81]Meester M. A. M., Schenk H., MacGillavry C. H., *Acta Crystallogr., Sect. B.*, 1971, **27**, 630.
[82]Ciganek E., *J. Am. Chem. Soc.*, 1965, **87**, 652, 1149; 1967, **89**, 1454; 1971, **93**, 2207.

Exercise 17 (D)

Predict the relative rates for the Cope rearrangement of bullvalene **A**, barbaralone **B**, protonated barbaralone **B′**, barbaralane **C** and octamethylsemibullvalene **D**.

Answer[83]

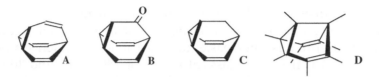

These compounds have a common motif:

When R and R′ are donors (acceptors), the 3–4 bond will be weakened (strengthened) and the Cope rearrangement facilitated (retarded). This allows four of the compounds to be put into order immediately. Rearrangement rates will increase:
B′ < **B** (One acceptor) < **C** (one donor) < **D** (two donors)
Any protonation of **B** will inhibit the rearrangement: a protonated carbonyl group is a powerful acceptor, whereas protonation at the double bond prevents its participation in the rearrangement. The only compound which is difficult to evaluate is **A**. A double bond can either release electrons from its π orbital or accept them into its π^* orbital. Cyclopropane has high-lying HOMOs, so the (deactivating) acceptor character of the double bond dominates. This stabilizes bullvalene, which is found to be less reactive experimentally than dihydrobullvalene. Hence **A** is clearly less reactive than **C** or **D**. However, why it should be less reactive than **B** is much less obvious (the activation energies are 12.8 kcal mol^{-1} for **A**, 9.6 kcal mol^{-1} for **B**, 7.8 kcal mol^{-1} for **C** and 6.4 kcal mol^{-1} for **D**), given that C=O is a *better* acceptor than C=C. The higher reactivity in barbaralone probably results from greater ring strain, hence it reflects a destabilized starting material rather than a stabilized transition state.

Exercise 18 (E)

Predict the changes in the following equilibrium as a function of R:

[83]Hoffmann R., Stohrer W. D., *J. Am. Chem. Soc.*, 1971, **93**, 6941.

Answer

The equilibrium will be move to the right if R is a donor, to the left for an acceptor.

7.6 Abnormal Valence Angles

Angular distortions do not require much energy so they are often fairly large. (Remember that infrared stretches are at much higher frequencies than bends). This is useful. It is much easier to notice abnormal valence angles (which may vary by up to 15° from the classical values of 180°, 120° and 109°28′) on a computer printout than unusual bond lengths, which are inherently more variable (1.09 Å for CH to 2.14 Å for CI). However it may also cause problems: large effects may be produced by small energy differences, so the utmost care should be taken when interpreting these results.

Applying rule 8 poses no real problems. On p. 208, for example, we saw how the stability of $CBFe(CO)_3$ results from interactions between the SOMOs of CB and the $Fe(CO)_3$ fragment. Any deformation which increases their overlap is favorable. Specifically, if the hydrogen atoms of the cyclobutadiene are pushed out of the CB plane and away from the iron, rehybridization causes the large lobe on carbon to be directed toward the metal. The C–Fe interaction increases and stability rises:

The validity of this interpretation is strengthened by an analysis of $TMMFe(CO)_3$. In this compound, the best way to increase stabilization is to draw the CH_2 groups closer to the $Fe(CO)_3$ by pyramidalizing the TMM. This is indeed observed. It is interesting to note that (a) this effect increases steric interactions between the TMM and $Fe(CO)_3$ fragments and (b) it is the reverse of the distortion observed in the cyclobutadiene complex.

Other kinds of deformation may occur (the cyclobutadiene may fold along a diagonal,[84] for example) but rule 8 holds in most cases.

---

[84]d'Angelo J., Ficini J., Martion S., Riche C., Sevin A., *J. Organomet. Chem.*, 1979, **117**, 265.

Exercise 19 (E)

Which of the following has (a) the smallest and (b) the largest barrier to rotation about the CHO group?

Answer

The strongest FO interaction occurring between the aryl and CHO fragments will involve the HOMO of X–Ph and the LUMO of CHO. As the HOMO energy rises, the interaction strengthens and the barrier to rotation increases. The HOMO is raised by donor ring substituents (such as MeO) and lowered by acceptors (NO$_2$).

   A similar rationale shows that the barrier to rotation of a methyl group should lower along the series: hydroxyacetone > acetone > fluoroacetone.[85] Here, the dominant interaction involves the carbonyl LUMO and the methyl group HOMO. As the LUMO energy rises, the barrier to rotation falls.

Exercise 20 (M)

Condensation of benzaldehyde with lithium enolates **53** and **55** gives two diastereomeric aldols **54** and four diastereomers of **56**. Why is only **56** obtained when M = K?

Answer[86]

For a C–C bond to cleave, there must be an acceptor substituent at one end and a donor at the other. The potassium aldolates form highly dissociated ion pairs and cleave easily. They give the thermodynamically controlled aldol products. When using lithium salts, the donor character of the O$^-$ function is reduced and the retroaldol process is retarded. Hence, the kinetic products are also observed.

---

[85]Radom L., Baker J., Gill P. M. W., Nobes R. H., Riggs N. V., *J. Mol. Struct.*, 1985, **126**, 271.
[86]Duhamel P., Cahard D., Quesnel Y., Poirier J. M., *Journées de Chimie Organique*, Palaiseau, 12–15 September 1995, Poster No. A241.

Exercise 21 (E)

The complex formed between benzene and Ag^+ adopts geometry **A** whereas the complex with I_2 has structure **B**. Offer an explanation.

Answer[87]

Ag$^+$ has a low-lying LUMO, so it interacts strongly with one of the relatively high-energy benzene HOMOs. $\sigma^*(I_2)$ lies at much higher energy, so it overlaps better with the Ψ_1 benzene orbital.

---

[87]Hudson R. F., *Angew. Chem. Int. Ed. Engl.*, 1973, **12**, 36.

8 Going Further

A theoretical chemist is not a mathematician, thinking mathematically, but a chemist, thinking chemically.

C. A. Coulson

Never do any calculations before you know the results.

J. A. Wheeler

We will say that we understand a molecular wave function when we can qualitatively predict the shape (sign and size of coefficients) of every molecular orbital of the molecule prior to doing the calculation.

R. Hoffmann

FO theory is only an *exploratory* method. A frontier orbital study should always be validated by experiments or theoretical investigations using more rigorous approaches. This is not really a problem: computer programs are increasingly user-friendly and computer power is rising exponentially. At the present time (2006), a program allowing fairly sophisticated calculations on a desktop computer costs only ~1000€. Naturally, an experimentalist does not become an expert in computational chemistry overnight, but a long training is not necessary either to benefit from this powerful tool. A similar situation exists in spectroscopy, where it takes many years to become a specialist, but 5 minutes to learn to distinguish between the IR spectra of a ketone, an ester, an acid and an aldehyde. Experimentalists have an advantage over computational chemists: they know their chemistry much better. And two essential steps in resolving a theoretical problem, the choice of an acceptable model and the interpretation of the results, require much more chemical than mathematical understanding.

So, our question should not be `can we go any further' but `when and how far do we wish to go?' To know *when*, we must define the limits of the FO approach.[1] To know *how*, we need more information concerning the available methods. Only some indications aimed at beginners are given here. For more detailed explanations and discussions, see the recommended books at the end of this chapter. Suffice it to say now that there is no one method that is ideal for all applications. The best way to learn computational chemistry is still to put one's shoulder to the wheel!

---

[1]Anh N. T., Maurel F., *New J. Chem.*, 1997, **21**, 861.

Frontier Orbitals   Nguyên Trong Anh
© 2007 John Wiley & Sons, Ltd

8.1 The Limits of Frontier Orbital Theory

8.1.1 The Simplifying Hypotheses of Frontier Orbital Theory

There are five of these:

1. All interactions occurring between two filled orbitals are neglected in studies of reactivity. HOMO–HOMO interactions are, however, crucial in conformational analysis.
2. The only interactions between filled and vacant orbitals which require consideration are those involving the FOs. In ionic reactions, the major interaction occurs between the nucleophile HOMO and the electrophile LUMO. In radical reactions, the SOMO plays the role of a HOMO, or a LUMO, or both.
3. It is essential that each reagent can be described accurately by a single electronic configuration. If not, the FOs cannot be determined unequivocally.
4. FO theory deals with the frontier orbitals of the transition state. However, in practice, the FOs of the starting materials are employed instead.
5. In principle, FO theory can only apply to bimolecular reactions.

The validity of these approximations and the subsequent limitations are discussed in details in the following sections. Suffice it here to say that:

- FO theory usually gives good predictions because *in the transition state*, the FOs are close in energy and possess large coefficients at the reacting sites.
- FO theory should be restricted to the comparison of the *same reaction* in *similar molecules*.
- Its predictions are generally more reliable for reactivity problems than for structural problems.

8.1.2 Consequences

Limitations Imposed by Approximation (1)

When studying reactivity, it is only possible to compare closely related compounds undergoing the same reaction (p. 114). The reason is that repulsive forces are larger than attractive forces (which is why activation energies are positive), so approximation (1) is vindicated only if, in the systems under study, the repulsive terms remain practically unchanged. At least their variations should be less than those of the frontier terms. This is no longer true if the molecules or the reactions are too different.

Conformational analysis is normally dominated by repulsions (which is why staggered conformations are energetically favored). The conformational energy profile of the HOMO usually has the same form as the total energy curve (Exercise 19, p. 153). This is probably because the overall change is dominated by the electrons having the highest energy.

Limitations Imposed by Approximation (2)

The chemistry of atoms is dominated by the outermost atomic orbitals, which are, so to speak, their `atomic frontier orbitals'. This is not true, however, of transition metals, where the $(n-1)$d orbitals must also be taken into account because they lie very close in energy to the ns and np orbitals. The same kind of problem occurs in molecules. The subjacent or superjacent MOs may determine the outcome of a reaction if they lie close in energy to the frontier orbitals (p. 113) or if they are nonbonding orbitals (Exercise 3, p. 59).

Charge control dominates (a) when the HOMO–LUMO gap is very large (p. 97; see also Exercise 21, p. 230); (b) in molecules whose rigidity prohibits good frontier orbital overlap (p. 156); (c) in very loose transition states (because the Coulomb interaction falls as a function of R^{-1} whereas the frontier orbital overlap falls exponentially with distance); (d) in the gas phase (p. 99).[2] Early transition states (which are not necessarily 'loose') can be governed by either charge or frontier orbital factors.

Conformational control is important (a) when transition state interactions between the reagents are weak (p. 166), (b) in the absence of highly polar substituents (e.g. propanal) and also (c) in reactions with very early (reactant-like) or very late (product-like) transition states.[3]

Limitations Imposed by Approximation (3)

Radical reactions have been discussed on p. 110. Salem suggested that *photochemical reactions* between a excited species A* and a ground-state molecule B are controlled by the interactions SO*(A)–LU(B) and SO(A)–HO(B) where SO* and SO refer to the two singly occupied orbitals generated by the excitation of one electron from the HOMO to the LUMO.[4] The reaction must involve the first singlet excited state and a single electronic configuration. This assumption is not always valid; the second excited state reacts in the electrocyclization of butadiene,[5] for example. Also, FO theory is not applicable to multistep reactions (see below). This means that photosensitized reactions can never be treated; they always involve intersystem crossing.

Frontier orbital theory can sometimes be remarkably efficient in explaining *organometallic chemistry* (pp. 207, 228; see also ref. 6). However, it is sometimes necessary to take the subjacent orbitals into account alongside the five d orbitals. Frontier orbital theory is difficult to operate in these cases because of frequent inversions of orbital

---

[2]Ionic reactions in the gas phase are under charge control. See, e.g.: Faigle J. F. G., Isolani P. C., Riveros J. M., *J. Am. Chem. Soc.*, 1976, **98**, 2049; Comisarow M., *Can. J. Chem.*, 1977, **55**, 171; Fukuda E. K., McIver R. T. Jr, *J. Am. Chem. Soc.*, 1979, **101**, 2498; Bohme D. K., Mckay G. I., Tanner S. D., *J. Am. Chem. Soc.*, 1980, **102**, 407; Jones M. E., Kass S. R., Filley J., Barkley R. M., Ellison G. B., *J. Am. Chem. Soc.*, 1985, **107**, 109; Johlman C. L., Wilkins C. L., *J. Am. Chem. Soc.*, 1985, **107**, 327; Houk K. N., Paddon-Row M. N., *J. Am. Chem. Soc.*, 1986, **108**, 2659; Sanchez Marcos E., Maraver J., Anguiano J., Bertrán J., *J. Chim. Phys.*, 1987, **84**, 765.
[3]Anh N. T., Maurel F., Lefour J. M., *New J. Chem.*, 1995, **19**, 353.
[4]Salem L., *J. Am. Chem. Soc.*, 1968, **90**, 543, 553.
[5]van der Lugt W. T. A. M., Oosterhoff L. J., *J. Am. Chem. Soc.*, 1969, **91**, 6042

energies and the large number of FO interactions involved. Correlation diagrams[7] are usually better adapted to organometallic chemistry and other multi-configurational systems.

The *isolobal analogy*[6,8] can be considered as an extension of FO theory. Rather than treating only the HOMO and LUMO, it focuses upon the more numerous (3–6) valence orbitals of the interacting fragments, thus allowing a more accurate description of transition metal complexes. The isolobal analogy underlines chemical relationships between metal fragments in much the same way as the periodic table highlights similarities between elements.

Limitations Imposed by Approximation (4)

As a reaction evolves from starting material to transition state, the FOs vary in form and energy. This may cause their relative reactivities to change (pp. 76–77), and may also have regiochemical consequences (p. 114).

Let us consider a two-step reaction, where the second is rate limiting. It is clear that the crucial FOs will be those of the intermediate rather than the starting materials. Hence a study of starting material FOs is unlikely to provide an understanding of *multistep reactions*, particularly those under thermodynamic control.

Limitations Imposed by Approximation (5)

In principle, FO theory can only be applied to bimolecular processes, so a trick is used to investigate structural problems. The molecule under study is formally divided into fragments whose recombination is treated as a bimolecular reaction. Results are usually less reliable than in problems of reactivity, for two reasons. The first is the assumption that reaction transition states and the recombination of fragments in an analysis of structure can be adequately described as perturbed forms of the starting materials. This assumption is more realistic when the perturbation involves only a *partial* formation of a bond, thus giving better results in reactivity studies. The second is that predicting a conformational preference requires accurately discerning energy differences of $1–2\,\mathrm{kcal\,mol^{-1}}$ (a typical energy gap between two conformers) when they are superimposed on a bond formation process liberating approximately $100\,\mathrm{kcal\,mol^{-1}}$ (a typical heat of formation for a σ bond). This requires a precision of 2% or better. The energy differences governing problems of regioselectivity ($1–2\,\mathrm{kcal\,mol^{-1}}$) and relative reactivity ($4–5\,\mathrm{kcal\,mol^{-1}}$) are only complicated by the heat liberated by the formation of a *partial* bond ($\sim 50\,\mathrm{kcal\,mol^{-1}}$). Hence the precision required to solve reactivity problems is 2–10 times lower. See also Exercise 3, p. 59.

[6]Albright T. A., Burdett J. K., Whangbo M. H., *Orbital Interactions in Chemistry*, John Wiley & Sons, Inc., New York, 1985.
[7]Salem L., *Electrons in Chemical Reactions*, John Wiley & Sons, Inc., New York, 1982, Chapter 5.
[8]Hoffmann R., *Angew. Chem. Int. Ed. Engl.*, 1982, **21**, 711.

8.2 The Capabilities of Computational Chemistry

8.2.1 Structural Problems

Quantum chemistry[9] can be used to calculate bond lengths and angles, dihedral angles, the equilibrium geometry, the relative stability of conformations, etc. It is equally easy to study isolated molecules or aggregates (acid–base complexes, charge-transfer complexes, hydrogen bonds), classical or nonclassical structures (e.g. trimethylenemethane, which cannot be represented by a Kekulé structure, or carboranes, where every boron or carbon atom is bound to six neighbours) or even molecules which have never been synthesized (e.g. fenestranes, which contain planar tetravalent carbon atoms). Quantum chemistry has many successes to its credit in these areas. To cite just one example, carbene had been assigned a linear structure on spectroscopic grounds. Theoretical studies gave bent structures and prompted further spectroscopic work, which confirmed the nonlinear structure.[10]

Semi-empirical methods[11] are usually less reliable for charged species and radicals.

8.2.2 Reactivity Problems

Probably the major contribution of computational chemistry is the determination of the structures and energies of transition states, which considerably expands our understanding of reaction mechanisms.

Reactants and products are stable molecules. They are easy to calculate, as efficient algorithms exist for localization of minima on potential surfaces.[12] On the other hand, transition states are harder to obtain. It is necessary to check that the structure obtained is really a saddle-point, i.e. it has one and only one imaginary vibration. If bond breaking and/or bond making occurs in the transition state, this imaginary frequency typically lies in the $400-2000 \, cm^{-1}$ range. Lower frequencies ($<200 \, cm^{-1}$) imply that the transition state corresponds to a deformation of the system, for example a conformational change. A graphical representation of this vibration can be a great help. Better still, the *intrinsic reaction coordinate*[13] should be used to show that this transition state does indeed connect the starting compound with the expected product.

The energy potential surface being rather flat at a saddle-point, transition state geometries obtained by various methods may differ markedly. Usually, this does not

---

[9]Molecular mechanics can be used for many common compounds and is particularly effective in exploring conformer distribution.

[10]Schaefer H. F. III, *The Electronic Structure of Atoms and Molecules*, Addison-Wesley, Reading, MA, 1972, p. 309.

[11]See Section 8.3 for the definitions of various methods.

[12]A potential surface is a graphical representation of the energy of the system as a function of its geometry. For a lucid account on potential energy surfaces, transition states, methods for calculating reaction paths, etc., see Chapter 2 in ref. 7.

[13]Intrinsic reaction coordinate is a steepest-descent pathway which is required to pass through reactant, transition state and product.

prevent effective comparison of closely related reactions. Low-level calculations, e.g. 3–21G, can therefore be used to calculate transition *structures*. However, to obtain reliable *energies*, methods including correlation are mandatory.

8.2.3 Beyond Potential Surfaces

Until now, computational studies of chemical reactions usually stop at the potential surface stage. This is sufficient for routine analyses. More thorough investigations involve first some refinements: kinetic energy is superimposed on the potential energy of the system to provide the internal energy, a PV term is introduced to give the free enthalpy and an entropic contribution is added to obtain the free energy.

$\Delta G^{\ddagger}$ only gives a static description of the reaction. A dynamic study is required to resolve many remaining questions. How do the initial conditions (relative positions and velocities of the reactants) influence the reaction? How should the reagents approach each other in order to achieve a reactive collision? How is the energy of the system divided between electronic, translational, rotational and vibrational components after the collision? Unfortunately, such calculations are difficult and only small systems can be treated by quantum dynamics at present. For more complicated structures, the potential surface is calculated using quantum mechanics and the dynamic aspects are treated using classical mechanics. To illustrate the kind of information that can be obtained from dynamic studies, let us consider the S_N2 reaction:

$$H^- + \overset{H}{\underset{H}{\underset{|}{}}}C-H \longrightarrow H-C\overset{H}{\underset{H}{\overset{|}{}}} + H^-$$

whose potential barrier is ~60 kcal mol⁻¹. The dynamic study shows that a *static* methane molecule will not react with an incoming hydride having an energy of 120 kcal mol⁻¹, even in a perfectly aligned system. However, reaction occurs at energies of 63–64 kcal mol⁻¹ *if at least 40% of this energy is localized within the methane molecule*.[14] The explanation is straightforward. In the first case, the hydride has enough energy to bring on the reaction, but 'rebounds' from the methane molecule before the Walden inversion can occur. In the second case, the vibrations in the methane skeleton are already relatively large, so a small input of energy is sufficient to drive the methane over the potential barrier.

Consider also a reaction having two possible transition states **A** and **B**, where **A** has the lower energy. This system may have a potential surface such as that shown below. Although minimum **B** is at higher energy, it may give rise to the principal product because its wider profile can tolerate a greater range of reactive trajectories, thus allowing more reaction 'events'.

---

[14]Leforestier C., *J. Chem. Phys.*, 1978, **68**, 4406.

8.3 The Methods of Quantum Chemistry

8.3.1 The Approximations

The Schrödinger equation is a second-order partial differential equation, involving a relation between the independent variables x, y, z and their second partial derivatives. This kind of equation can be solved only in some very simple cases (for example, a particle in a box). Now, chemical problems are N-body problems: the motion of any electron will depend on those of the other $N - 1$ particles of the system, because all the electrons and all the nuclei are mutually interacting. Even in classical mechanics, these problems must be solved numerically.

Electrons are much lighter than nuclei (a proton is 1836 times heavier than an electron) and thereby move much faster. In the time it takes the electrons of a molecule to explore the entire space around all the nuclei, these have essentially remained at standstill. This means that when calculating the energies of the electrons, we may treat the nuclei as fixed. This *Born–Oppenheimer* approximation greatly reduces the problem complexity by removing the motion of the nuclei.

The self-consistent field (SCF or *Hartree–Fock*) method replaces the calculation of a multi-electron system by a series of simpler calculations. Let us first take a system containing only the fixed nuclei and *one* electron. This allows us to compute a set of molecular orbitals (which are, by definition, monoelectronic wavefunctions) and distribute the electrons among them. Now choose one electron and find the average field provided by all the others.[15] It is assumed then that our electron will move in this effective electric field. In such a case, the electron will be described by a wave equation containing only its own coordinates, and not the coordinates of any other electron. Solving it gives a first-improved MO for this electron, which can be used to work out a first-improved MO for a second, then a third electron, etc. In the same way, starting with these, we obtain a second-improved set of MOs. This process is repeated until self-consistency, i.e. until the MOs undergo no further changes upon iteration.

The Hartree–Fock equations are the coupled differential equations of the SCF procedure. The *LCAO approximation* transforms these differential equations into an ensemble of algebraic equations, which are substantially easier to solve.

These SCF MOs incorporate average interelectronic repulsion effects, but do not include *instantaneous electron correlation*, which is the tendency for electrons to stay at any time as far removed as possible from each other. Without this correlation, electrons 'get in each other's way' to a greater extent than they should. This is the reason why SCF calculations give energies, especially transition energies, which are much too high.

---

[15]This average field is simply the field that would be provided if each of these electrons was a charge cloud and molecular orbitals allow us to calculate these clouds precisely.

8.3.2 The Principal Theoretical Models

Ab Initio and Semi-empirical Models

Pople refers to a specific set of approximations as defining a *theoretical model*. Hence the *ab initio* or *Hartree–Fock models* employ the Born–Oppenheimer, LCAO and SCF approximations. If the system under study is a *closed-shell* system (even number of electrons, singlet state), the constraint that each spatial orbital should contain two electrons, one with α and one with β spin, is normally made. Such wavefunctions are known as *restricted Hartree-Fock* (RHF). Open-shell systems are better described by *unrestricted Hartree-Fock* (UHF) wavefunctions, where α and β electrons occupy different spatial orbitals. We have seen that Hartree–Fock (HF) models give rather unreliable energies.

Solving the Schrödinger equation using LCAO methods involves the calculation of about n^4 integrals, where n is the number of AOs of the system.[16] Consequently, the number of calculations rises very rapidly with increasing molecular complexity. Every one of these integrals has to be evaluated in *ab initio* calculations, which therefore requires substantial computer time. *Semi-empirical* models (of which AM1 and PM3 are currently the most popular) incorporate two further approximations, reducing the value of n considerably. The first involves taking into account only the valence shells. The second neglects most of the $\langle \varphi_i \varphi_j | r_{12} | \varphi_k \varphi_l \rangle$ integrals. These integrals correspond to the repulsion between the portion of electron 1 localized in the overlapping region of φ_i and φ_j with the fraction of electron 2 in the $\varphi_k \varphi_l$ overlap region, r_{12} being the distance separating these electron clouds. Calculations are facilitated by neglecting three-and four-center integrals (we talk of k-center integrals when the AOs belong to k different atoms) of this type, which are simultaneously the most numerous, the most difficult to calculate and also the smallest. The errors resulting from these approximations are compensated by empirical parameters.

Semi-empirical models, like force field methods, perform best when their parameters are good, that is, for well-known systems. They cannot predict completely new structures. Even for known compounds, their results are not always dependable. Thus, semi-empirical calculations tend to overstress diradical mechanisms in cycloadditions.[17] AM1 and PM3 underestimate frontier interactions with respect to steric repulsions. Basicities of anions are overestimated and their nucleophilicities underestimated. Optimizations may give unreasonable structures. Usually, but not systematically, PM3 gives more reliable structures and AM1 more realistic energies.[18]

Basis Sets

If each AO is represented by only a single mathematical function, we talk of a *minimal* basis set; when it is represented by several functions, we have an *extended* basis set.

---

[16]More exactly, n is the number of basis functions. Therefore, a calculation using extended basis sets is more time consuming than the same calculation with a minimal basis set.

[17]Caramella P., Houk K. N., Domelsmith L. N., *J. Am. Chem. Soc.*, 1977, **99**, 4511; Dewar M. J. S., *J. Am. Chem. Soc.*, 1984, **106**, 209; Brown F. K., Houk K. N., *Tetrahedron Lett.*, 1984, 4609.

[18]Anh N. T., Frison G., Solladié-Cavallo A., Metzner P., *Tetrahedron*, 1998, **54**, 12841.

Thus, each AO has only one coefficient in the LCAO when using a minimal basis set, but several for an extended set. Extended basis set calculations are more accurate (because they have more parameters to adjust), but they are more difficult to interpret, for precisely the same reason.

The minimal STO-3G basis set was used for many years. It employs Slater orbitals (STO = Slater Type Orbital) of the form $P(r)\exp(-\zeta r)$, where $P(r)$ is a polynomial in r. With STO functions, two-electron integrals are difficult to evaluate numerically, so they are replaced each by a sum of three Gaussian functions[19] (hence the acronym STO-3G). The combination of the three Gaussians is treated as a single entity, so each AO has only one coefficient in the LCAO and the base remains minimal in character.

Minimal basis sets can only be used for ʻnormalʼ molecules. They are not flexible enough to describe correctly more special situations such as small-ring compounds or transition states. Increasing its size is an obvious way to improve a basis set. In a *double zeta* (DZ) basis set, each AO is represented by two functions, one ʻinternalʼ function with a large ζ exponent to describe the region near the nucleus, one ʻexternalʼ function with a smaller ζ for the outer region. Using the relative values of the ʻinternalʼ and ʻexternalʼ coefficients, we can depict with equal ease a tight or loose electron cloud. A DZ basis set uses two s functions (1s and 1sʼ) for hydrogen; four s functions (1s, 1sʼ, 2s and 2sʼ) and two p functions (2p, 2pʼ) for first-row elements; six *s* functions and four *p* functions for second-row elements. *Triple zeta* (TZ), *quadruple zeta* (QZ) and even *quintuple zeta* (5Z) basis sets have been used.

These basis sets are fairly expensive. As the computer time rises as n^4, n being the number of basis functions, a double zeta or triple zeta calculation will require 16 or 81 times more computer time, respectively, than a minimal basis set calculation. Now it is clear that doubling the core functions will not significantly improve the description of bond making or bond breaking, which involves only valence functions. Hence a good compromise would be the use of *split-valence* sets, where only the number of valence functions is augmented. For example, in the 3–21G basis set, each orbital is represented by three Gaussians, but the valence orbitals are split into two groups: two ʻinternalʼ and one ʻexternalʼ Gaussians. In the LCAO expression, each core AO has one coefficient whereas each valence AO has two. These 3–21G calculations are no more expensive to use than STO-3G calculations and are usually more accurate, because the valence orbital coefficients are optimized separately. This gives greater flexibility in the description of the charge cloud. In the 6–31G basis set, each core electron is described by six Gaussians and each valence electron by four: three internal and one external. In the even more extended 6–311G basis set, each valence orbital has five Gaussians, which are subdivided into three groups.

Standard basis sets are not suitable for calculating anions. The reason is that the exponents ζ (which give the size of AOs) are optimized for neutral molecules. With these exponents, the electrons of anions are still compelled to move in the same volumes, so their repulsion is unduly exaggerated. A small basis set is not flexible enough to alleviate this effect. In particular, the 3–21G basis set gives unreliable activation energies and geometries for ionic reactions. Take a gas-phase S_N2 reaction. It is

---

[19]i.e. functions of the form $\exp(-\zeta r^2)$.

generally accepted that its reaction profile shows a double minimum.[20] The first step involves the interaction of the nucleophile with the substrate to give an ion–molecule complex. This initial complex rearranges, and the rearranged complex dissociates into the products. In a 3–21G calculation, the initial ion–molecule complex is very tight, to maximize charge transfer and reduce the nucleophile's instability. However, the rearrangement of this complex requires very little energy, so the calculated energy barrier is too low.[21]

This flaw can be corrected by employing *diffuse orbitals* which increase the volume of the charge cloud. *Polarization orbitals* have a principal quantum number which is greater by one than the valence orbitals: thus p orbitals are polarization functions for s orbitals; d orbitals are polarization functions for p orbitals. They can describe much better the electronic cloud and are particularly useful for transition states. Consider, for example, the ene reaction:

If a hydrogen is described by an s orbital, the electron cloud in its neighbourhood will have spherical symmetry. Thus the circled hydrogen can be described reasonably well with s-type orbitals in the reagent or in the product. In the transition state, however, this hydrogen is simultaneously bonded to C_1 and C_5 and loses its spherical symmetry. Its electronic density is mostly localized along the C_1H *and* the C_5H axes. Such a broken line is impossible to represent with s orbitals, but poses no problem to p orbitals. In conclusion, to describe correctly the transition state of the ene reaction, polarization orbitals are mandatory, at least for the hydrogen to be transferred.

The presence of diffuse orbitals is indicated by a + sign and polarization orbitals by an asterisk. For example, 3–21+G denotes a 3–21G base having diffuse orbitals on the heavy atoms; 6–31++G* is a 6–31G base having diffuse orbitals on heavy *and* hydrogen atoms, with by polarization orbitals only on the heavy atoms; 6–31G** is a 6–31G base having no diffuse orbitals but with polarization orbitals on all atoms. Notation such as 6–31G*//3–21G implies that the energies have been calculated with a 6–31G* base, using a geometry optimized at the 3–21G level.

Correlation Models

Instantaneous correlation can be taken into account either by mixing ground state with excited state wavefunctions (configuration interaction and Møller–Plesset models) or by introducing explicitly approximate correction terms [density functional theory (DFT) models].

[20]Olmstead W. N., Brauman J. I., *J. Am. Chem. Soc.*, 1977, **99**, 4219; Pellerite M. J., Brauman J. I., *J. Am. Chem. Soc.*, 1980, **102**, 5993; 1983, **105**, 2672.
[21]Anh N. T., Maurel F., Thanh B. T., Thao H. H., N'Guessan Y. T., *New J. Chem.*, 1994, **18**, 473; Anh N. T., Maurel F., Thao H. H., N'Guessan Y. T., *New J. Chem.*, 1994, **18**, 483.

Configuration Interaction Models

Configuration interaction (CI) methods improve on HF models by making a 'resonance' between several electronic configurations. The trial wavefunction is a linear combination of ground state and excited state configurations, whose coefficients are determined by a variational calculation. Excited states are generated by exciting electrons from occupied MOs to vacant MOs.[22] As their number is very large, it is necessary to limit severely the number of electron promotions. Although CISDTQ (CI with singly, doubly, triply and quadruply excited states) has been employed for small molecules, the only general applicable methods are CID and CISD.[23] CI calculations are significantly more costly than HF calculations.

MCSCF (multi-configuration SCF) methods are CI in which not only the coefficients of the linear combination are optimized but also the MOs from which the excited configurations are constructed. Usually, the MOs are divided into *active* (FOs and some other nearby MOs) and *inactive*. If all the configurations generated from the active orbitals are taken into consideration, the model is called CASSCF (complete active space SCF).[24]

Møller–Plesset Models

In the Møller–Plesset perturbation method, the correlated system is considered to be a perturbation of the Hartree–Fock system. Consequently, each correlated configuration may be expressed as a linear combination of HF configurations. The acronyms MP2, MP3, MP4, etc. indicate Møller–Plesset methods truncated at the second, third or fourth order of perturbation. Only the MP2 method is of common use. In higher order MP models, geometry must be optimized numerically.

DFT Models

Density functional theory (DFT) is based on the Hohenberg–Kohn theorem, which states that there is a functional which gives the *exact* ground-state energy for the exact electron density. DFT models have become very popular because they are not more costly than Hartree–Fock models. The reason is that in the HF, CI and MP models, a wavefunction for an N-electron system depends on $3N$ coordinates, whereas in the DFT approach, the electron density depends on only three coordinates, irrespective of the number of electrons. The problem is that the exact functional would be the Schrödinger equation itself! Several approximate functionals have been developed by many authors (Becke, Parr, Perdew, and others) and different forms of the functional can yield slightly different results. Some of the most common DFT models are:

---

[22]These are the Hartree–Fock MOs and are kept fixed in the CI calculation.

[23]CIS does not improve on the Hartree–Fock energy, as there is no interaction between the fundamental state and the singly excited states (Brillouin theorem). In CISD, the singly excited states can interact with the fundamental state through the intermediacy of the doubly excited states. We have met a similar situation in three orbital interactions (p. 27): two MOs belonging to the same molecule can interact if perturbed by a third MO belonging to another molecule.

[24]The method is also called FORS (full optimized reaction space).

- *Local density models* = density functional models in which the electron density is assumed to vary only slowly throughout all space.
- *Non-local density functional models* or *gradient corrected density functional models* = density functional models which take explicitly into account the nonuniformity in electron distributions, thus improving on the local density model.
- *Hybrid density functional models* = density functional models which incorporate the exchange energy from the Hartree–Fock model.
- *BLYP (Becke–Lee–Yang–Parr) model* = a non-local density functional model.
- *B3LYP model* = both a hybrid density functional model and a non-local density functional model, it incorporates three adjustable parameters.
- *BP (Berke–Perdew) model* = a non-local density functional model.
- *EDF1 model* = a non-local density functional model.

Solvent Effects

We have seen (p. 99) that reactions in the gas phase may be completely different from solution chemistry. An accurate description of solvent effects is therefore critical. There are currently two general approaches, neither entirely satisfactory, to the treatment of molecules in solution:

The explicit approach uses a Monte Carlo simulation of the solute immersed in a box containing a large number of solvent molecules. Results are usually good but the computational cost is rather high.

In the implicit approach, as in the SCRF (self-consistent field reaction) method,[25] the solvent is treated as a continuum whose principal characteristic is its dielectric constant. The solute is placed in a cavity within the 'solvent', which becomes polarized by its presence. In turn, this creates an electric field within the cavity. Subsequently, the free energy of solvation is calculated by a multipolar development. This method does not require much computer time. It ignores, however, specific solvent–solute interactions such as hydrogen bonds.

A reliable and inexpensive treatment of solvent effects would certainly represent major progress in computational chemistry.

8.3.3 A Few Technical Points

Choosing the Model

A chemist *never* calculates a reaction, only a model for it.[26] Any reaction in solution involves trillions of molecules. A calculation generally takes into account two reagents,

---

[25]Tapia O., Goscinski O., *Mol. Phys.*, 1975, **29**, 1653 ; Miertus S., Scrocco E., Tomasi J., *J. Chem. Phys.*, 1981, **55**, 117 ; Rivail J. L., Terryn B., Rinaldi D., Ruiz-Lopez M. F., *Theochem*, 1985, **120**, 387 ; Karelson M. M., Katritzky A. R., Zerner M. C., *Int. J. Quantum Chem. Symp.*, 1986, **20**, 521; Mikkelson K. V., Agren H., Jensen H. J. A., Helkager T., *J. Chem. Phys.*, 1988, **89**, 3086.
[26]See also p. 104.

along with maybe a few solvent molecules! When trying to strike a balance between computing time and chemical accuracy, it is best to *simplify the reagents, not the reaction*. Never be tempted into the worst solution: modelling the reagents 'realistically', which usually means incorporating a mass of useless details. This generates large systems and encourages temptation to 'simplify' the reaction trajectory, which renders the calculation worthless. For instance, we could impose a symmetry plane upon the trajectory for the reaction of CCl_2 with norbornadiene and this would probably 'simplify' the calculation sufficiently to allow us to use the real reagents in our model. The only trouble is that we would come out with the wrong transition state. It is reasonable to replace norbornadiene with cyclopentene (even ethylene) and CCl_2 by CH_2 to save computer time, but not to impose a particular reaction trajectory: we would almost certainly miss the nonlinear cheletropic profile.

If a molecule contains a single substituent, it is unwise to represent it by Me or F. The inductive and hyperconjugative effects of a methyl group are small, so the calculation will often be determined by conformational effects (pp. 166, 167). Fluorine exaggerates dipolar influences and tends to underplay frontier orbital contributions. Sometimes, it may be interesting, however, to amplify the influence of a substituent as an aid to analyzing its effects. NH_2 and BH_2 are seductive models for donor and acceptor groups, respectively; their effects are substantial and a simple rotation through $90°$ is enough to 'switch off' their conjugation with a neighboring π system.

At first sight, LiCN, which takes into account the counterion, may seem to be a more realistic nucleophile model than CN^-. However, if the solvent is not introduced explicitly into the calculation, the cation will tend to chelate to the cyanide and the carbonyl oxygen, thus promoting a four-centerd cyclic transition state, which is unlikely to be present in solution. It would be better to omit the cation altogether than to introduce it in the absence of solvent. If the solvent acts purely as a Lewis base, computing time can be saved by replacing ethers with water (even when modelling Grignard reagents!).

Here is a caricatural example illustrating how an unsound model may lead to absurd results. Consider a summer day when on the highway cars are staying as close to one another as possible. Reducing the car speed will then increase the security and also render the traffic more fluid. Indeed, as the braking distance is proportional to the square of the speed v, if v is divided by 2, the safety distance separating two successive cars is divided by 4, so the number of cars passing by a given point each hour is doubled. If v is divided by 4, the number of passing cars is 4 times larger: clearly, if the speed is zero, the hourly number of passing cars is infinite! So, where is the error? Be assured that the mathematics are perfectly sound![27]

Choosing a Basis Set

The best basis set is the one which gives the best answers, not the one which is the most expensive! This rather obvious statement can be illustrated by two examples. Minimal

---

[27]The error comes from modeling cars by material points. If their lengths are taken into account, the speed cannot be reduced beyond a certain limit.

basis sets predict that the protonation of ethylene gives an open cation; some extended bases give a structure wherein the proton bridges the double bond. As acids do not add to alkenes stereospecifically, the open cation is in better agreement with experiment. The bridged structure is an artifact which results from having diffuse and polarization orbitals upon *every* atom; their presence makes the proton so large that it can interact with both of the carbon atoms simultaneously.

In a study of the addition of a hydrogen radical H$^\bullet$ to ethylene, the error in the activation energy *increases* with the size of the basis set![28] In fact, the basis set was augmented only for the carbon atoms, thus improving the model for the ethylene while ignoring the H$^\bullet$. This lack of balance induced an artificially high activation energy. Excellent results were obtained with much less refined descriptions for the carbon, provided that p orbitals were added to the incoming hydrogen radical.

Attempts to achieve perfection may give poor results. `Normal' AOs are optimized for neutral molecules, so they are too small to give good descriptions of anions, particularly in the gas phase. A better model is obtained by a *slight* contraction of the Gaussian exponents. However, if these exponents had been completely optimized, the anions would become so stable that they would lose all reactivity: for example, a hydride ion H$^-$ will arrive at a carbonyl group, bounce on it and go away without reacting!

8.4 To Dig Deeper

1. Hehre W. J., *A Guide to Molecular Mechanics and Quantum Chemical Calculations*, Wavefunction, Irvine, CA, 2003.
   Plenty of practical information and useful advice. Highly recommended. This book (which can be acquired separately) is part of the Spartan package.
2. Jensen F., *Introduction to Computational Chemistry*, John Wiley & Sons, Inc., New York, 1999.
   The theory is covered in more details than in Hehre's book.
3. Koch W., Holthausen M. C., *A Chemist's Guide to Density Functional Theory*, 2nd edn, Wiley-VCH, Weinheim, 2002.

---

[28]Delbecq F., Ilavsky D., Anh N. T., Lefour J. M., *J. Am. Chem. Soc.*, 1985, **107**, 1623.

Appendix: MO Catalog

A1 Organization of the Catalog

These Hückel calculations were made using the parameters on p. 22. The unusual numbering of some molecules is due to the Hückel software.

The numbers in **bold italics** give the orbital energies in units of β, with a reference value α of zero. Thus, the frontier orbital energies of butadiene are given as **0.618** and **−0.618**. The orbital energy appears above a list of its coefficients descending in a column from atom 1 to atom n. Orbitals lying above the LUMO have been omitted for some of the larger alternant hydrocarbons. They may be found easily by using the pairing theorem (p. 36).

When Hückel parameters are not available or reliable (e.g. for silicon or sulfur compounds), SCF calculations are used instead. In these cases, only pertinent MOs are given. Energies are then given in eV. *Do not forget that FO is an approximate theory, to be used only for preliminary studies*. Spending too much time calculating the frontier orbitals would be futile. Avoid sophisticated methods and use only simple ones (Hückel, MNDO, AM1, PM3, STO-3G or 3–21G).

A2 Chapter 3

Exercise 2

$CH_2=CH_2$

1.000	*−1.000*
0.707	0.707
0.707	−0.707

$CH_2=CH−CH=CH_2$

1.618	*0.618*	*−0.618*	*−1.618*
0.37	0.60	0.60	0.37
0.60	0.37	−0.37	−0.60
0.60	−0.37	−0.37	0.60
0.37	−0.60	0.60	−0.37

Frontier Orbitals   Nguyên Trong Anh
© 2007 John Wiley & Sons, Ltd

2.110	*1.000*	*0.618*	*−0.254*	*−1.618*	*−1.861*
0.25	0.50	0.00	0.75	0.00	0.36
0.52	0.50	0.00	−0.19	0.00	−0.66
0.43	0.00	−0.60	−0.35	−0.37	0.44
0.39	−0.50	−0.37	0.28	0.60	−0.15
0.39	−0.50	0.37	0.28	−0.60	−0.15
0.43	0.00	0.60	−0.35	0.37	0.44

Exercise 3

2.310	*1.652*	*0.356*	*0.887*	*0.477*	*−0.400*	*−0.738*	*−1.579*	*−1.869*	*−2.095*
0.32	0.27	0.22	0.26	0.54	0.06	0.30	0.44	0.25	0.26
0.28	0.32	0.00	0.58	0.00	−0.32	0.00	−0.55	−0.27	0.00
0.32	0.27	−0.22	0.26	−0.54	0.06	−0.30	0.44	0.25	−0.26
0.29	−0.19	−0.48	−0.22	0.16	−0.47	0.36	−0.08	0.32	−0.34
0.20	−0.43	−0.36	0.16	0.34	−0.10	−0.48	0.27	−0.41	0.16
0.17	−0.53	0.00	0.36	0.00	0.51	0.00	−0.34	0.43	0.00
0.20	−0.43	0.36	0.16	−0.34	−0.10	0.48	0.27	−0.41	−0.16
0.29	−0.19	0.48	−0.22	−0.16	−0.47	−0.36	−0.08	0.32	0.34
0.47	0.12	0.30	−0.35	0.26	0.29	−0.22	−0.14	−0.20	−0.54
0.47	0.12	−0.30	−0.35	−0.26	0.29	0.22	−0.14	−0.20	0.54

2.303	*1.618*	*1.303*	*1.000*	*0.618*	*−0.618*	*−1.000*
0.46	0.00	0.35	0.41	0.00	0.00	0.41
0.30	−0.26	0.40	0.00	0.43	−0.43	0.00
0.23	−0.43	0.17	−0.41	0.26	0.26	−0.41
0.23	−0.43	−0.17	−0.41	−0.26	0.26	0.41
0.30	−0.26	−0.40	0.00	−0.43	−0.43	0.00
0.46	0.00	−0.35	0.41	0.00	0.00	−0.41
0.30	0.26	−0.40	0.00	0.43	0.43	0.00
0.23	0.43	−0.17	−0.41	0.26	−0.26	0.41
0.23	0.43	0.17	−0.41	−0.26	−0.26	−0.41
0.30	0.26	0.40	0.00	−0.43	0.43	0.00

Use the pairing theorem for the remaining MOs

2.414	2.000	1.414	1.414	1.000	1.000	0.414
0.22	0.29	0.28	0.17	0.41	0.00	0.31
0.15	0.29	0.12	0.41	0.20	0.35	0.22
0.15	0.29	−0.12	0.41	−0.20	0.35	−0.22
0.22	0.29	−0.28	0.17	−0.41	0.00	−0.31
0.22	−0.29	−0.28	0.17	0.41	0.00	−0.31
0.15	−0.29	−0.12	0.41	0.20	−0.35	−0.22
0.15	−0.29	0.12	0.41	−0.20	−0.35	0.22
0.22	−0.29	0.28	0.17	−0.41	0.00	0.31
0.30	0.00	0.40	−0.24	0.00	0.00	−0.44
0.30	0.00	−0.40	−0.24	0.00	0.00	0.44
0.37	0.29	−0.29	−0.17	−0.20	−0.35	0.09
0.37	0.29	0.29	−0.17	0.20	−0.35	−0.09
0.37	−0.29	0.29	−0.17	−0.20	0.35	−0.09
0.37	−0.29	−0.29	−0.17	0.20	0.35	0.09

Use the pairing theorem for the remaining MOs

A3 Chapter 4

Exercise 1

2.247	0.802	0.555	−0.555	−0.802	−2.247
0.23	0.52	0.42	0.42	0.52	0.23
0.52	0.42	0.23	−0.23	−0.42	−0.52
0.42	0.23	−0.52	−0.52	0.23	0.42
0.42	−0.23	−0.52	0.52	0.23	−0.42
0.52	−0.42	0.23	0.23	−0.42	0.52
0.23	−0.52	0.42	−0.42	0.52	−0.23

2.194	*1.295*	*1.194*	*0.295*	*−0.295*	*−1.194*	*−1.295*	*−2.194*
0.22	0.26	0.30	0.54	0.54	0.30	0.26	0.22
0.48	0.34	0.36	0.16	−0.16	−0.36	−0.34	−0.48
0.36	−0.16	0.48	−0.34	−0.34	0.48	−0.16	0.36
0.30	−0.54	0.22	−0.26	0.26	−0.22	0.54	−0.30
0.30	−0.54	−0.22	0.26	0.26	−0.22	−0.54	0.30
0.36	−0.16	−0.48	0.34	−0.34	0.48	0.16	−0.36
0.48	0.34	−0.36	−0.16	−0.16	−0.36	0.34	0.48
0.22	0.26	−0.30	−0.54	0.54	0.30	−0.26	−0.22

2.000	*0.000*	*0.000*	*−2.000*
0.50	0.50	0.50	0.50
0.50	−0.50	0.50	−0.50
0.50	−0.50	−0.50	0.50
0.50	0.50	−0.50	−0.50

2.000	*1.000*	*1.000*	*−1.000*	*−1.000*	*−2.000*
0.41	0.00	0.58	−0.58	0.00	0.41
0.41	0.50	0.29	0.29	0.50	−0.41
0.41	0.50	−0.29	0.29	−0.50	0.41
0.41	0.00	−0.58	−0.58	0.00	−0.41
0.41	−0.50	−0.29	0.29	0.50	0.41
0.41	−0.50	0.29	−0.29	−0.50	−0.41

2.356	*1.477*	*1.095*	*0.262*	*−0.262*	*−1.095*	*−1.477*	*−2.356*
0.36	0.34	0.16	0.48	0.48	0.16	0.34	0.36
0.48	0.16	0.34	0.36	0.36	−0.34	−0.16	−0.48
0.30	−0.26	0.54	−0.22	−0.22	0.54	−0.26	0.30
0.22	−0.54	0.26	−0.30	−0.30	−0.26	0.54	−0.22
0.22	−0.54	−0.26	0.30	0.30	−0.26	−0.54	0.22
0.30	−0.26	−0.54	0.22	0.22	0.54	0.26	−0.30
0.48	0.16	−0.34	−0.36	−0.36	−0.34	0.16	0.48
0.36	0.34	−0.16	−0.48	−0.48	0.16	−0.34	−0.36

Exercise 2

2.349	1.393	0.570	−0.643	−1.000
0.18	0.29	0.65	0.56	0.37
0.43	0.40	0.37	−0.36	−0.62
0.22	0.60	−0.31	−0.40	0.58
0.09	0.43	−0.55	0.62	−0.35
0.85	−0.46	−0.18	0.10	0.12

2.286	1.530	0.509	−0.686	−1.638
0.05	0.41	0.60	0.58	0.36
0.11	0.62	0.31	−0.40	−0.59
0.21	0.55	−0.45	−0.31	0.60
0.37	0.21	−0.53	0.61	−0.40
0.90	−0.32	0.25	−0.16	0.08

Exercise 4

1.618	−0.618
0.85	0.53
0.53	−0.85

2.315	1.435	−0.750
0.30	0.82	0.48
0.39	0.36	−0.85
0.87	−0.44	0.22

2.633	1.618	0.314	−0.618	−0.948
0.20	0.60	0.60	0.37	0.31
0.33	0.37	0.19	−0.60	−0.60
0.84	0.00	−0.44	0.00	0.32
0.33	−0.37	0.19	0.60	−0.60
0.20	−0.60	0.60	−0.37	0.31

2.792	2.315	2.000	1.435	1.271	−0.750	−1.063
0.31	0.62	0.60	0.31	0.16	0.15	0.14
0.35	0.28	0.00	−0.25	−0.17	−0.60	−0.59
0.70	0.00	−0.53	0.00	0.37	0.00	0.31
0.35	−0.28	0.00	0.25	−0.17	0.60	−0.59
0.31	−0.62	0.60	−0.31	0.16	−0.15	0.14
0.19	0.21	0.00	−0.58	−0.62	0.34	0.29
0.19	−0.21	0.00	0.58	−0.62	−0.34	0.29

Exercise 5

2.385	1.404	−0.789
0.30	0.83	0.47
0.41	0.34	−0.85
0.86	−0.45	0.24

2.792	2.241	1.628	1.256	−0.917
0.19	0.22	0.39	0.75	0.44
0.35	0.27	0.25	0.19	−0.84
0.31	0.78	−0.46	−0.18	0.20
0.70	−0.20	0.42	−0.48	0.24
0.50	−0.47	−0.63	0.36	−0.05

2.524	2.000	1.350	−0.874
0.29	0.00	0.84	0.45
0.45	0.00	0.30	−0.84
0.60	−0.71	−0.32	0.21
0.60	0.71	−0.32	0.21

2.524	2.000	1.331	−0.911
0.29	0.00	0.85	0.44
0.46	0.00	0.28	−0.84
0.55	−0.75	−0.29	0.20
0.63	0.66	−0.34	0.23

O 1
‖
H 2 —C— O 3 —CH3 4

2.728	1.786	1.282	−0.796
0.16	0.53	0.69	0.47
0.28	0.41	0.19	−0.84
0.75	0.27	−0.55	0.25
0.58	−0.69	0.43	−0.05

Exercise 6

1 N — CH3 5
‖
3 H3C — 2 — CH3 4

2.566	2.224	2.000	0.755	−1.045
0.35	0.26	0.00	0.71	0.56
0.42	−0.12	0.00	0.46	−0.78
0.51	−0.37	0.71	−0.26	0.18
0.51	−0.37	−0.71	−0.26	0.18
0.43	0.80	0.00	−0.40	−0.13

1 N — OH 5
‖
3 H3C — 2 — CH3 4

2.600	2.262	2.000	0.698	−1.060
0.39	0.24	0.00	0.69	0.57
0.40	−0.16	0.00	0.47	−0.77
0.46	−0.44	0.71	−0.25	0.18
0.46	−0.44	−0.71	−0.25	0.18
0.51	0.73	0.00	−0.42	−0.15

1 N — NH2 5
‖
3 H3C — 2 — CH3 4

2.536	2.000	1.954	0.584	−1.074
0.31	0.00	0.45	0.61	0.57
0.44	0.00	0.02	0.48	−0.76
0.57	−0.71	−0.29	−0.24	0.17
0.57	0.71	−0.29	−0.24	0.17
0.24	0.00	0.79	−0.53	−0.18

Exercise 9

$$\overset{\overset{\displaystyle O\ 1}{\|}}{\underset{2}{\overset{3}{H_2N}-C}}\diagdown H$$

2.067	1.257	−0.824
0.48	0.75	0.46
0.51	0.19	−0.84
0.72	−0.63	0.29

$$\overset{\overset{\displaystyle O\ 1}{\|}}{\underset{\overset{3}{H_3C}\quad\underset{2}{\quad}\quad\overset{4}{NH_2}}{\diagup\ \diagdown}}$$

2.450	1.759	1.234	−0.943
0.32	0.29	0.79	0.43
0.47	0.22	0.19	−0.84
0.72	−0.64	−0.17	0.20
0.39	0.68	−0.56	0.27

$$\overset{\overset{\displaystyle O\ 1}{\|}}{\underset{\overset{3}{H_2N}\ \ \underset{2}{\ }\ \overset{4}{O}\underset{5}{\diagdown}CH_3}{\diagup\ \diagdown}}$$

2.777	2.044	1.460	1.203	−0.984
0.19	0.39	0.06	0.80	0.42
0.34	0.40	0.03	0.16	−0.83
0.21	0.59	−0.59	−0.43	0.27
0.73	−0.05	0.56	−0.32	0.23
0.52	−0.58	−0.58	0.22	−0.04

Some Other Carbonyl Compounds

$$\overset{\overset{\displaystyle O\ 1}{\|}}{\underset{2}{\overset{3}{Cl}-C}}\diagdown H$$

2.129	1.533	−0.662
0.26	0.82	0.51
0.30	0.44	−0.85
0.92	−0.37	0.13

$$\overset{\overset{\displaystyle O\ 1}{\|}}{\underset{\overset{3}{H_3C}\quad\underset{2}{\quad}\quad\overset{4}{Cl}}{\diagup\ \diagdown}}$$

2.389	2.000	1.402	−0.791
0.30	0.00	0.83	0.47
0.42	0.00	0.33	−0.85
0.75	0.50	−0.39	0.21
0.43	−0.87	−0.22	0.12

Exercise 13

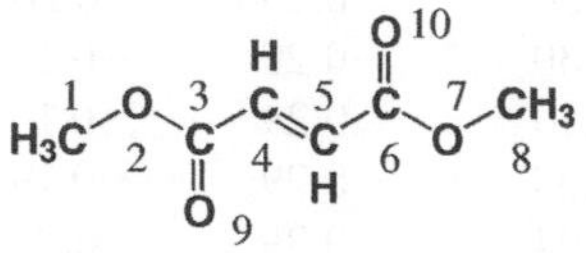

$$\underset{1}{H_3C}\overset{}{\underset{2}{O}}\overset{H}{\underset{3}{C}}\overset{}{\underset{4}{C}}\overset{5}{\underset{6}{C}}\overset{O\,10}{\underset{}{C}}\overset{7}{\underset{8}{O}}CH_3$$

(structure with labels: H_3C—1—O—2—C—3 (H above 4), C—4 (=O 9 below, H below), C—5, C—6 (=O 10 above), O—7, CH_3—8)

2.784	2.752	2.000	1.851	1.359	1.298	1.000	−0.108	−1.143	−1.792
0.37	0.39	0.45	0.48	0.40	0.33	0.00	0.04	0.03	0.01
0.51	0.53	0.00	−0.13	−0.46	−0.41	0.00	−0.16	−0.15	−0.09
0.25	0.22	−0.31	−0.32	0.09	0.13	0.00	0.39	0.58	0.43
0.14	0.06	−0.31	−0.11	0.24	0.06	0.50	0.44	−0.27	−0.54
0.14	−0.06	−0.31	0.11	0.24	−0.06	0.50	−0.44	−0.27	0.54
0.25	−0.22	−0.31	0.32	0.09	−0.13	0.00	−0.39	0.58	−0.43
0.51	−0.53	0.00	0.13	−0.46	0.41	0.00	0.16	−0.15	0.09
0.37	−0.39	0.45	−0.48	0.40	−0.33	0.00	−0.04	0.03	−0.01
0.14	0.13	−0.31	−0.37	0.24	0.45	−0.50	−0.35	−0.27	−0.15
0.14	−0.13	−0.31	0.37	0.24	−0.45	−0.50	0.35	−0.27	0.15

(structure with labels: H_3C—1—C—2 (=O 6 above), O—3 (labeled 3 with C), C—3, O—4, C—5 (H above), CH_2)

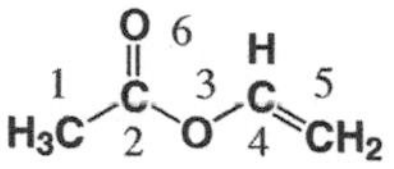

2.715	2.130	1.386	0.787	−0.848	−1.171
0.40	0.82	0.36	0.03	0.19	0.08
0.41	0.15	−0.32	−0.05	−0.75	−0.38
0.74	−0.48	0.16	−0.34	0.13	0.26
0.25	−0.23	0.20	0.56	0.30	−0.66
0.09	−0.11	0.14	0.72	−0.36	0.56
0.24	0.14	−0.83	0.22	0.41	0.17

Exercise 15

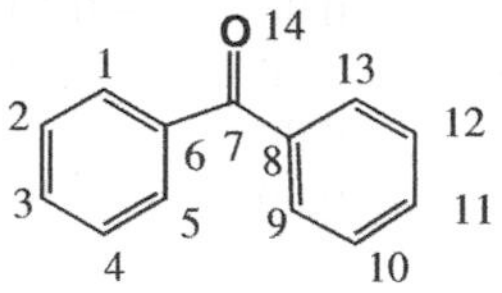

(structure: two benzene rings connected by C=O, benzophenone, with atom numbers 1–14, O 14 at top)

2.375	2.000	1.689	1.000	1.000	1.000	1.000
0.23	0.29	0.14	0.26	0.44	0.14	0.19
0.15	0.29	0.28	−0.10	0.54	−0.05	−0.11
0.13	0.29	0.33	0.37	0.10	−0.18	−0.30
0.15	0.29	0.28	−0.26	−0.44	−0.14	−0.19
0.23	0.29	0.14	0.10	−0.54	0.05	0.11
0.40	0.29	−0.04	0.37	−0.10	0.18	0.30
0.48	0.00	−0.35	0.00	0.00	0.00	0.00
0.40	−0.29	−0.04	−0.36	0.04	−0.16	0.33
0.23	−0.29	0.14	−0.40	0.02	0.37	0.15
0.15	−0.29	0.28	−0.04	−0.03	0.53	−0.18
0.13	−0.29	0.33	0.36	−0.04	0.16	−0.33
0.15	−0.29	0.28	0.40	−0.02	−0.37	−0.15
0.23	−0.29	0.14	0.04	0.03	−0.53	0.18
0.35	0.00	−0.51	−0.01	0.06	−0.02	−0.63

−0.294	−1.000	−1.000	−1.000	−1.487	−2.000	−2.283
0.29	0.02	0.51	0.17	0.04	0.29	0.26
0.04	−0.13	−0.48	0.22	−0.30	−0.29	−0.19
−0.30	0.11	−0.03	−0.39	0.40	0.29	0.16
0.04	0.02	0.51	0.17	−0.30	−0.29	−0.19
0.29	−0.13	−0.48	0.22	0.04	0.29	0.26
−0.13	0.11	−0.03	−0.39	0.23	−0.29	−0.41
−0.53	0.00	0.00	0.00	−0.43	0.00	0.42
−0.13	−0.11	0.03	0.39	0.23	0.29	−0.41
0.29	−0.42	0.05	−0.34	0.04	−0.29	0.26
0.04	0.53	−0.08	−0.05	−0.30	0.29	−0.19
−0.30	−0.11	0.03	0.39	0.40	−0.29	0.16
0.04	−0.42	0.05	−0.34	−0.30	0.29	−0.19
0.29	0.53	−0.08	−0.05	0.04	−0.29	0.26
0.41	0.00	0.00	0.00	0.17	0.00	−0.13

2.463	2.000	1.504	1.000	1.000	−0.458	−1.000	−1.376	−2.133
0.18	0.35	0.04	0.56	0.06	0.35	0.50	0.03	0.40
0.11	0.35	0.23	0.29	0.48	0.09	−0.50	0.36	−0.33
0.09	0.35	0.30	−0.27	0.42	−0.39	0.00	−0.53	0.31
0.11	0.35	0.23	−0.56	−0.06	0.09	0.50	0.36	−0.33
0.18	0.35	0.04	−0.29	−0.48	0.35	−0.50	0.03	0.40
0.33	0.35	−0.17	0.27	−0.42	−0.25	0.00	−0.40	−0.51
0.46	0.00	−0.33	0.00	0.00	−0.58	0.00	0.50	0.30
0.32	0.00	−0.66	−0.27	0.42	0.40	0.00	−0.21	−0.10
0.70	−0.50	0.47	0.00	0.00	0.17	0.00	−0.10	−0.05

Exercise 17

1.879	1.000	− 0.347	−1.532
0.67	0.58	0.43	0.23
0.58	0.00	−0.58	−0.58
0.43	−0.58	−0.23	0.66
0.23	−0.58	0.66	−0.43

2.529	2.351	2.000	1.545	0.791	−0.585	−1.631
0.18	0.27	0.00	0.67	0.52	0.36	0.21
0.28	0.36	0.00	0.36	−0.11	−0.58	−0.56
0.26	0.08	0.00	0.28	−0.66	−0.14	0.63
0.39	−0.18	0.00	0.08	−0.41	0.66	−0.46
0.37	0.72	0.00	−0.56	0.06	0.16	0.11
0.52	−0.35	−0.71	−0.12	0.24	−0.18	0.09
0.52	−0.35	0.71	−0.12	0.24	−0.18	0.09

Exercise 20

AM1 calculations. Energies given in eV.

	HO-3	HO-1	HO	LU
	−12.46	−10.45	−8.94	−0.56
$C_2\,2p_z$	0.28	0.33	-	−0.74
$S_3\,2p_z$	0.19	0.75	-	0.54
$2p_y$	-	-	0.95	-
$O_4\,2p_z$	0.46	−0.45	-	0.27

AM1 calculations. Energies given in eV

	HO-3	HO-1	HO	LU
	−11.86	−9.06	−8.75	−0.94
$C_2\,2p_z$	0.44	0.06	-	0.69
$S_3\,2p_z$	0.49	−0.52	-	−0.54
$2p_y$	-	-	0.91	-
$O_4\,2p_z$	0.30	0.74	-	−0.30

Exercise 21

2.713	1.665	0.734	−1.112
0.10	0.21	0.72	0.66
0.26	0.35	0.53	−0.73
0.76	0.47	−0.41	0.19
0.59	−0.78	0.18	−0.04

$$\overset{2}{\underset{N}{1}}\overset{3}{\diagdown}4$$

1.712	0.812	− 0.460	−1.565
0.50	0.63	0.52	0.29
0.60	0.20	−0.50	−0.59
0.54	−0.47	−0.29	0.64
0.31	−0.58	0.63	−0.41

$$\overset{1}{\diagup}\overset{3}{\underset{2}{\diagdown}}\overset{5}{\underset{4}{\diagup}}\overset{CH_3}{\underset{O}{\diagdown}}\overset{}{\underset{6}{}}$$

2.718	1.750	1.425	0.465	−0.711	−1.646
0.02	0.20	0.38	−0.60	0.58	−0.35
0.05	0.35	0.54	−0.28	−0.41	0.58
0.12	0.41	0.39	0.47	−0.29	−0.60
0.26	0.37	0.02	0.50	0.61	0.41
0.75	0.30	−0.46	−0.30	−0.19	−0.09
0.59	−0.66	0.45	0.11	0.04	0.01

$$\diagup\diagdown\diagup\diagdown\diagup$$

1.802	1.247	0.445	−0.445	−1.247	−1.802
0.23	0.42	0.52	0.52	0.42	0.23
0.42	0.52	0.23	−0.23	−0.52	−0.42
0.52	0.23	−0.42	−0.42	0.23	0.52
0.52	−0.23	−0.42	0.42	0.23	−0.52
0.42	−0.52	0.23	0.23	−0.52	0.42
0.23	−0.42	0.52	−0.52	0.42	−0.23

$$\overset{1}{\diagup}\overset{3}{\underset{2}{\diagdown}}\overset{5}{\underset{4}{\diagup}}\overset{O}{\underset{6}{\diagdown}}$$

1.940	1.497	0.709	−0.241	−1.136	−1.771
0.13	−0.37	0.52	−0.55	0.46	−0.26
0.26	−0.55	0.37	0.13	−0.52	0.46
0.37	−0.46	−0.26	0.52	0.13	−0.55
0.46	−0.13	−0.55	−0.26	0.37	0.52
0.52	0.26	−0.13	−0.46	−0.55	−0.37
0.55	0.52	0.46	0.37	0.26	0.13

$$\overset{1}{\diagup}\overset{3}{\underset{2}{\diagdown}}\overset{5}{\underset{4}{\diagup}}\overset{6}{\underset{N}{\diagdown}}$$

1.843	1.357	0.589	−0.325	−1.178	−1.782
0.20	−0.39	−0.52	0.54	0.44	−0.25
0.37	−0.53	−0.30	−0.17	−0.52	0.44
0.48	−0.33	0.34	−0.48	0.17	−0.54
0.51	0.08	0.50	0.33	0.32	0.50
0.47	0.44	−0.05	0.37	−0.55	−0.39
0.35	0.51	−0.53	−0.45	0.33	0.17

Exercise 22

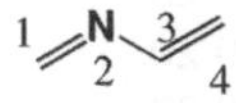

1.828	0.683	− 0.547	−1.464
0.38	0.51	0.69	0.35
0.69	0.35	−0.38	−0.51
0.54	−0.44	−0.30	0.65
0.30	−0.65	0.54	−0.44

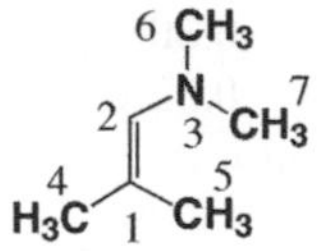

2.710	2.351	2.000	2.000	1.298	−0.318	−1.274
0.19	0.38	0.00	0.00	0.18	0.57	0.68
0.25	0.10	0.00	0.00	0.47	0.51	−0.67
0.01	−0.18	0.00	0.00	0.55	−0.50	0.21
0.18	0.60	−0.17	−0.69	−0.18	−0.24	−0.15
0.18	0.60	0.17	0.69	−0.18	−0.24	−0.15
0.48	−0.22	−0.69	−0.17	−0.44	0.17	−0.04
0.48	−0.22	0.69	0.17	−0.44	0.17	−0.04

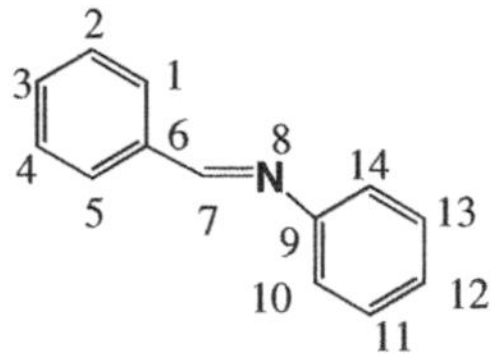

2.279	2.070	1.598	1.171	1.000	1.000	0.600
0.21	0.32	0.09	0.14	0.50	0.03	0.24
0.15	0.29	0.25	−0.27	0.50	0.03	−0.09
0.13	0.28	0.32	−0.45	0.00	0.00	−0.29
0.15	0.29	0.25	−0.27	−0.50	−0.03	−0.09
0.21	0.32	0.09	0.14	−0.50	−0.03	0.24
0.33	0.38	−0.11	0.43	0.00	0.00	0.23
0.34	0.13	−0.35	0.22	0.00	0.00	−0.34
0.43	−0.10	−0.45	−0.17	0.00	0.00	−0.43
0.43	−0.29	−0.14	−0.34	0.00	0.00	0.29
0.27	−0.25	0.11	−0.11	0.03	−0.50	0.30
0.20	−0.23	0.32	0.21	0.03	−0.50	−0.11
0.17	−0.22	0.40	0.35	0.00	0.00	−0.37
0.20	−0.23	0.32	0.21	−0.03	0.50	−0.11
0.27	−0.25	0.11	−0.11	−0.03	0.50	0.30

−0.413	*−1.000*	*−1.000*	*−1.139*	*−1.436*	*−2.058*	*−2.173*
0.30	0.01	0.50	0.13	0.01	0.25	0.31
0.07	−0.01	−0.50	0.20	−0.30	−0.23	−0.24
−0.33	0.00	0.00	−0.36	0.42	0.20	0.22
0.07	0.01	0.50	0.20	−0.30	−0.23	−0.24
0.30	−0.01	−0.50	0.13	0.01	0.25	0.31
−0.19	0.00	0.00	−0.35	0.28	−0.29	−0.42
0.53	0.00	0.00	0.14	−0.43	0.09	0.30
0.41	0.00	0.00	0.18	0.34	0.11	−0.24
0.15	0.00	0.00	−0.44	−0.22	−0.37	0.33
−0.24	−0.50	0.01	0.16	−0.01	0.32	−0.24
−0.05	0.50	−0.01	0.26	0.24	−0.30	0.19
0.26	0.00	0.00	−0.46	−0.33	0.29	−0.18
−0.05	−0.50	0.01	0.26	0.24	−0.30	0.19
−0.24	0.50	−0.01	0.16	−0.01	0.32	−0.24

1.947	*0.683*	*− 1.129*
0.24	0.72	0.65
0.47	0.49	−0.73
0.85	−0.48	0.22

Exercise 23

1.281	*−0.781*
0.79	0.62
0.62	−0.79

2-Azabutadiene MOs are given p. 257. Acrylonitrile MOs (p. 256) may be taken as approximate MOs for 1-azabutadiene. The AM1 energies of the azabutadienes and butadiene FOs are given below.

1-Azabutadiene	HO: −10.37 eV	LU: 0.21 eV	Difference: 10.58 eV
2-Azabutadiene	HO: −9.78 eV	LU: 0.24 eV	Difference: 10.02 eV
Butadiene	HO: −9.33 eV	LU: 0.45 eV	Difference: 9.78 eV

Exercise 24

4—2—N—5 / 3—1—O—6

2.761	**1.831**	**1.425**	**0.602**	**−0.538**	**−1.584**
0.33	0.43	0.02	0.55	0.56	0.31
0.14	0.41	0.39	0.35	−0.43	−0.59
0.06	0.32	0.54	−0.33	−0.33	0.63
0.02	0.18	0.38	−0.55	0.60	−0.40
0.75	0.21	−0.47	−0.37	−0.18	−0.07
0.55	−0.68	0.45	0.15	0.04	0.01

4—2—N—5—10 / 3—1 9 / 6 8 / 7

2.232	**1.721**	**1.220**	**1.000**	**0.575**	**−0.381**	**−1.000**
0.45	0.39	0.26	0.00	0.45	0.41	0.00
0.27	0.46	−0.20	0.00	0.31	−0.50	0.00
0.15	0.40	−0.50	0.00	−0.27	−0.22	0.00
0.07	0.23	−0.41	0.00	−0.47	0.58	0.00
0.51	0.01	0.39	0.00	−0.28	0.14	0.00
0.35	−0.18	0.11	0.50	−0.30	−0.23	0.50
0.26	−0.33	−0.26	0.50	0.10	−0.05	−0.50
0.23	−0.38	−0.42	0.00	0.36	0.25	0.00
0.26	−0.33	−0.26	−0.50	0.10	−0.05	0.50
0.35	−0.18	0.11	−0.50	−0.30	−0.23	−0.50

−1.159	**−1.596**	**−2.111**
0.23	0.30	0.26
0.10	−0.54	−0.17
−0.35	0.56	0.10
0.30	−0.35	−0.05
−0.49	−0.10	−0.50
0.17	−0.07	0.40
0.29	0.22	−0.34
−0.51	−0.27	0.32
0.29	0.22	−0.34
0.17	−0.07	0.40

2.782	2.048	1.605	1.297	0.858	−0.251	−1.154	−1.684
0.17	0.47	0.27	0.07	0.56	0.29	0.31	0.42
0.08	0.33	0.46	−0.12	0.14	−0.56	0.13	−0.55
0.03	0.21	0.47	−0.23	−0.44	−0.15	−0.46	0.51
0.01	0.10	0.29	−0.18	−0.51	0.60	0.40	−0.30
0.35	0.39	−0.16	−0.18	0.06	0.34	−0.64	−0.37
0.73	−0.05	−0.33	−0.55	−0.06	−0.13	0.17	0.08
0.52	−0.56	0.46	0.44	0.03	0.03	−0.03	−0.01
0.19	0.37	−0.27	0.60	−0.45	−0.28	0.30	0.14

2.772	2.177	1.983	1.584	1.379	1.000	0.983
0.34	0.12	0.31	0.21	0.01	0.00	0.22
0.18	0.26	0.45	−0.17	0.08	0.00	0.30
0.08	0.28	0.24	−0.22	−0.24	0.00	0.34
0.03	0.34	0.02	−0.18	−0.41	0.00	0.03
0.08	0.16	0.35	−0.26	0.34	0.00	−0.26
0.03	0.10	0.23	−0.24	0.39	0.00	−0.56
0.02	0.47	−0.20	−0.06	−0.33	0.00	−0.31
0.01	0.34	−0.21	0.04	−0.02	0.50	−0.16
0.00	0.27	−0.21	0.12	0.30	0.50	0.15
0.00	0.25	−0.21	0.16	0.44	0.00	0.31
0.00	0.27	−0.21	0.12	0.30	−0.50	0.15
0.00	0.34	−0.21	0.04	−0.02	−0.50	−0.16
0.74	−0.06	0.01	0.49	−0.09	0.00	−0.25
0.54	−0.20	−0.45	−0.66	0.08	0.00	0.13

0.479	−0.298	−0.708	−1.000	−1.321	−1.853	−2.168
0.49	0.59	0.06	0.00	0.14	0.26	0.09
0.23	−0.29	−0.06	0.00	−0.23	−0.58	−0.21
−0.38	−0.19	0.47	0.00	−0.14	0.40	0.26
−0.41	0.35	−0.27	0.00	0.41	−0.16	−0.35
0.01	−0.31	−0.49	0.00	0.30	0.40	0.12
−0.22	0.39	0.40	0.00	−0.16	−0.17	−0.05
0.19	0.09	−0.27	0.00	−0.41	−0.10	0.49
0.25	−0.19	0.23	0.50	0.06	0.17	−0.36
−0.07	−0.03	0.11	−0.50	0.32	0.22	0.29
−0.28	0.20	−0.31	0.00	−0.49	0.24	−0.27
−0.07	−0.03	0.11	0.50	0.32	−0.22	0.29
0.25	−0.19	0.23	−0.50	0.06	0.17	−0.36
−0.30	−0.22	−0.02	0.00	−0.04	−0.06	−0.02
0.11	0.05	0.00	0.00	0.01	0.01	0.00

2.315	2.136	1.896	1.400	1.325	1.000	1.000	0.920	0.484
0.46	0.13	0.17	0.22	0.33	0.00	0.00	0.06	0.44
0.41	0.00	0.41	0.04	−0.05	0.00	0.00	0.28	0.20
0.25	−0.13	0.26	−0.29	0.03	0.00	0.00	0.39	−0.34
0.18	−0.28	0.09	−0.44	0.09	0.00	0.00	0.08	−0.36
0.23	0.00	0.35	0.12	−0.43	0.00	0.00	−0.19	0.00
0.13	0.00	0.25	0.13	−0.52	0.00	0.00	−0.46	−0.19
0.17	−0.47	−0.09	−0.33	0.09	0.00	0.00	−0.32	0.17
0.10	−0.36	−0.13	−0.01	0.01	0.50	−0.03	−0.18	0.22
0.07	−0.30	−0.16	0.32	−0.07	0.50	−0.03	0.15	−0.06
0.06	−0.28	−0.17	0.45	−0.11	0.00	0.00	0.32	−0.25
0.07	−0.30	−0.16	0.32	−0.07	−0.50	0.03	0.15	−0.06
0.10	−0.36	−0.13	−0.01	0.01	−0.50	0.03	−0.18	0.22
0.42	0.21	−0.18	0.16	0.32	0.00	0.00	−0.25	−0.20
0.26	0.16	−0.25	0.00	0.05	−0.03	−0.50	−0.15	−0.27
0.18	0.14	−0.30	−0.16	−0.26	−0.03	−0.50	0.12	0.07
0.15	0.13	−0.31	−0.22	−0.39	0.00	0.00	0.26	0.30
0.18	0.14	−0.30	−0.16	−0.26	0.03	0.50	0.12	0.07
0.26	0.16	−0.25	0.00	0.05	0.03	0.50	−0.15	−0.27

−0.171	−0.706	−1.000	−1.000	−1.139	−1.341	−1.822	−2.107	−2.192
0.45	0.03	0.00	0.00	0.20	0.19	0.19	0.20	0.21
−0.36	−0.06	0.00	0.00	0.16	−0.18	−0.51	−0.12	−0.30
−0.11	0.48	0.00	0.00	−0.08	−0.17	0.37	−0.02	0.28
0.38	−0.28	0.00	0.00	−0.06	0.41	−0.17	0.15	−0.31
−0.27	−0.47	0.00	0.00	−0.30	0.22	0.37	0.07	0.16
0.41	0.39	0.00	0.00	0.18	−0.12	−0.16	−0.03	−0.06
0.05	−0.28	0.00	0.00	0.16	−0.37	−0.06	−0.30	0.40
−0.19	0.24	0.50	−0.01	−0.06	0.05	0.14	0.25	0.28
−0.02	0.11	−0.50	0.01	−0.09	0.31	−0.20	−0.21	0.22
0.20	−0.32	0.00	0.00	0.16	−0.47	0.22	0.20	−0.20
−0.02	0.11	0.50	−0.01	−0.09	0.31	−0.20	−0.21	0.22
−0.19	0.24	−0.50	0.01	−0.06	0.05	0.14	0.25	−0.28
0.06	0.03	0.00	0.00	−0.49	−0.17	0.07	−0.40	−0.27
−0.23	−0.02	0.01	0.50	0.18	0.02	−0.16	0.32	0.20
−0.02	−0.01	−0.01	−0.50	0.29	0.14	0.22	−0.27	−0.15
0.23	0.03	0.00	0.00	−0.51	−0.22	−0.24	0.26	0.14
−0.02	−0.01	0.01	0.50	0.29	0.14	0.22	−0.27	−0.15
−0.23	−0.02	−0.01	−0.50	0.18	0.02	−0.16	0.32	0.20

2.785	2.222	2.084	1.784	1.387	1.318	1.056	1.000
0.20	0.33	0.31	0.00	0.10	0.17	0.38	0.00
0.10	0.40	0.26	0.32	0.07	−0.02	0.24	0.00
0.04	0.31	0.05	0.25	−0.27	0.01	0.19	0.00
0.02	0.30	−0.15	0.13	−0.44	0.04	−0.04	0.00
0.04	0.24	0.18	0.32	0.26	−0.21	−0.32	0.00
0.02	0.14	0.11	0.25	0.30	−0.26	−0.57	0.00
0.01	0.35	−0.37	−0.02	−0.34	0.04	−0.23	0.00
0.00	0.24	−0.31	−0.09	−0.02	0.01	−0.10	−0.50
0.00	0.18	−0.27	−0.13	0.32	−0.03	0.12	−0.50
0.00	0.16	−0.26	−0.14	0.46	−0.04	0.23	0.00
0.00	0.18	−0.27	−0.13	0.32	−0.03	0.12	0.50
0.00	0.24	−0.31	−0.09	−0.02	0.01	−0.10	0.50
0.35	0.17	0.24	−0.32	0.02	0.16	−0.02	0.00
0.72	−0.12	−0.05	−0.21	−0.15	−0.58	0.03	0.00
0.52	−0.29	−0.34	0.53	0.14	0.48	−0.02	0.00
0.20	0.14	0.22	−0.40	0.05	0.51	−0.43	0.00

0.679	−0.056	−0.704	−1.000	−1.119	−1.364	−1.902	−2.172
0.47	0.31	0.01	0.00	0.28	0.24	0.33	0.11
−0.02	−0.43	−0.06	0.00	0.19	−0.13	0.55	−0.24
−0.49	−0.04	0.49	0.00	−0.11	−0.22	0.35	0.27
−0.31	0.43	−0.29	0.00	−0.07	0.42	−0.11	−0.35
0.01	−0.25	−0.45	0.00	−0.39	0.15	0.37	0.13
0.03	0.44	0.38	0.00	0.24	−0.08	−0.15	−0.05
0.28	0.02	−0.29	0.00	0.19	−0.36	−0.13	0.49
0.25	−0.22	0.24	−0.50	−0.07	0.03	0.18	−0.35
−0.11	−0.01	0.11	0.50	−0.11	0.32	−0.22	0.28
−0.32	0.22	−0.32	0.00	0.19	−0.46	0.23	−0.26
−0.11	−0.01	0.11	−0.50	−0.11	0.32	−0.22	0.28
0.25	−0.22	0.24	0.50	−0.07	0.03	0.18	−0.35
0.11	0.26	0.05	0.00	−0.65	−0.32	−0.24	−0.07
−0.08	−0.11	−0.02	0.00	0.17	0.08	0.05	0.01
0.03	0.03	0.00	0.00	−0.03	−0.01	−0.01	0.00
−0.33	−0.24	−0.03	0.00	0.31	0.14	0.08	0.02

2.446	1.538	0.630	−0.534	−1.581
0.47	0.22	0.57	0.56	0.31
0.24	0.54	0.34	−0.43	−0.59
0.12	0.60	−0.36	−0.32	0.63
0.05	0.39	−0.57	0.61	−0.40
0.84	−0.38	−0.33	−0.18	−0.07

Exercise 25

Hückel parameters for sulfur are not very reliable, so AM1 calculations have been used for this exercise. All values in eV.

HO: −8.56   LU: 0.61

HO: −9.47   LU: 0.29

HO: −9.44   LU: −1.13

HO: −9.68   LU: −1.34

HO: −9.79   LU: −1.51

HO: −10.16   LU: −1.40

HO: −9.99   LU: 1.36

HO: −10.69   LU: −0.14

A4 Chapter 5

Exercise 1

For butadiene, see p. 245; isoprene and pentadiene, p. 249; vinyl acetate, p. 253; acrolein, p. 254; acrylonitrile, p. 256; eneamine, p. 258.

2.268	*0.814*	*−1.083*
0.92	0.35	0.16
0.35	−0.59	−0.72
0.16	−0.73	0.67

2.007	*1.437*	*0.434*	*−0.727*	*−1.650*
0.48	0.04	0.47	0.61	0.42
0.36	0.45	0.48	−0.27	−0.60
0.24	0.61	−0.26	−0.42	0.58
0.12	0.42	−0.60	0.57	−0.35
0.76	−0.49	−0.35	−0.22	−0.11

2.336	*0.773*	*−1.108*
0.91	−0.37	0.19
0.38	0.57	−0.73
0.16	0.74	0.66

2.136	*1.414*	*1.000*	*0.662*	*−0.662*	*−1.000*	*−1.414*	*−2.136*
0.14	0.35	0.00	0.60	0.60	0.00	0.35	0.14
0.31	0.50	0.00	0.39	−0.39	0.00	−0.50	−0.31
0.51	0.35	0.00	−0.33	−0.33	0.00	0.35	0.51
0.39	0.00	−0.50	−0.31	0.31	−0.50	0.00	−0.39
0.33	−0.35	−0.50	0.13	0.13	0.50	−0.35	0.33
0.31	−0.50	0.00	0.39	−0.39	0.00	0.50	−0.31
0.33	−0.35	0.50	0.13	0.13	−0.50	−0.35	0.33
0.39	0.00	0.00	−0.31	0.31	0.50	0.00	−0.39

2.117	1.211	0.525	−0.658	−1.646
0.26	0.18	0.69	0.54	0.37
0.55	0.22	0.36	−0.36	−0.63
0.33	0.57	−0.26	−0.41	0.57
0.16	0.47	−0.50	0.63	−0.34
0.71	−0.61	−0.30	0.13	0.16

2.746	1.805	1.242	0.545	−0.642	−1.686
0.11	0.25	0.19	0.67	0.55	0.37
0.31	0.44	0.23	0.37	−0.36	−0.63
0.13	0.35	0.53	−0.28	−0.41	0.57
0.05	0.20	0.42	−0.52	0.62	0.34
0.75	0.25	−0.54	−0.24	0.11	0.14
0.56	−0.72	0.40	0.09	−0.02	−0.02

2.047	1.387	0.656	−0.328	−0.863	−1.899
0.27	0.22	0.52	0.68	0.18	0.34
0.55	0.31	0.34	−0.22	−0.16	−0.64
0.35	0.47	−0.40	−0.08	−0.53	0.47
0.17	0.34	−0.60	0.25	0.61	−0.25
0.50	−0.26	0.10	−0.52	0.48	0.41
0.48	−0.67	−0.28	0.39	−0.26	−0.14

Exercise 2

H-CHS

AM1 calculations

Me-CHS
AM1 calculations

S 0.72
0.51 ‖
H₃C—C—H
−11.40 eV

S 0.63
−0.75 ‖
H₃C—C—H
−0.78 eV

O=⟋—S

0.31 C··−0.64 S
O=⟍−0.28⟋0.65
AM1
−1.69 eV

0.30 C··−0.51 S
O=⟍−0.19⟋0.55
3–21G
0.14 eV

0.46 C··−0.59 S
O=⟍−0.31⟋0.72
STO-3G
4.39 eV

−0.38　0.59 S
O=⟍−0.14⟋0.70
AM1
−12.39 eV

−0.22　0.35 S
O=⟍−0.11⟋0.34
3–21G
−11.62 eV

−0.36　0.56 S
O=⟍−0.17⟋0.66
STO-3G
−8.56 eV

AM1 calculations

O
Si—O
−0.36
0.20　　−0.54
　　　　0.32
−8.99 eV

O
Si—O
−0.33
−0.29　　0.44
　　　　0.37
0.49 eV

O=⟋—⟍=O

AM1 calculations

0.60　−0.38 O
O=⟍　　⟋−0.60
　　0.38
−14.28 eV

−0.47　0.53 O
O=⟍　　⟋−0.47
　　0.53
−0.76 eV

Exercise 3

AM1 calculations

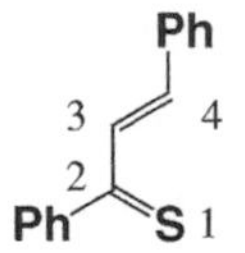

Ph
3 4
2
Ph S 1

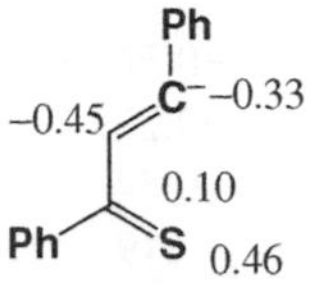

Ph
−0.45 C −0.33
0.10
Ph S 0.46
−9.04 eV

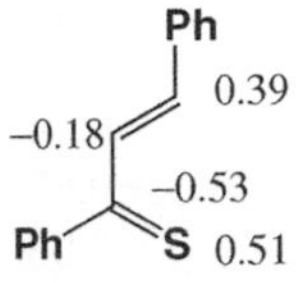

Ph
0.39
−0.18
−0.53
Ph S 0.51
−1.53 eV

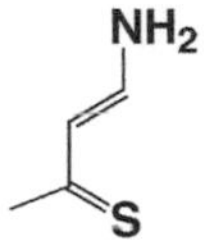

NH₂

S

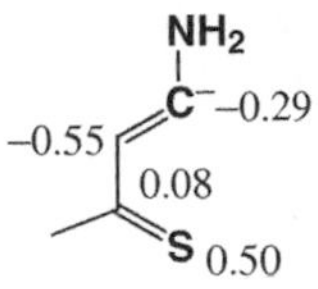

NH₂
−0.55 C −0.29
0.08
S 0.50
−8.62 eV

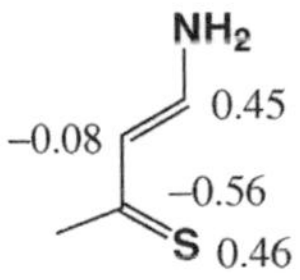

NH₂
0.45
−0.08
−0.56
S 0.46
−0.59 eV

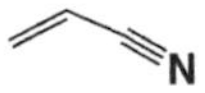

N

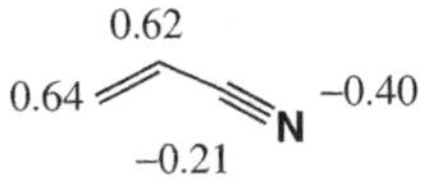

0.62
0.64 −0.40
N
−0.21
−10.86 eV

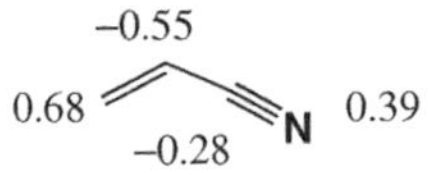

−0.55
0.68 N 0.39
−0.28
0.05 eV

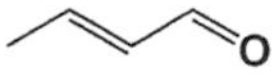

O

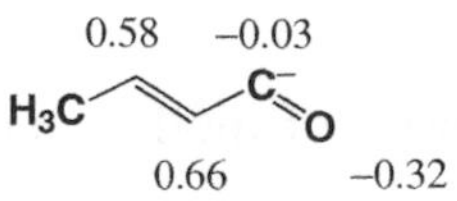

0.58 −0.03
H₃C C O
0.66 −0.32
−10.45 eV

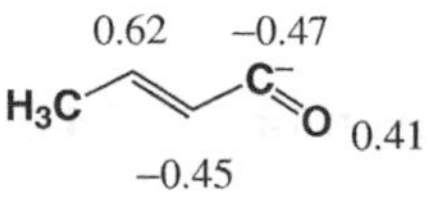

0.62 −0.47
H₃C C O 0.41
−0.45
−0.13 eV

Exercise 4

Hückel MOs of pentadiene are given on p. 249 and those of acrolein on p. 254.

2.784	2.325	1.830	1.310	0.856	−0.453	−1.652
0.19	0.11	0.48	0.57	0.44	0.39	0.22
0.34	0.15	0.40	0.18	−0.06	−0.56	−0.59
0.19	0.38	0.10	0.14	−0.54	−0.29	0.65
0.07	0.16	0.05	0.11	−0.63	0.64	−0.39
0.72	−0.18	0.19	−0.60	0.06	0.19	0.13
0.51	−0.32	−0.63	0.49	−0.03	−0.04	−0.02
0.17	0.81	−0.40	−0.14	0.33	0.08	−0.12

2.367	1.724	0.859	−0.358	−1.593
0.19	0.71	0.47	0.44	0.21
0.26	0.51	−0.07	−0.60	−0.56
0.43	0.18	−0.53	−0.23	0.67
0.18	0.10	−0.62	0.63	−0.42
0.82	−0.44	0.33	0.07	−0.13

2.767	1.930	1.336	1.000	−0.436	−1.597
0.19	0.48	0.45	0.58	0.38	0.24
0.33	0.45	0.15	0.00	−0.54	−0.61
0.14	0.32	0.26	−0.58	−0.29	0.63
0.05	0.17	0.19	−0.58	0.67	−0.39
0.74	0.08	−0.63	−0.00	0.19	0.14
0.54	−0.65	0.53	0.00	−0.04	−0.02

Exercise 6

Hückel MOs of acrolein are given on p. 254 and those of methyl acrylate in the previous exercise.

2.409	1.584	1.000	−0.413	−1.580
0.33	0.59	0.58	0.39	0.23
0.46	0.35	0.00	−0.55	−0.60
0.23	0.36	−0.58	−0.27	0.64
0.10	0.23	−0.58	0.67	−0.40
0.79	−0.58	0.00	0.16	0.12

Exercise 7

AM1 calculations
PhCHO

The HOMO (−10 eV) and the HOMO −1 (−10.04 eV) are mainly localized on the benzene fragment. It is the HOMO −2 (−10.71 eV) which corresponds to the 2p oxygen lone pair.

The HOMO (−3.11 eV), the HOMO −1 (−5.81 eV) and the HOMO −2 (−6.09 eV) are mainly localized on the styrene fragment. The HOMO −3 (−6.61 eV) is an oxygen lone pair and the HOMO −4 (−7.15 eV) a chlorine lone pair.

Exercise 8

Hückel MOs of fulvene are given on p. 246, those of 1-aminobutadiene on p. 264.

2.228	1.360	0.618	0.186	−1.618	−1.775
0.45	0.67	0.00	0.55	0.00	0.24
0.55	0.24	0.00	−0.45	0.00	−0.67
0.39	−0.17	0.60	−0.32	0.37	0.47
0.32	−0.47	0.37	0.39	−0.60	−0.17
0.32	−0.47	−0.37	0.39	0.60	−0.17
0.39	−0.17	−0.60	−0.32	−0.37	0.47

Exercise 9

2.811	2.670	1.979	1.591	1.475	1.276	0.716
0.24	0.11	0.37	0.22	0.24	0.03	0.48
0.12	0.09	0.22	0.09	0.52	0.18	0.12
0.09	0.13	0.06	−0.07	0.53	0.21	−0.41
0.14	0.25	−0.10	−0.21	0.27	0.08	−0.39
0.37	0.68	−0.33	−0.32	−0.18	−0.13	0.16
0.24	0.32	0.11	0.37	−0.15	0.03	0.14
0.13	0.19	0.12	0.63	−0.31	0.12	−0.50
0.32	−0.10	0.40	−0.12	−0.03	−0.18	0.10
0.18	−0.06	0.41	−0.20	−0.05	−0.66	−0.35
0.61	−0.41	0.02	−0.26	−0.28	0.51	−0.08
0.42	−0.34	−0.58	0.36	0.29	−0.39	0.03

−0.369	−0.800	−1.297	−2.053
0.23	0.00	0.25	0.58
−0.57	−0.15	0.28	−0.43
−0.02	0.12	−0.60	0.31
0.58	0.06	0.51	−0.19
−0.24	−0.20	−0.07	0.12
0.12	0.65	−0.22	−0.39
−0.09	−0.36	0.10	0.13
0.36	−0.51	−0.37	−0.37
−0.26	0.28	0.16	0.12
−0.13	0.15	0.09	0.08
0.03	−0.03	−0.02	−0.01

2.694	1.648	1.482	0.660	−0.500	−1.138	−1.846
0.19	0.41	0.25	0.52	0.31	0.35	0.50
0.13	0.28	0.55	0.17	−0.54	0.16	−0.51
0.16	0.06	0.57	−0.41	−0.04	−0.53	0.44
0.29	−0.19	0.30	−0.44	0.56	0.44	−0.30
0.79	−0.46	−0.17	0.16	−0.30	0.03	0.15
0.39	0.39	−0.19	0.18	0.39	−0.56	−0.41
0.23	0.60	−0.39	−0.53	−0.26	0.26	0.15

2.520	2.193	2.000	0.518	−1.230
0.42	0.08	0.00	0.56	0.71
0.26	−0.24	0.00	0.66	−0.66
0.56	0.30	−0.71	−0.27	−0.15
0.56	0.30	0.71	−0.27	−0.15
0.36	−0.87	0.00	−0.31	0.14

2.562	2.471	2.000	2.000	1.341	0.224	−0.893	−1.705
0.20	0.38	0.00	0.00	0.21	0.51	0.59	0.39
0.17	0.15	0.00	0.00	0.60	0.40	0.33	−0.57
0.22	−0.01	0.00	0.00	0.59	−0.42	−0.30	0.57
0.41	−0.18	0.00	0.00	0.19	−0.49	0.60	−0.41
0.25	0.56	−0.10	0.70	−0.23	−0.20	−0.14	−0.07
0.25	0.56	0.10	−0.70	−0.23	−0.20	−0.14	−0.07
0.50	−0.27	0.75	0.10	−0.21	0.19	−0.14	0.08
0.58	−0.31	−0.65	−0.09	−0.24	0.22	−0.17	0.09

Exercise 10

2.771	2.102	1.791	1.303	0.674	0.215	−1.022	−1.833
0.03	0.25	0.18	0.03	0.33	0.71	0.43	0.32
0.08	0.53	0.33	0.04	0.22	0.15	−0.44	−0.59
0.15	0.38	−0.01	−0.11	0.50	−0.49	−0.20	0.55
0.33	0.27	−0.35	−0.18	0.11	−0.26	0.64	−0.42
0.73	−0.07	−0.21	0.59	−0.08	0.13	−0.18	0.09
0.53	−0.40	0.57	−0.47	0.04	−0.04	0.03	−0.01
0.05	0.48	0.41	0.13	−0.67	−0.20	0.22	0.21
0.19	0.24	−0.44	−0.60	−0.35	0.33	−0.32	0.15

Exercise 11

The MOs of cyclopentadienone are shown p. 269.

2.760	2.000	1.497	0.618	0.436	−1.618	−1.692
0.32	0.38	0.09	0.00	0.57	0.35	0.65
0.15	0.38	−0.16	0.60	0.26	0.16	−0.49
0.08	0.38	−0.33	0.37	−0.46	−0.53	0.18
0.08	0.38	−0.33	−0.37	−0.46	0.44	0.18
0.15	0.38	−0.17	−0.60	0.26	0.03	−0.19
0.74	0.00	0.57	0.00	−0.33	−0.56	−0.15
0.54	−0.54	−0.63	0.00	0.12	0.26	0.02

Exercise 12

2.320	1.189	0.618	−1.008	−1.618
0.74	0.56	0.00	0.37	0.00
0.38	−0.11	−0.60	−0.59	0.37
0.29	−0.58	−0.37	0.29	−0.60
0.29	−0.58	0.37	0.29	0.60
0.38	−0.11	0.60	−0.59	−0.37

Exercise 13

2.549	2.279	2.000	1.386	0.305	−0.833	−1.686
0.56	0.09	0.75	0.20	0.20	0.15	0.08
0.44	0.04	0.00	−0.17	−0.47	−0.61	−0.43
0.22	−0.07	0.00	−0.57	−0.46	0.26	0.59
0.11	−0.19	0.00	−0.61	0.33	0.37	−0.57
0.07	−0.36	0.00	−0.28	0.56	−0.59	0.36
0.09	−0.90	0.00	0.32	−0.23	0.15	−0.07
0.64	0.11	−0.66	0.23	0.22	0.17	0.09

2.497	2.261	2.177	1.216	0.346	−0.776	−1.720
0.40	0.31	0.79	0.20	0.25	0.14	0.07
0.28	0.12	0.20	−0.22	−0.60	−0.56	−0.39
0.42	0.04	−0.12	−0.41	−0.39	0.33	0.62
0.23	−0.12	−0.03	−0.61	0.32	0.37	−0.56
0.15	−0.32	0.05	−0.33	0.50	−0.62	0.36
0.22	−0.87	0.20	0.30	−0.21	0.16	−0.07
0.68	0.14	−0.53	0.42	0.19	−0.10	−0.13

Exercise 14

The MOs of butadiene and of acrolein are shown pp. 245 and 254, respectively.

Exercise 15

Hückel MOs of naphthalene and azulene are given p. 246.

2.484	1.773	1.325	0.856	0.600	−0.826	−1.105	−1.458	−2.150
0.24	0.25	0.42	0.39	0.27	0.41	0.27	0.36	0.34
0.16	0.40	0.41	−0.16	0.37	−0.39	0.26	−0.42	−0.31
0.15	0.45	0.12	−0.52	−0.04	−0.09	−0.56	0.26	0.32
0.21	0.40	−0.25	−0.29	−0.40	0.46	0.36	0.05	−0.39
0.37	0.27	−0.45	0.27	−0.20	−0.29	0.16	−0.33	0.51
0.44	0.05	0.14	0.49	−0.21	0.05	−0.56	−0.10	−0.42
0.60	−0.54	0.28	−0.31	−0.25	−0.20	0.23	0.15	0.07
0.30	−0.32	−0.21	−0.24	0.49	0.52	−0.19	−0.45	0.11
0.27	0.02	−0.49	0.04	0.49	−0.27	0.02	0.53	−0.29

2.721	1.898	1.353	0.893	0.621	−0.808	−1.083	−1.449	−2.148
0.18	0.28	0.40	0.43	0.23	0.37	0.32	0.36	0.34
0.10	0.37	0.45	−0.11	0.39	−0.40	0.21	−0.44	−0.31
0.08	0.42	0.21	−0.53	0.01	−0.05	−0.55	0.28	0.33
0.13	0.43	−0.17	−0.37	−0.38	0.44	0.39	0.04	−0.39
0.27	0.39	−0.44	0.21	−0.24	−0.31	0.14	−0.34	0.51
0.39	0.17	0.10	0.50	−0.24	0.10	−0.56	−0.09	−0.42
0.76	−0.45	0.20	−0.24	−0.17	−0.17	0.19	0.12	0.06
0.30	−0.11	−0.26	−0.16	0.54	0.54	−0.17	−0.43	0.11
0.21	0.15	−0.51	0.05	0.47	−0.29	0.03	0.53	−0.29

Exercise 16

2.795	1.944	1.891	1.313	1.000	0.781	0.705
0.71	−0.54	0.00	0.05	0.00	0.35	0.00
0.35	0.02	0.27	−0.02	0.35	−0.27	−0.30
0.16	0.11	0.32	0.31	0.35	−0.37	0.14
0.08	0.19	0.34	0.43	0.00	−0.02	0.40
0.07	0.26	0.32	0.25	−0.35	0.35	0.14
0.12	0.32	0.27	−0.10	−0.35	0.30	−0.30
0.26	0.36	0.19	−0.38	0.00	−0.12	−0.35
0.26	0.36	−0.19	−0.38	0.00	−0.12	0.35
0.12	0.32	−0.27	−0.10	0.35	0.30	0.30
0.07	0.26	−0.32	0.25	0.35	0.35	−0.14
0.08	0.19	−0.34	0.43	0.00	−0.02	−0.40
0.16	0.11	−0.32	0.31	−0.35	−0.37	−0.14
0.35	0.02	−0.27	−0.02	−0.35	−0.27	0.30

−0.762	−1.000	−1.123	−1.317	−1.948	−2.278
0.11	0.00	0.21	0.00	0.13	0.00
−0.20	−0.35	−0.40	0.06	−0.32	−0.29
−0.26	0.35	0.24	0.29	0.33	0.21
0.39	0.00	0.13	−0.44	−0.32	−0.18
−0.04	−0.35	−0.39	0.29	0.29	0.21
−0.36	0.35	0.30	0.06	−0.25	−0.29
0.32	0.00	0.05	−0.37	0.19	0.46
0.32	0.00	0.05	0.37	0.19	−0.46
−0.36	−0.35	0.30	−0.06	−0.25	0.29
−0.04	0.35	−0.39	−0.29	0.29	−0.21
0.39	0.00	0.13	0.44	−0.32	0.18
−0.26	−0.35	0.24	−0.29	0.33	−0.21
−0.20	0.35	−0.40	−0.06	−0.32	0.29

Exercise 20

2.660	1.525	0.676	−0.895	−1.467
0.34	0.20	0.53	0.68	0.32
0.26	0.69	0.31	−0.34	−0.50
0.22	0.50	−0.47	−0.20	0.66
0.33	0.08	−0.63	0.52	−0.47
0.81	−0.48	0.06	−0.33	0.03

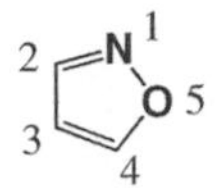

2.699	1.340	0.811	−0.782	−1.568
0.40	0.27	0.63	0.55	0.27
0.22	0.61	0.22	−0.44	−0.58
0.20	0.55	−0.45	−0.20	0.64
0.31	0.12	−0.59	0.60	−0.43
0.81	−0.48	−0.03	−0.33	0.04

Exercise 24

Shown below are the HOMO coefficients at the α- and γ-positions of three enolates. calculated by the AM1 and STO-3G methods. STO-3G values are indicated in italics.

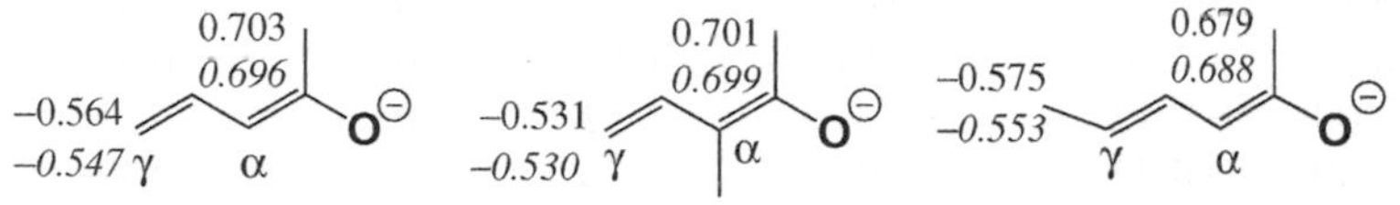

Exercise 25

For acrolein, styrene and methyl acrylate, see pp. 254, 264 and 268, respetively. 1-Hexene can be modeled by propene (p. 264). Ethyl vinyl ether can be modeled by methyl vinyl ether, enol or even propene.

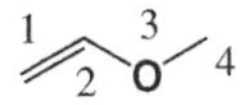

2.713	1.665	0.734	−1.112
0.10	0.21	0.72	0.66
0.26	0.35	0.53	−0.73
0.76	0.47	−0.41	0.19
0.59	−0.78	0.18	−0.03

A5   Chapter 6

Exercise 11

For hexatriene, see p. 256.
Beware: the numbering here is *different* from the tropone numbering in Exercise 26 of
Chapter 5.

2.197	1.618	1.247	0.714	−0.445	−0.618	−1.802	1.912
0.45	0.49	0.00	0.65	0.00	0.30	0.00	0.20
0.54	0.30	0.00	−0.19	0.00	−0.49	0.00	−0.59
0.37	0.00	−0.42	−0.39	−0.52	0.00	−0.23	0.46
0.27	−0.30	−0.52	−0.09	0.23	0.49	0.42	−0.29
0.22	−0.49	−0.23	0.33	0.42	−0.30	−0.52	0.10
0.22	−0.49	0.23	0.33	−0.42	−0.30	0.52	0.10
0.27	−0.30	0.52	−0.09	−0.23	0.49	−0.42	−0.29
0.37	0.00	0.42	−0.39	0.52	0.00	0.23	0.46

Exercise 17

1   2   3   4   5
Me–CH=CH–OMe

2.721	2.256	1.635	0.569	−1.181
0.13	0.91	0.20	0.31	0.15
0.13	0.33	−0.10	−0.63	−0.68
0.27	0.11	−0.31	−0.58	0.70
0.75	−0.09	−0.50	0.38	−0.18
0.58	−0.21	0.77	−0.15	0.03

A6 Summary Table

p. 245 $CH_2{=}CH_2$ $CH_2{=}CH{-}CH{=}CH_2$

p. 246

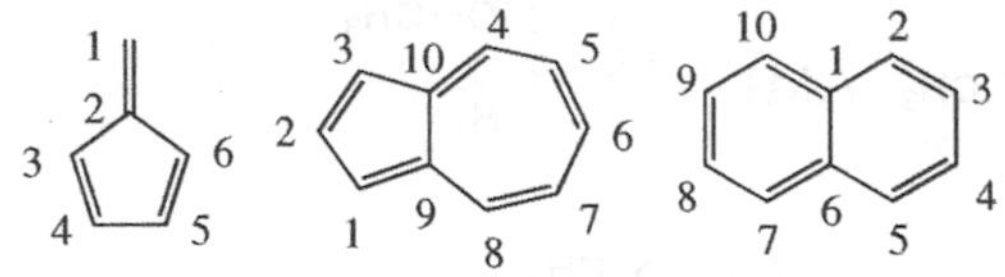

p. 247

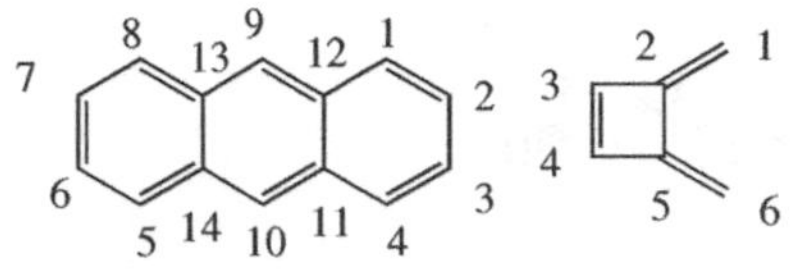

p. 248

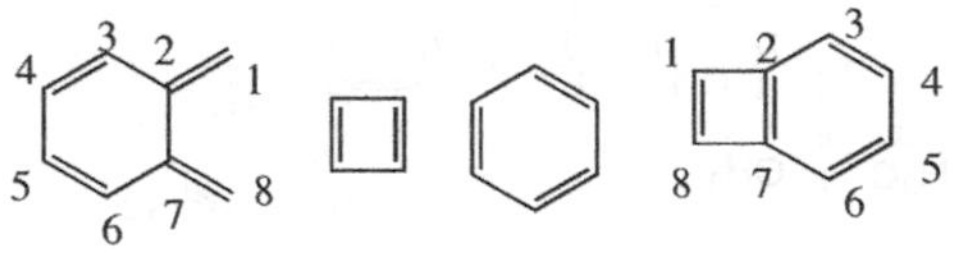

p. 249

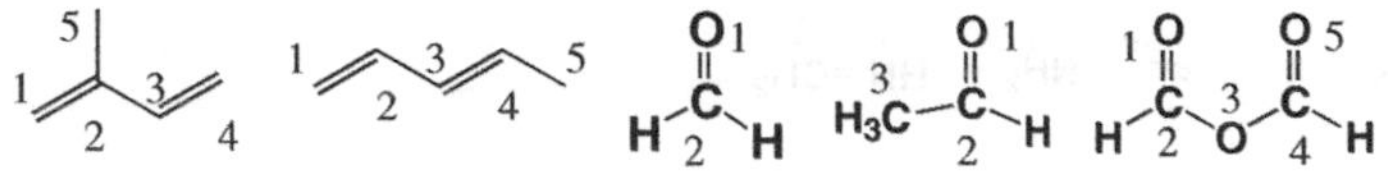

p. 250

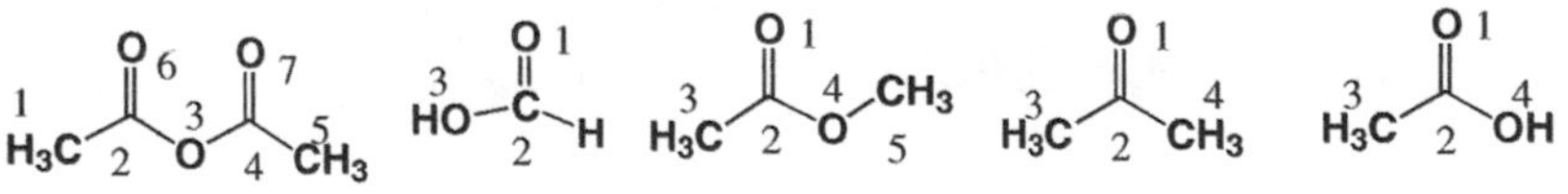

p. 251

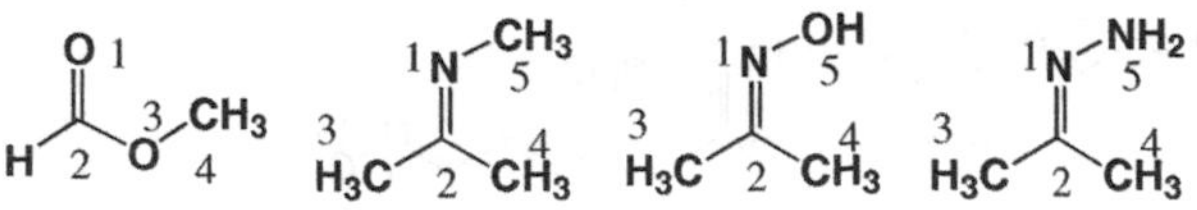

p. 252

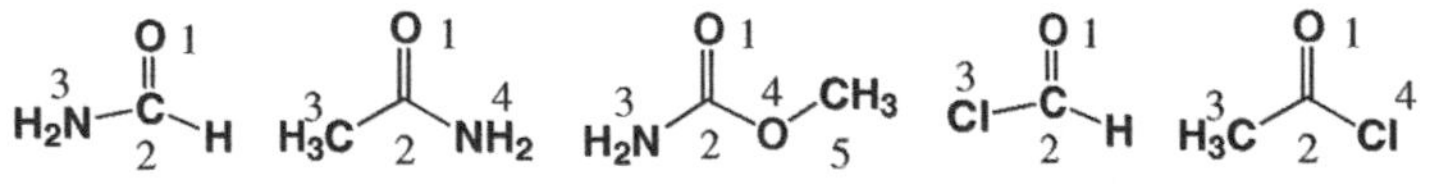

p. 253

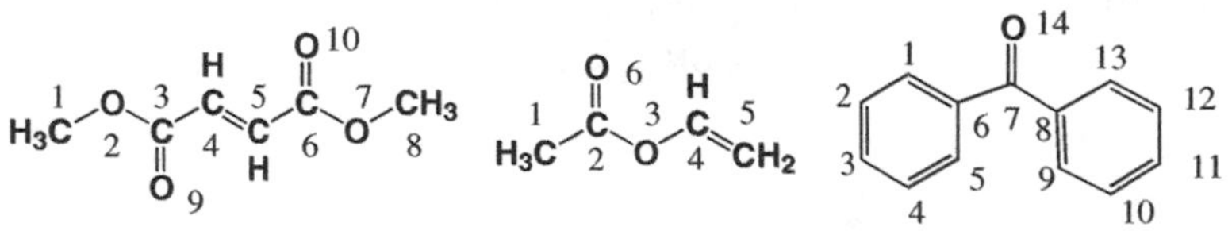

p. 254

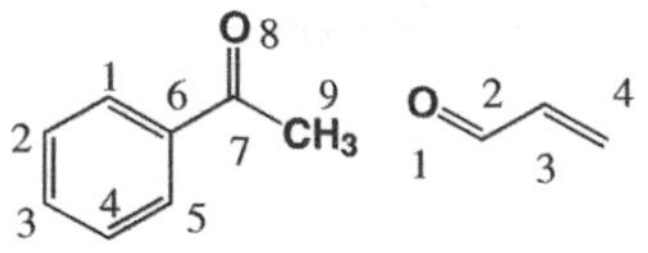

p. 255

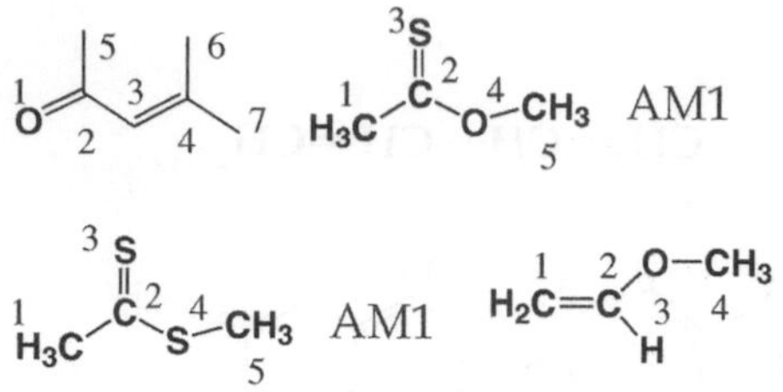

p. 256

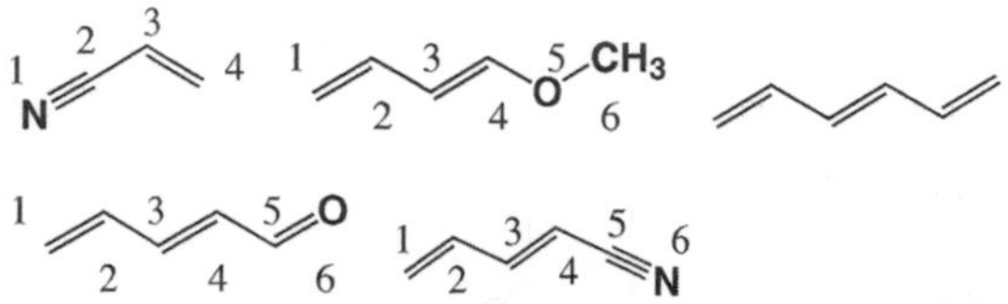

p. 257

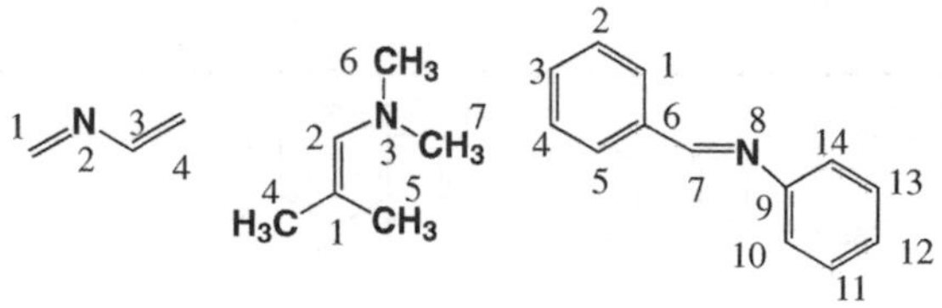

p. 258

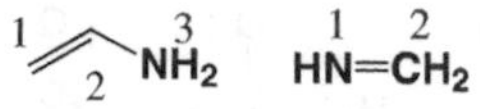

p. 259

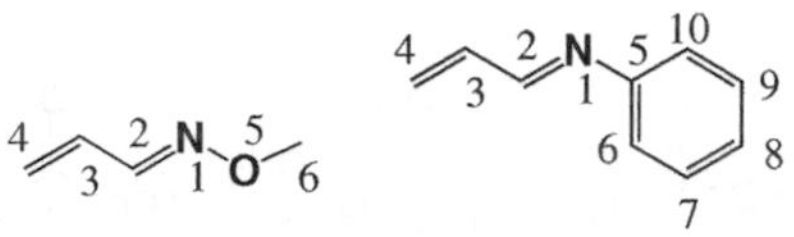

p. 260

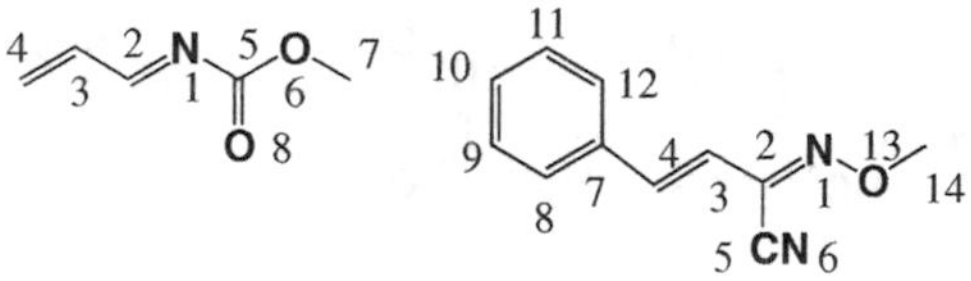

p. 261

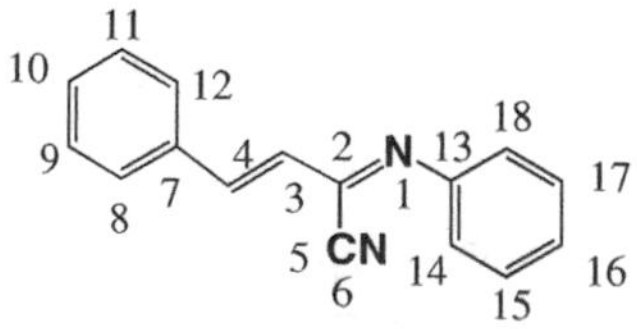

p. 262

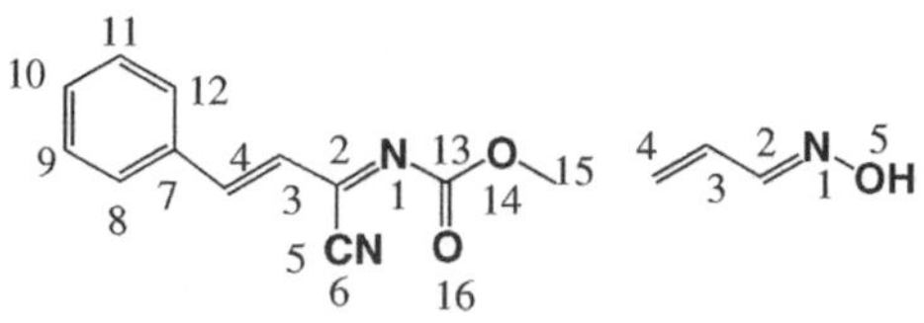

p. 263 AM1 calculations

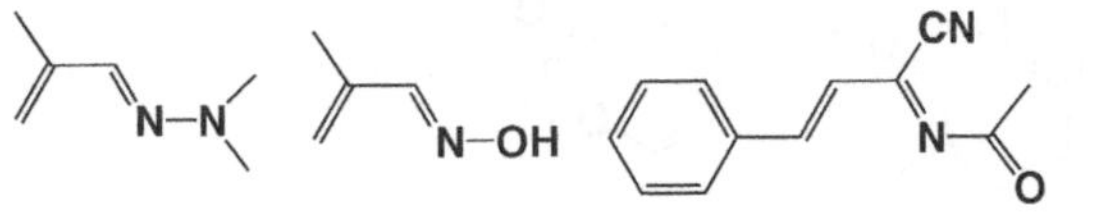

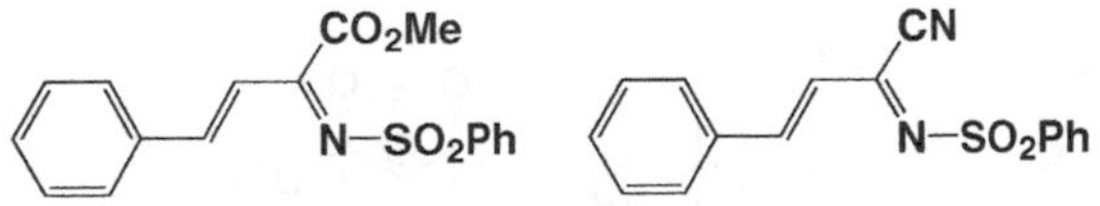

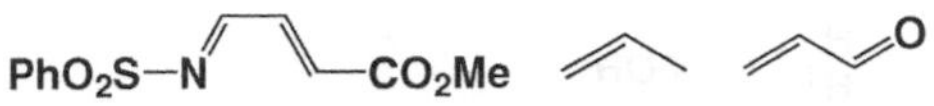

p. 264 Hückel

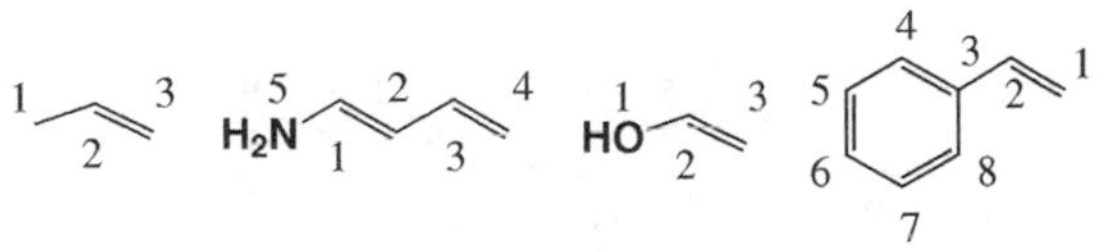

p. 265

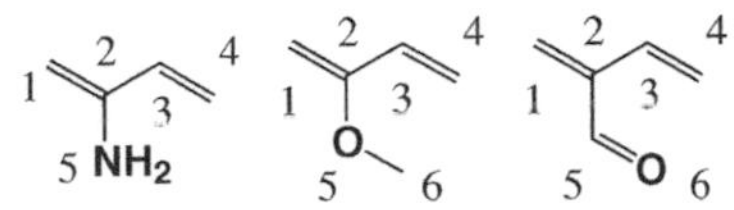

H-CHS (AM1)

p. 266 **Me-CHS**

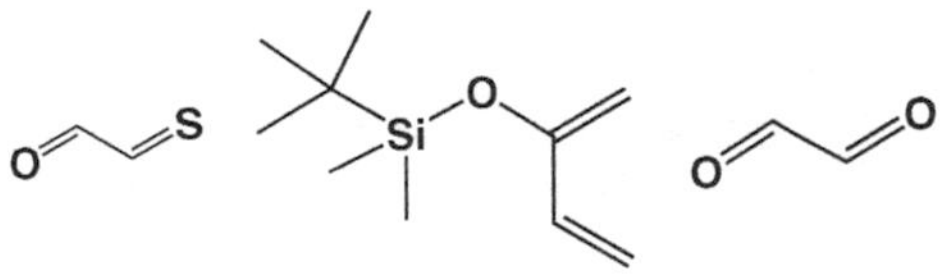

p. 267

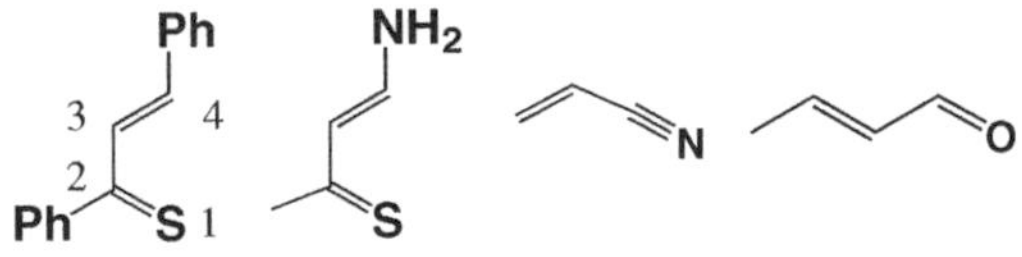

p. 268

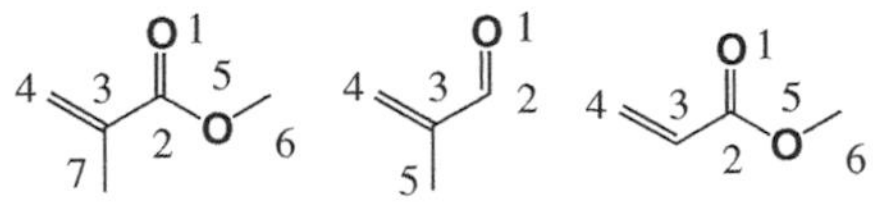

p. 269 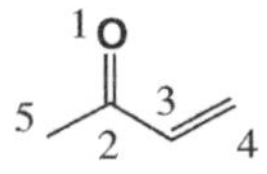 **Ph-CHO**

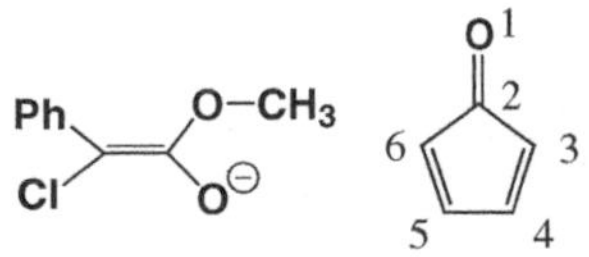

p. 270

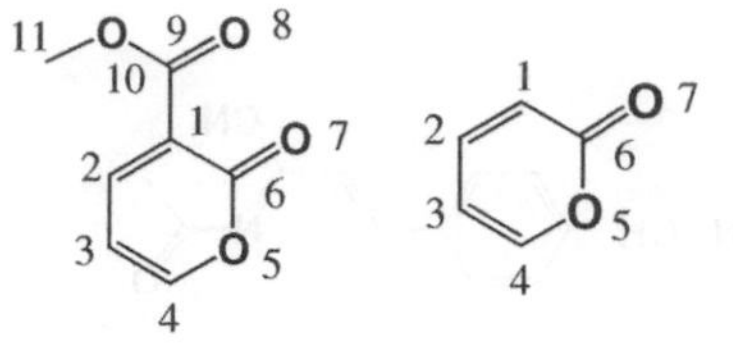

p. 271

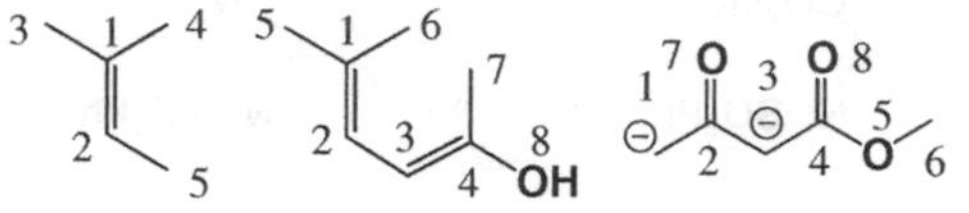

p. 272

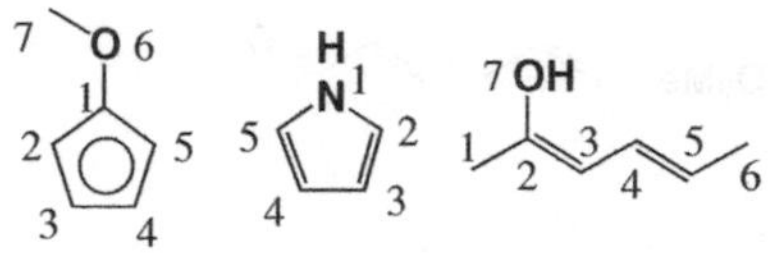

p. 273

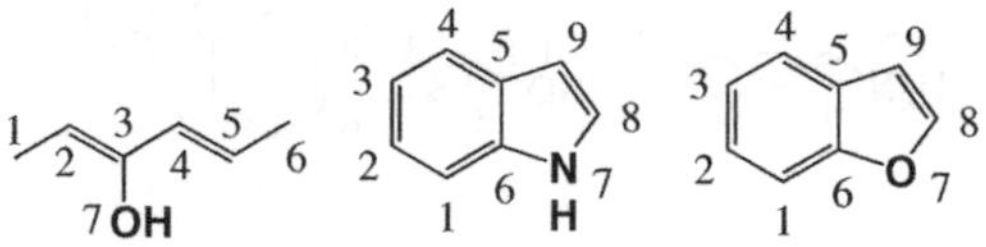

p. 274

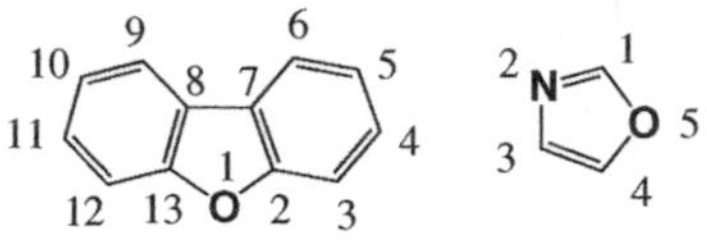

p. 275

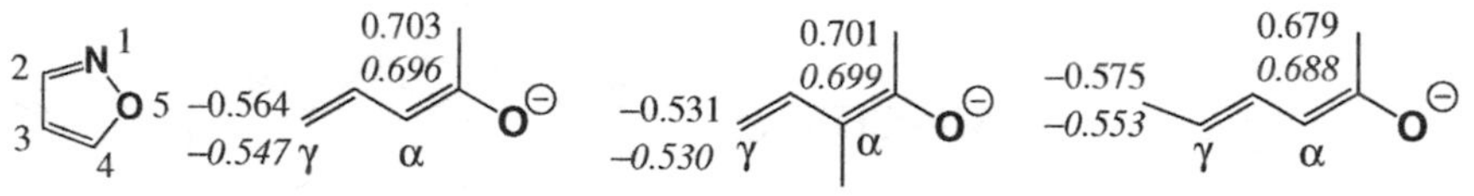
0.703
0.696
–0.564
–0.547 γ α
0.701
0.699
–0.531
–0.530 γ α
0.679
0.688
–0.575
–0.553 γ α

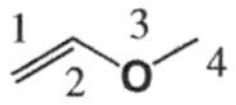

p. 276

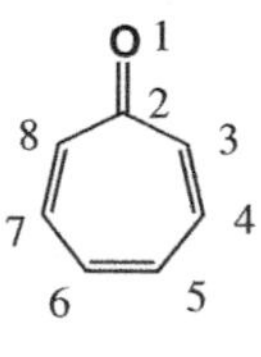

1 2 3 4 5
Me–CH=CH–OMe

Index

Printed and bound by CPI Group (UK) Ltd, Croydon, CR0 4YY

27/10/2024

14580171-0005